MATHEMATICAL LOGIC

about the editors . . .

AYDA I. ARRUDA is Full Professor in the Department of Mathematics at the State University of Campinas in Brazil. She received her Ph.D. in mathematics from the University of Paraná in 1964, where she taught until 1968, and has had visiting appointments in Chile and France. Her principal research interests lie in paraconsistent logics, relevant logics, and set theory. She is the author or coauthor of numerous articles and the coeditor of one book. She is a member of the Association of Symbolic Logic.

NEWTON C. A. DA COSTA is Full Professor of Mathematics at the University of São Paulo, Brazil. Already the recipient of a degree in civil engineering, he received a Ph.D. degree in mathematics from the University of Paraná in 1960. He has taught at the University of Paraná, the University of Campinas, and has held numerous visiting positions in Brazil, Argentina, France, United States, Chile, and Australia. He has written many articles for mathematics journals and has written two books and is coeditor of a third. His research interests involve modal logic, paraconsistent logic, and lattice theory. He is a member of the Brazilian Mathematical Society, the American Mathematical Society, and the Association for Symbolic Logic, and is a member of the Academy of Sciences of the State of São Paulo, Brazil.

ROLANDO CHUAQUI is Full Professor of Mathematics and Head of the Graduate Program in Mathematics at the Catholic University of Chile in Santiago. Having received an M.D. degree from the University of Chile School of Medicine in 1960, he went on to study at the University of California at Berkeley, where he received a Ph.D. degree in logic and the methodology of science in 1965. He has taught at the University of Chile, and has held visiting positions at the University of California at Berkeley, the Institute for Advanced Study, the University of California at Los Angeles, the State University of Campinas, and the University of São Paulo. His research interests lie in foundations of probability, cardinal algebras with applications to measure theory, and foundations of set theory, and he has published widely in the literature. He is President of the Chilean Mathematical Society, and a member of the American Mathematical Society and the Association for Symbolic Logic. Dr. Chuaqui is a member of the Chilean Academy of Sciences.

PURE AND APPLIED MATHEMATICS

A Program of Monographs, Textbooks, and Lecture Notes

Contributions to *Lecture Notes in Pure and Applied Mathematics* are reproduced by direct photography of the author's typewritten manuscript. Potential authors are advised to submit preliminary manuscripts for review purposes. After acceptance, the author is responsible for preparing the final manuscript in camera-ready form, suitable for direct reproduction. Marcel Dekker, Inc. will furnish instructions to authors and special typing paper. Sample pages are reviewed and returned with our suggestions to assure quality control and the most attractive rendering of your manuscript. The publisher will also be happy to supervise and assist in all stages of the preparation of your camera-ready manuscript.

LECTURE NOTES
IN PURE AND APPLIED MATHEMATICS

1. *N. Jacobson,* Exceptional Lie Algebras
2. *L.-Å. Lindahl and F. Poulsen,* Thin Sets in Harmonic Analysis
3. *I. Satake,* Classification Theory of Semi-Simple Algebraic Groups
4. *F. Hirzebruch, W. D. Newmann, and S. S. Koh,* Differentiable Manifolds and Quadratic Forms
5. *I. Chavel,* Riemannian Symmetric Spaces of Rank One
6. *R. B. Burckel,* Characterization of C(X) Among Its Subalgebras
7. *B. R. McDonald, A. R. Magid, and K. C. Smith,* Ring Theory: Proceedings of the Oklahoma Conference
8. *Y.-T. Siu,* Techniques of Extension of Analytic Objects
9. *S. R. Caradus, W. E. Pfaffenberger, and B. Yood,* Calkin Algebras and Algebras of Operators on Banach Spaces
10. *E. O. Roxin, P.-T. Liu, and R. L. Sternberg,* Differential Games and Control Theory
11. *M. Orzech and C. Small,* The Brauer Group of Commutative Rings
12. *S. Thomeier,* Topology and Its Applications
13. *J. M. López and K. A. Ross,* Sidon Sets
14. *W. W. Comfort and S. Negrepontis,* Continuous Pseudometrics
15. *K. McKennon and J. M. Robertson,* Locally Convex Spaces
16. *M. Carmeli and S. Malin,* Representations of the Rotation and Lorentz Groups: An Introduction
17. *G. B. Seligman,* Rational Methods in Lie Algebras
18. *D. G. de Figueiredo,* Functional Analysis: Proceedings of the Brazilian Mathematical Society Symposium
19. *L. Cesari, R. Kannan, and J. D. Schuur,* Nonlinear Functional Analysis and Differential Equations: Proceedings of the Michigan State University Conference
20. *J. J. Schäffer,* Geometry of Spheres in Normed Spaces
21. *K. Yano and M. Kon,* Anti-Invariant Submanifolds
22. *W. V. Vasconcelos,* The Rings of Dimension Two
23. *R. E. Chandler,* Hausdorff Compactifications
24. *S. P. Franklin and B. V. S. Thomas,* Topology: Proceedings of the Memphis State University Conference
25. *S. K. Jain,* Ring Theory: Proceedings of the Ohio University Conference
26. *B. R. McDonald and R. A. Morris,* Ring Theory II: Proceedings of the Second Oklahoma Conference
27. *R. B. Mura and A. Rhemtulla,* Orderable Groups
28. *J. R. Graef,* Stability of Dynamical Systems: Theory and Applications
29. *H.-C. Wang,* Homogeneous Banach Algebras
30. *E. O. Roxin, P.-T. Liu, and R. L. Sternberg,* Differential Games and Control Theory II
31. *R. D. Porter,* Introduction to Fibre Bundles
32. *M. Altman,* Contractors and Contractor Directions Theory and Applications
33. *J. S. Golan,* Decomposition and Dimension in Module Categories
34. *G. Fairweather,* Finite Element Galerkin Methods for Differential Equations
35. *J. D. Sally,* Numbers of Generators of Ideals in Local Rings
36. *S. S. Miller,* Complex Analysis: Proceedings of the S.U.N.Y. Brockport Conference
37. *R. Gordon,* Representation Theory of Algebras: Proceedings of the Philadelphia Conference
38. *M. Goto and F. D. Grosshans,* Semisimple Lie Algebras
39. *A. I. Arruda, N. C. A. da Costa, and R. Chuaqui,* Mathematical Logic: Proceedings of the First Brazilian Conference

Other Volumes in Preparation

MATHEMATICAL LOGIC

Proceedings of the First Brazilian Conference

EDITED BY

Ayda I. Arruda
Department of Mathematics
State University of Campinas, Brazil

Newton C. A. da Costa
Department of Mathematics
University of São Paulo, Brazil

Rolando Chuaqui
Institute of Mathematics
Catholic University of Chile, Chile

MARCEL DEKKER, INC. New York and Basel

Library of Congress Cataloging in Publication Data

Brazilian Conference on Mathematical Logic, 1st,
Universidad Estadual de Campinas, 1977.
Mathematical logic.

(Lecture notes in pure and applied mathematics ;
v. 39)
"Sponsored by the Center of Logic, Epistemology,
and History of Science, and by the Institute of
Mathematics, Statistics, and Computing Science of
the State University of Campinas."
1. Logic, Symbolic and mathematical--Congresses.
I. Arruda, Ayda I. II. da Costa, Newton C.A.
III. Chaqui, R. IV. Universidade Estual de
Campinas. Centro de Lógica, Epistemologia, e
História de Ciência. V. Universidade Estadual de
Campinas. Instituto de Matemática, Estatística,
e Ciencia da Computação. VI. Title.
QA9.A1B73 1977 511'.3 78-14488
ISBN 0-8247-6772-1

MARCEL DEKKER, INC.

270 Madison Avenue, New York, New York 10016

Current printing (last digit)

10 9 8 7 6 5 4 3 2 1

PRINTED IN THE UNITED STATES OF AMERICA

PREFACE

This volume constitutes the Proceedings of the First Brazilian Conference on Mathematical Logic held at the State University of Campinas, Brazil, from 4 to 6 July 1977, and sponsored by the Center of Logic, Epistemology, and History of Science, and by the Institute of Mathematics,Statistics and Computing Science of the State University of Campinas. The Organizing Committee consisted of A. I. Arruda, N. C. A. da Costa, O. Porchat Pereira, and E. H. Alves. This is the first of a series of meetings that the Center of Logic, Epistemology and History of Science is planning to organize each year.

The papers included in this volume are the texts of the lectures delivered at the meeting and also expanded and revised texts of short communications.

The Editors would like to acknowledge the financial support given to the meeting by the State Univesity of Campinas and by the Fundação de Âmparo ã Pesquisa do Estado de São Paulo (FAPESP).

The Editors.

Centro de Lógica, Epistemologia
e História de Ciência

and

Instituto de Matemática, Estatística,
e Ciências da Computação.

November 1977

CONTENTS

CONTRIBUTORS

ELIAS H. ALVES
The Center of Logic, Epistemology, and History of Science, Universidade Estadual de Campinas, Campinas, SP., Brazil.

AYDA I. ARRUDA
Department of Mathematics, Universidade Estadual de Campinas, Campinas, SP., Brazil.

ROLANDO CHUAQUI
Institute of Mathematics, Universidad Católica de Chile, Santiago, Chile.

ROBERTO CIGNOLI
Department of Mathematics, Universidade Estadual de Campinas, Campinas, SP., Brazil.

MANUEL CORRADA
Institute of Mathematics, Universidad Católica de Chile, Santiago, Chile.

NEWTON C. A. DA COSTA
Department of Mathematics, Universidade de São Paulo, São Paulo, SP., Brazil

JOSÉ EDUARDO DE ALMEIDA MOURA
Department of Philosophy, Universidade Federal do Rio Grande do Norte, Natal, RN., Brazil.

MANUEL M. FIDEL
Institute of Mathematics, Universidad Nacional del Sur, Bahía Blanca, Argentina.

MARCEL GUILLAUME
Department of Mathematics, Université de Clermont-Ferrand, Clermont-Ferrand, France.

JERZY KOTAS
Institute of Mathematics, Uniwersytet Mikołaja Kopernika, Toruń, Poland.

ANDRÉA LOPARIĆ
The Center of Logic, Epistemology, and History of Science, Universidade Estadual de Campinas, Campinas, SP., Brazil.

LUIS HENRIQUE LOPES DOS SANTOS
The Center of Logic, Epistemology, and History of Science, Universidade Estadual de Campinas, Campinas, SP., Brazil.

CARLOS A. LUNGARZO
The Center of Logic, Epistemology, and History of Science, Universidade Estadual de Campinas, Campinas, SP., Brazil.

MARIA VICTORIA MARSHALL
Institute of Mathematics, Universidad Católica de Chile, Santiago, Chile.

IRENE MIKENBERG
Institute of Mathematics, Universidad Católica de Chile, Santiago, Chile.

CHARLES C. PINTER
Department of Mathematics, Bucknell University, Lewisburg, Pa., U.S.A.

ANDRÉS R. RAGGIO
Department of Mathematics, Universidade Estadual de Campinas, Campinas, SP., Brazil.

H. P. SANKAPPANAVAR
Institute of Mathematics, Universidade Federal da Bahía, Salvador, BA., Brazil.

ANTONIO MÁRIO SETTE
Institute of Mathematics, Universidade Federal de Pernambuco, Recife, PE., Brazil

J. SERGIO SETTE
Institute of Mathematics, Universidade Federal de Pernambuco, Recife, PE., Brazil.

INTRODUCTION

The Center of Logic, Epistemology, and History of Science at the State University of Campinas.

In 1976, the Center of Logic, Epistemology, and History of Science (CLE) was created at the State University of Campinas (UNICAMP), Brazil, through the efforts of Dr. Oswaldo Porchat Pereira who is its present Coordinator. The faculty of the CLE consists of professors and researchers from several areas of scientific knowledge who belong to the various Departments of the University. Researchers from other institutions are also admitted to the CLE.

The field of activities of the CLE includes the disciplines which appear in its name. Its main objectives are the organization of Colloquia, seminars, and lectures; the coordination of research work; the publishing of specialized literature; and the creation of conditions for the establishing interdisciplinary graduate courses and degrees.

The CLE receives its financial support from UNICAMP and other Brazilian organizations that support scientific research. Thanks to these resources, the CLE has had the opportunity of inviting several local and foreign professors, mainly from the U.S.A. and France.

The main activities of the CLE during 1976 and 1977 have been:

- A graduate program in Logic and Philosophy of Science, leading to Masters and Doctors degrees. This program is conducted in collaboration with the Department of Philosophy at UNICAMP.

- Three permanent Seminars which meet weekly since the beginning of 1976: Logic, Philosophy of Language, and Epistemology of Natural Sciences.

- Publication of a journal of Philosophy called *Manuscrito* to appear twice a year starting in November 1977.

- Organization of the First and Second Colloquia of Philosophy dealing with Hume (1976), and Kant (1977).

- Organization of the First Brazilian Conference in Mathematical Logic.

One of the most developed sectors of the CLE is Logic, which is coordinated by Professor Newton C. A. da Costa, from the University of São Paulo. This development is due mainly to the fact that the Institute of Mathematics, Statistics and Conputing Science at UNICAMP already had a group working in Logic which joined the CLE. This group which counts among its members Professors Ayda I. Arruda and Rolando Chuaqui, was mainly responsible for the organization in July 1976 at UNICAMP of the Third Latin-American Symposium on Mathematical Logic. The Proceedings of this meeting were published by North-Holland Pub. Co. under the tithe Non-Classical, Logics, Model Theory and Computability.

A periodical publication in Logic is planned under the title *Cadernos de Lógica*, the first issue will appear in 1978.

The Logic Sector of the CLE has programmed the Brasilian Conferences in Mathematical Logic to be held every year at UNICAMP to assemble Brazilian logicians and stimulate research in the area. The first of this Conferences, whose Proceedings are contained in this volume, was held in July 1977, co-sponsored by the Institute of Mathematics, Statistics and Computing Science of UNICAMP. Its organization has been possible due to the financial support of UNICAMP and FAPESP (Fundação de Âmparo à Pesquisa do Estado de São Paulo).

The program for this conference was the following:

July 4.

9:30 A.M.- Opening session.

10:00-11:00 A.M. - *On Bernays' class theory*, Rolando Chuaqui, Universidad Católica de Chile and Universidade Estadual de Campinas.

2:00-2:30 P.M. - *Internal models for any finite subset of the axioms for the impredicative theory of classes*, Manuel Corrada, Universidad de Chile.

2:30-3:00 P.M. - *From total to partial algebras*, Irene Mikenberg, Universidad Católica de Chile.

3:00-3:30 P.M. - *The real numbers in D**, Matias F. Dias, Universidade

Federal da Paraíba, Brasil (presented by N. C. A. da Costa).

4:00-5:00 P.M. - *On the logic of negationless intuitionistic mathematics*, Ayda I. Arruda, Universidade Estadual de Campinas, Brasil.

JULY 5.

9:30-10:30 A.M. - *Decision problems:history and methods*, H. P. Sankappanavar, Universidade Federal da Bahía, Brasil.

11:00-12:00 A.M. - *Functorialization of predicate calculus*, A. M. Sette, Universidade Federal de Pernambuco, Brasil

2:00-2:30 P.M. - *An algebraic study of a propositional system of Nelson*, Manuel M. Fidel, Universidad Nacional del Sur, Argentina (presented by L. Tapia).

2:30-3:00 P.M. - *On a paraconsistent system of higher-order*, J. E. Moura Universidade Federal do Rio grande do Norte, Brasil.

3:30-4:30 P.M. - *An introduction to the logic of quantum mechanics*, R. L. Cignoli, Universidade Estadual de Campinas, Brasil.

4:30-5:00 P.M. - *Superposition of states in quantum logic*, C. A. Lungarzo, Universidade Estadual de Campinas, Brasil.

JULY 6.

9:00-10:00 A.M. - *The method of valuation in modal logic*, Andréa Loparić, Universidade Estadual de Campinas, Brasil.

10:00-10:30 A.M. - *Discussive versions of T, B, S4 and S5*, L. H. Lopes dos Santos, Universidade Estadual de Campinas, Brasil.

11:00-12:00 A.M. - A *proof-theoretic analysis of da Costa's* C_{ω}^{*}, A. R. Raggio, Universidade Estadual de Campinas, Brasil, and Universitaet Erlangen, B.D.R.

2:00-3:00 P.M. - *On the problem of Jaśkowski and the logics of Łukasiewicz*, Newton C. A. da Costa, Universidade de São Paulo, Brasil.

3:00-3:30 P.M. - Closing session.

The following communications where presented by title:

Some remarks in set theory, Marcel Guillaume, Université de Clermont-Ferrand, France.

Constructibility in the impredicative theory of classes, M. V. Marshall and R. Chuaqui, Universidad Católica de Chile and Universidade de Estadual de Campinas, Brasil.

Cylindric algebras with a property of Rasiowa and Sikorski, Charles C. Pinter, Bucknell University, U.S.A.

There were 46 participants of whom 22 were students.

The Organizing Committee.

On some Higher-Order Paraconsistent Predicate Calculi.

by *E. H. ALVES and J. E. de A. MOURA.*

ABSTRACT. In this paper we present a hierarchy of higher-order predicate calculi based on the calculi $\mathcal{C}_n$, $1 \leqslant n < \omega$, of da Costa (see da Costa, *On the theory of Inconsistent formal systems*, Notre Dame Journal of Formal Logic, vol. XV. nº 4 (1974), 497-510, and Alves, *Lógica e Inconsistência, um estudo dos cálculos* $\mathcal{C}_n$, $1 \leqslant n \leqslant \omega$, Universidade de São Paulo, 1976). We denote these calculi by $\mathbf{P}_n^m$, $1 \leqslant n < \omega$, $1 \leqslant m \leqslant \omega$, where n indicates which of the calculi C_n is being considered and the superscript m indicates the order. We give details only concerning the calculus $\mathbf{P}_1^2$, that is, the second order calculus based on $\mathcal{C}_1$. The generalization to other cases is immediate. We introduce also a semantics for these calculi, which is a generalization of the semantics of $\mathcal{C}_1^=$, presented in A.I. Arruda and N. C. A. da Costa, *Une sémantique pour le calcul* $\mathcal{C}_1^=$, C. R. Acad. Sc. Paris, 284 A (1977), 279-282. We show also that these calculi are weakly complete relatively to this semantics. Finally, we show how the classical arithmetic can be obtained from $\mathbf{P}_1^m$, $m > 2$.

1. The calculi $\mathbf{P}_1^m$, $1 \leqslant m \leqslant \omega$.

One defines *types* as follows: (I) ι is a type; (II) if $t_1, t_2, \ldots, t_n$ $(1 \leqslant n < \omega)$ are types, then $\langle t_1, t_2, \ldots, t_n \rangle$ is a type; (III) the only types are those given by (I)-(II).

The primitive symbols of $\mathbf{P}_1^m$, $1 \leqslant m \leqslant \omega$, are: 1) the propositional logical operators: $\supset$, $\&$, $\vee$, $\neg$; 2) the quantifiers: $\forall$ and $\exists$; 3) parentheses: (,); 4) for each type, a denumerably infinite set of variables of this type; 5) for each type, an arbitrary set of constants of this type.

The terms of type t are the constants and variables of type t.

The formulae of $\mathbf{P}_1^m$, $1 \leqslant m \leqslant \omega$, are defined as follows: (I) if F is a term of type $\langle t_1, t_2, \ldots, t_n \rangle$ $(1 \leq n \leq \omega)$ and $T_1, T_2, \ldots, T_n$ are terms of types $t_1, t_2, \ldots, t_n$, respectively, then $F(T_1\ T_2\ \ldots\ T_n)$ is a formula; (II) if A and B are formulae, so are $(A \supset B)$, $(A \,\&\, B)$, $(A \vee B)$, $\neg A$; (III) if A is a formula and X is a variable, $(\forall X)A$ and $(\exists X)A$ are formulae; (IV) the only formulae are given by (I)-(III).

The notions of term free for another in a formula, sentence (formula without free variables), etc., are introduced as usual. $(A \equiv B)$ is an abbreviation for $(A \supset B) \,\&\, (B \supset A)$, and A^o for $\neg(A \,\&\, \neg A)$. We will omit parentheses, according to usual conventions.

The postulates (axioms schemata and primitive rules) of $\mathbf{P}_1^m$, $1 \leq m \leq \omega$, are the following, with the usual restrictions:

(1) $A \supset (B \supset A)$.

(2) $(A \supset B) \supset ((A \supset (B \supset C)) \supset (A \supset C))$.

(3) $A,\ A \supset B \,/\, B$.

(4) $A \,\&\, B \supset A$.

(5) $A \,\&\, B \supset B$.

(6) $A \supset (B \supset A \,\&\, B)$.

(7) $A \supset A \vee B$.

(8) $B \supset A \vee B$.

(9) $(A \supset C) \supset ((B \supset C) \supset (A \vee B \supset C))$.

(10) $A \vee \neg A$

(11) $\neg\neg A \supset A$

(12) $B^o \supset ((A \supset B) \supset (A \supset \neg B) \supset \neg A))$.

(13) $A^o \,\&\, B^o \supset (A \,\&\, B)^o$.

(14) $A^o \,\&\, B^o \supset (A \supset B)^o$.

(15) $A^o \,\&\, B^o \supset (A \vee B)^o$.

(16) $(\forall X)A(X) \supset A(T)$.

(17) $A(T) \supset (\exists X)A(X)$.

(18) $(\forall X)(A(X))^o \supset ((\forall X)\, A(X))^o$

(19) $(\forall X)(A(X))^{o} \supset ((\exists X)A(X))^{o}$.

(20) $A \equiv B$, *if either* A *and* B *are congruent formulae, or one differs from the other by elimination of vacuous quantifiers.*

(21) $C \supset A(X) / C \supset (\forall X)A(X)$.

(22) $A(X) \supset C / (\exists X)A(X) \supset C$.

(23) $(\exists F)(\forall X_1)\ldots(\forall X_n)(F(X_1 \ldots X_n) \equiv A(X_1, \ldots, X_n))$.

(24) $(\forall X_1)\ldots(\forall X_n)(F(X_1 \ldots X_n) \equiv G(X_1 \ldots X_n)) \supset F = G$.

(25) $X = Y \supset (A(X) \supset A(Y))$.

The symbol $\vdash$ (of deduction) and the usual syntactical notions are introduced as in classical higher-order predicate calculus; we also employ the usual definition for identity.

THEOREM 1. *All schemata and rules of classical positive logic of higher order are valid in* $\mathbf{P}_1^m$, $1 \leqslant m \leqslant \omega$.

DEFINITION 1. $\neg^* A =_{df} \neg A \ \& \ A^{o}$.

THEOREM 2. *The strong negation,* $\neg^*$, *has all properties of the classical one.*

REMARK. In general, several of the theorems of [2] and [4] are valid in $\mathbf{P}_1^m$, $1 \leqslant m \leqslant \omega$. It will be easier to demonstrate some of these results by applying the semantics which follows.

2. A SEMANTICS FOR THE SECOND ORDER CALCULUS $\mathbf{P}_1^2$.

DEFINITION 2. *Let* D *be a non empty set. A structure for* $\mathbf{P}_1^2$ *is a function* S, *such that:*

(1) $S(\iota) = D$;

(2) $S(\langle t_1, \ldots, t_n \rangle) \subseteq \mathbb{P}(S(t_1) \times \ldots \times S(t_n)) \times \mathbb{P}(S(t_1) \times \ldots \times S(t_n))$,

$1 \leq n < \omega$, t_i $(1 \leq i \leq n)$ *is a type, and provided that the ordered pairs* (α, β) *belonging to* $S(\langle t_1, \ldots, t_n \rangle)$ *are such that* $\alpha \cup \beta$ *is equal to* $S(t_1) \times \ldots \times S(t_n)$.

DEFINITION 3. *An interpretation for* $\mathbf{P}_1^2$, *in a structure* S, *is a function* $\mathcal{J}$ *such that:* (1) *to each constant of type* ι *of* $\mathbf{P}_1^2$, $\mathcal{J}$ *associates an element of* D; *and* (2) *to each constant of* $\mathbf{P}_1^2$, *of type* $\langle t_1, \ldots, t_n \rangle$, $1 \leq n < \omega$, $\mathcal{J}$ *associates an ordered pair of* $S(\langle t_1, \ldots, t_n \rangle)$.

DEFINITION 4. $S\mathbf{P}_1^2$ *denotes the laguage obtained by adding to* $\mathbf{P}_1^2$ *a new constant for each element of the sets which are the values of* S (see the analogous for first order logic in [5]).

DEFINITION 5. *A valuation for* $\mathbf{P}_1^2$, *based on an interpretation* $\mathcal{J}$, *is a function* ϑ *from the set of sentences of* $S\mathbf{P}_1^2$ *on* $\{0,1\}$ *such that* ($F(T_1 \ldots T_n)$ *is any atomic sentence,* A *and* B *are any sentences whatsoever of* $S\mathbf{P}_1^2$, *and the other notations have clear meanings*):

(1) $\vartheta(F(T_1 \ldots T_n) = 1 \iff \langle \mathcal{J}(T_1), \ldots, \mathcal{J}(T_n) \rangle \in \alpha$, *where* $\mathcal{J}(F) = (\alpha, \beta)$;

(2) $\vartheta(\neg F(T_1 \ldots T_n)) = 1 \iff \langle \mathcal{J}(T_1), \ldots, \mathcal{J}(T_n) \rangle \in \beta$, *where* $\mathcal{J}(F) = (\alpha, \beta)$;

(3) $\vartheta(A \supset B) = 1 \iff \vartheta(A) = 0$ *or* $\vartheta(B) = 1$;

(4) $\vartheta(A \,\&\, B) = 1 \iff \vartheta(A) = 0 = \vartheta(B) = 1$;

(5) $\vartheta(A \vee B) = 1 \iff \vartheta(A) = 1$ *or* $\vartheta(B) = 1$;

(6) $\vartheta(A) = 0 \Rightarrow \vartheta(\neg A) = 1$;

(7) $\vartheta(\neg\neg A) = 1 \Rightarrow \vartheta(A) = 1$;

(8) $\vartheta(B^o) = \vartheta(A \supset B) = \vartheta(A \supset \neg B) = 1 \Rightarrow \vartheta(A) = 0$;

(9) $\vartheta(A^o) = \vartheta(B^o) = 1 \Rightarrow \vartheta(A \supset B)^o = \vartheta(A \,\&\, B)^o = \vartheta(A \vee B)^o = 1$;

(10) $\vartheta((\forall X) A(X)) = 1 \iff \vartheta(A(C)) = 1$ *for each constant* C *of* $S\mathbf{P}_1^2$;

(11) $\vartheta((\exists X) A(X)) = 1 \iff \vartheta(A(C)) = 1$ *for a constant* C *of* $S\mathbf{P}_1^2$;

(12) $\vartheta(((\forall X)(A(X))^{o}) = 1 \Rightarrow \vartheta(((\forall X)A(X))^{o}) = \vartheta(((\exists X)A(X))^{o}) = 1$;

(13) *If A and B are congruent sentences, or one differs from the other by elimination of vacuous quantifiers, then $\vartheta(A) = \vartheta(B)$;*

(14) $\vartheta(C = C') = \vartheta(A(C)) = 1 \Rightarrow \vartheta(A(C') = 1$, *for C and C' in* $S\mathbf{P}^2_1$.

One defines the notions of principal and secondary interpretation, sound interpretation, valid formula, secondarily valid formula, etc., in analogy with classical calculus (see [3]).

One says that a valuation ϑ is a model for Γ (Γ is a set of sentences of $\mathbf{P}^2_1$) if $\vartheta(A) = 1$ for each sentence A belonging to Γ and ϑ is based on a sound interpretation. If the interpretation is a principal one, we call the model a principal model; otherwise, we say that the model is secondary. If one has $\vartheta(A) = 1$ for any valuation ϑ which is a model for Γ, we write $\Gamma \models A$ (in particular, $\models A$ if A is secondarily valid).

The definition of sound interpretation makes it trivially true that $\vdash A \Rightarrow \models A$ and that $\Gamma \vdash A \Rightarrow \Gamma \models A$. The soundness of $\mathbf{P}^2_1$ is thus proved.

Let Γ be a set of sentences of $\mathbf{P}^2_1$. One says that Γ is trivial, if for each sentence A of $\mathbf{P}^2_1$ $\Gamma \vdash A$; otherwise, it is non-trivial. Γ is inconsistent if there is a sentence A of $\mathbf{P}^2_1$ such that $\Gamma \vdash A$ and $\Gamma \vdash \neg A$; otherwise, Γ is consistent. One says that Γ is a Henkin set if, for each formula $A(X)$, where X is the only free variable, there exists a constant C_A (of $\mathbf{P}^2_1$) such that $\Gamma \vdash (\exists X)A(X) \supset A(C_A)$.

It is immediate to extend the notion of validity and the notion of semantic consequence to the set of formulae of $\mathbf{P}^2_1$.

THEOREM 3. $\Gamma \models A \Rightarrow \Gamma \vdash A$.

PROOF: In analogy with Henkin's proof for the correspondent weak completeness theorem of classical calculus, using the notion of maximal non trivial set, instead of maximal consistent set. It is necessary to remark that the characteristic function ϑ over the Henkin maximal non trivial set obtained, will be defined within an interpretation $\mathcal{J}$ which differs essentially from the classical one only with respect to the atomic formulae. That is: if F is a constant of type $\langle t_1, \ldots, t_n \rangle$, $1 \leqslant n < \omega$, where t_i, $1 \leqslant i \leqslant n$, is of type ι, $\mathcal{J}$ associates to F the ordered pair $(\alpha, \beta) \in S(\langle t_1, \ldots, t_n \rangle)$, $1 \leqslant n < \omega$, such that α is the set of n-uples $(A_1, \ldots, A_n)$, $(B_1, \ldots, B_n)$, ... defined on D such that $F(A_1 \ldots A_n)$, $F(B_1 \ldots B_n)$ ... belong to Γ, and $\mathcal{J}$ associates to β the set of n-uples $(A'_1, \ldots, A'_n)$, $(B'_1, \ldots, B'_n)$, ... defined on D such that $\neg F(A'_1 \ldots A'_n)$, $\neg F(B'_1 \ldots B'_n)$, ... belong to Γ. ■

3. Aplications of the semantics.

As an application of the semantics, it is easy to prove the following:

THEOREM 4. *The following schemata are not valid in* $\mathbf{P}_1^2$:

(1) $A \,\&\, \neg A \supset B$;

(2) $A \supset (\neg A \supset B)$;

(3) $\neg((\forall X)\, F(X) \,\&\, \neg(\forall X)\, F(X))$;

(4) $(\exists X)\,(\exists F)\,(F(X) \,\&\, \neg F(X)) \supset B$

(5) $(\forall F)\,((\exists X) \neg (F(X))^o \supset ((\forall X)\, F(X) \,\&\, (\;X) \neg F(X)))$.

REMARK. One can evidently build higher-order calculi based on the calculi C_n, $1 \leqslant n < \omega$. The above semantics can be generalized to these calculi.

4. The arithmetic in $\mathbf{P}_1^\omega$.

In analogy with the presentation in [6], it is possible to develop in $\mathbf{P}_1^\omega$ an arithmetic which preserves the basic characteristics of the one founded on the classical calculus. Introducing the Russelian contextual definition of the descriptor (the symbol ι) it is easy to define the Peano's primitive concepts and prove Peano's Axioms, as follows (we shall not make explicit the types and other restrictions to which are subjected the other notations employed):

DEFINITION 6.

$$F(\iota X\, A(X)) =_{df} (\exists X)(F(X) \;\&\; (\forall Y)(A(Y) \equiv X = Y)).$$

DEFINITION 7.

$$\hat{X}_1 \ldots \hat{X}_n A =_{df} \iota P((\forall X_1) \ldots (\forall X_n)(P(X_1 \ldots X_n) \equiv A)).$$

DEFINITION 8.

$$0 =_{df} \iota F\, \neg^*(\exists X) F(X).$$

DEFINITION 9.

$$n' =_{df} \hat{F}((\exists G)(n(G) \;\&\; (\exists H)(\neg^* G(H) \;\&\; (\forall X)(F(X) \equiv (G(X) \vee X = H))))).$$

DEFINITION 10.

$$N =_{df} \hat{n}((\forall F)((F(0) \;\&\; (\forall m)(F(m) \supset F(m'))) \supset F(n)).$$

THEOREM 5. $\vdash N(0)$.

THEOREM 6. $\vdash (\forall n)(N(n) \supset N(n'))$.

THEOREM 7. $\vdash (\forall n)(N(n) \supset \neg^*(0 = n'))$.

THEOREM 8. $\vdash (\forall F)((F(0) \;\&\; (\forall n)(N(n) \supset (F(n) \supset F(n')))) \supset (\forall n)(N(n) \supset F(n)))$.

These theorems are the postulates 1, 2, 4 and 5 of Peano's Arithmetic. As in the classical calculus, Peano's postulate 3, as a form of the Axiom of Infinity, cannot be proved. We will adopt the Axiom of Infinity essentially in the form used in Principia Matematica (II, 203, *120.03):

(26) $(\forall F)(N(F) \supset (\exists G) F(G))$.

The results of this Note will be developed and published elsewhere.

REFERENCES.

[1] E. H. Alves, Lógica e Inconsistência: um estudo dos cálculos $\mathcal{C}_n$, $1 \leqslant n \leqslant \omega$, Master Thesis, University of São Paulo, 1976.

[2] A. I. Arruda and N. C. A. da Costa, *Une sémantique pour le calcul* $\mathcal{C}_1^=$, C. R. Acad. Sc. Paris, 284 A (1977), 279-282.

[3] A. Church, Introduction to Mathematical Logic, I, Princeton University Press, 1956.

[4] N. C. A. da Costa, *On the theory of inconsistent formal systems*, Notre Dame Journal of Formal Logic, vol. XV, nº 4 (1974), 497-510.

[5] J. R. Shoenfield, Mathematical Logic, Addison-Wesley, 1967.

[6] A. N. Whitehead and B. Russell, Principia Mathematica, volumes I, II and III, Cambridge University Press, 1950.

Centro de Lógica e Epistemologia
Universidade Estadual de Campinas
13.100 Campinas, S.P., Brazil

and

Departamento de Filosofia
Universidade Federal do Rio Grande
do Norte
59.000 Natal, R.N., Brazil.

SOME REMARKS ON GRISS' LOGIC OF NEGATIONLESS INTUITIONISTIC MATHEMATICS.

by *AYDA I. ARRUDA*.

ABSTRACT. This is a partially expository paper on Griss' logic of negationless intuitionistic mathematics, in which we develop some of our results on the subject, announced in *On Griss' propositional calculus*, The Journal of Symbolic Logic, 36, nº 3 (1971), p.579.

After an introduction on Griss' concepts about negantionless intuitionistic mathematics, we develop the Griss propositional calculus, $\mathcal{G}$, and discuss his logic of species. Then, we study a modification of $\mathcal{G}$, $\mathcal{G}_\vartheta$, proposed by Vredenduin, and present an algebrization of $\mathcal{G}$ (by Lindenbaum algebra associated to $\mathcal{G}$) which is the dual of the Griss' albegra as defined by Imai and Isēki in *On Griss algebra, I*, Proceedings of the Japan Academy, vol. 42, nº 3 (1966), pp. 213-216.

1. INTRODUCTION.

In 1944 Griss made the first attempt to construct a negationless intuitionistic mathematics [4] which was afterwards developed by him in a more complete and systematic way (see [5]-[7], [10] and [11]). After this, many people tried to obtain a logic for negationless intuitionistic mathematics; in this paper we restrict ourselves to the Griss' logic of negationless intuitionistic mathematics [9].

To explain the main ideas that brought Griss to the construction of a negationless intuitionistic mathematics, and his version of a corresponding logic, we quote some parts of his papers.

"In 1947 Prof. L. E. J. Brouwer gave a formulation of the directives of intuitionistic mathematics (*Richtlijnen der in-*

tuïtionistische Wiskunde, Proc. Kon. Ned. Akad. v. Wetensch., 50, 1947). It is remarkable that negation does not occur in an explicit way, so one might be inclined to believe negationless mathematics to be a consequence of this formulation. The notion of species, however, is introduced in this way (translated from the Dutch text): 'Finally in this construction of mathematics at any stage properties that can be supposed to hold for mathematical conceivabilities already obtained are allowed to be added as new mathematical conceivabilities under the name of species'. By this formulation it is possible that there are properties that can be supposed to hold for mathematical conceivabilities but are not known to be true. With it, negation and null-species are introduced simultaneously but at the cost of evidence. What are the properties that can be supposed? What other criterium could there be than 'to hold for mathematical conceivabilities already obtained'? In the definition of the notion of species the words 'can be supposed' should be replaced by 'are known'. *One should restrict oneself in intuitionistic mathematics to mathematical conceivabilities and properties of those mathematical conceivabilities, and one should not make suppositions of which one does not know whether it is possible to fulfil them.* (The well-known turn in mathematics: "suppose *ABC* to be retangular" seems to be a supposition, but mostly means: "Consider a retangular triangle *ABC*".)"(cf. [7], p. 108.)

"A property of natural numbers defines the species of natural numbers having this property. The species may consist of a single element a, e.g., the species of natural numbers identical to a. From this definition it follows that the species contain at least one element". ([7], p. 110.)

"... In this way, the *disjunction* is defined in a particular case. It is evident the disjunction a or b in the usual meaning (the assertion a is true or the assertion b is true) does not occur in negationless mathematics, because there is no question of assertions that are not true. In general, our definition of disjunction runs as follows: *a or b is true for all elements of the set V means that the property a holds for a subspecies V' and the property b holds for a subspecies V'', V being the sum of V' and V''.*" ([7], p. 109.)

"In propositional logic p represents a proposition that is either true or false. In intuitionistic mathematics it is not certain whether p is true or false. In negationless intuitionistic mathematics false propositions do not occur. The *implication* $p \to q$ will have its natural meaning: q follows from p, p is true, so q is also true. The *conjunction* p & q retains its ordinary meaning". ([9], p. 41.)

The logic of negationless intuitionistic mathematics "will differ much from classic logic and even from Heyting's logic. If, for example, from propositional logic (of classic or intuitionistic mathematics) only the negation is omitted, this does not give us the propositional logic of negationless mathematics, because in the latter the classic interpretation of the formula $p \to p \vee q$ is rejected". ([9], p. 41.)

By these words we see that in Griss' propositional calculus the only primitive connectives are $\to$ (*implication*) and & (*conjunction*). The *disjunction* may appear in the predicate calculus (as in Vredenduin's [14]), or in a logic of species, but in this last case interpreted as *union*. The negation as well as empty species are rejected, but negation reappears in Griss' logic of species interpreted as *complement* and submitted to some restrictions.

All Griss' results in [9], except for the last section, are based on Heyting's propositional calculus, $\mathcal{H}$; so it will be worthwhile to remember the axiomatics of $\mathcal{H}$: if p, q, r,... are propositional variables, then the axioms of $\mathcal{H}$ are the following:

1. $p \to (p.p)$.
2. $(p.q) \to (q.p)$.
3. $(p \to q) \to ((p.r) \to (q.r))$.
4. $((p \to q).(q \to r)) \to (p \to r)$.
5. $q \to (p \to q)$.
6. $(p.(p \to q)) \to q$.
7. $p \to (p \vee q)$.
8. $(p \vee q) \to (q \vee p)$.
9. $((p \to r).(q \to r)) \to ((p \vee q) \to r)$.
10. $\sim p \to (p \to q)$.
11. $((p \to q).(p \to \sim q)) \to \sim p$.

The deduction rules are *substitution* and *modus ponens*.

It deserves to be remarked that axioms 10 and 11 can be replaced, respectively, by 10' and 11':

10' $(p \vee q).(\sim p \vee q)) \to q$. 11' $((p.q) \to \sim r) \to ((p.r) \to \sim q)$.

2. THE PROPOSITIONAL CALCULUS $\mathcal{G}$.

To obtain the axiomatics for $\mathcal{G}$ Griss considers only the deduction rules and axioms 1-6 of $\mathcal{H}$, but remarking that: "The axioms we accept must represent reasonings we actually use in mathematics. The axiom $q \to (p \to q)$ does not represent such a reasoning. We replace it by $p \,\&\, q \to p$. As in $p \to q$, q never represents a false proposition, we can omit the axiom $p \,\&\, (p \to q) \to q$." ([9], p. 42.) Nonetheless, Griss' justification for the elimination of axiom 6 of $\mathcal{H}$, is not in agreement with his interpretation of implication. Anyhow, this axiom may be dispensed with since we have the rule of modus ponens which can be used instead of it.

Now we will formulate the axiomatics of $\mathcal{G}$, using syntactical variables for formulas instead of propositional variables (as Griss does in [9]):

A1. $A \to A \,\&\, A$.

A2. $A \,\&\, B \to B \,\&\, A$.

A3. $A \,\&\, B \to A$

A4. $(A \to B) \to (A \,\&\, C \to B \,\&\, C)$.

A5. $(A \to B) \,\&\, (B \to C) \to (A \to C)$.

R1. $\dfrac{A \quad A \to B}{B}$

R2. $\dfrac{C \quad A \to B}{A \to B \,\&\, C}$

In the original formulation of Griss there is also the rule $A, B / A \,\&\, B$, but it is not necessary because it is derivable from the axioms and rules mentioned above; evidently the rule of substitution is not necessary.

DEFINITION 2.1. $A \leftrightarrow B =_{def} (A \to B) \,\&\, (B \to A)$.

THEOREM 2.1. *In $\mathcal{G}$ we prove, among others, the following schemata and rules:*

1. $A \leftrightarrow A$.
2. $(A \leftrightarrow B) \,\&\, (B \leftrightarrow C) \to (A \leftrightarrow C)$.
3. $(A \leftrightarrow B) \to (B \leftrightarrow A)$
4. $(A \to B) \,\&\, (C \to D) \to (A \,\&\, C \to B \,\&\, D)$.
5. $A \,\&\, B \to B$.
6. $(A \to B) \,\&\, (A \to C) \to (A \to B \,\&\, C)$.
7. $(A \to B) \to (C \,\&\, A \to C \,\&\, B)$.
8. $(A \to B \,\&\, C) \to (A \to B) \,\&\, (A \to C)$.
9. $(A \to B) \to (A \to A \,\&\, B)$.
10. $(A \to B) \,\&\, (A \,\&\, B \to C) \to (A \to C)$.

11. $(A \leftrightarrow B) \rightarrow (C \,\&\, A \leftrightarrow C \,\&\, B)$.

12. $(A \leftrightarrow B) \rightarrow (A \,\&\, C \leftrightarrow B \,\&\, C)$.

13. $(A \,\&\, B) \,\&\, C \leftrightarrow A \,\&\, (B \,\&\, C)$.

14. $A \leftrightarrow B \vdash (A \rightarrow C) \leftrightarrow (B \rightarrow C)$.

15. $A \,\&\, B \vdash A \leftrightarrow B$.

16. $A \leftrightarrow B \vdash (C \rightarrow A) \leftrightarrow (C \rightarrow B)$.

17. $A \vdash B \rightarrow A$.

18. $A, B \vdash A \,\&\, B$.

THEOREM 2.2. *In $\mathcal{G}$ the rules* R2 *and* R2', $A / B \rightarrow A$, *are equivalent.*

PROOF: By Theorem 2.1, item 17, we see that R2' can be obtained from **R2**. On the other hand, from C and $A \rightarrow B$ we obtain $A \rightarrow C$ and $A \rightarrow B$; so, by Axioms A4, A1, A5, and rule R1, we have, $A \rightarrow C \,\&\, A$ and $A \,\&\, C \rightarrow B \,\&\, C$, consequently, by Axioms A2, A5 and rule R1, we obtain $A \rightarrow B \,\&\, C$. ∎

COROLLARY 2.1. *The system $\mathcal{G}$ and the system $\mathcal{G}'$, obtained from $\mathcal{G}$ substituting* R2' *for* R2, *are equivalent.*

THEOREM 2.3. *The following schemata and rules are not valid, in general, in $\mathcal{G}$:*

1. $A \rightarrow (B \rightarrow A)$.

2. $(A \rightarrow B) \rightarrow ((A \rightarrow C) \rightarrow (B \rightarrow C))$.

3. $((A \rightarrow B) \rightarrow A) \rightarrow A$.

4. $(A \rightarrow B) \rightarrow ((C \rightarrow A) \rightarrow (C \rightarrow B))$.

5. $A \,\&\, B \rightarrow C \vdash A \rightarrow (B \rightarrow C)$.

6. $(A \rightarrow B) \rightarrow ((A \rightarrow (B \rightarrow C)) \rightarrow (A \rightarrow C))$.

7. $A \,\&\, B \rightarrow (A \rightarrow B)$.

8. $(A \rightarrow B) \rightarrow ((B \rightarrow C) \rightarrow (A \rightarrow C))$.

9. $A \rightarrow (B \rightarrow A \,\&\, B)$.

10. $(A \,\&\, B \rightarrow C) \rightarrow (A \rightarrow (B \rightarrow C))$.

11. $A \,\&\, (A \rightarrow B) \rightarrow B$.

12. $(A \rightarrow (A \rightarrow B)) \rightarrow (A \rightarrow B)$.

13. $A \rightarrow (A \rightarrow B) \vdash A \rightarrow B$.

14. $A \rightarrow (B \rightarrow C) \vdash A \,\&\, B \rightarrow C$.

15. $(A \rightarrow (B \rightarrow C)) \rightarrow (A \,\&\, B \rightarrow C)$.

PROOF: For 1-10, use the matrix $M_1 = \langle N, I, \dot{\rightarrow}, \dot{\&} \rangle$, where N is the set of natural numbers different of zero, I is the set of odd natural numbers, and the functions $\dot{\rightarrow}$ and $\dot{\&}$ and the valuation ϑ (i. e., a function from the set of formulas of $\mathcal{G}$ in N) are defined as follows (the other notations have clear meanings) :

$\vartheta(A \rightarrow B) = \vartheta(A) \dot{\rightarrow} \vartheta(B) = \vartheta(B) + 2$ if $\vartheta(A) \in I$ and $\vartheta(B) \in N - I$, or

$\vartheta(A) \in N - I$ and is less than $\vartheta(B) \in N - I$;

$= 1$ in the remaining cases.

$\vartheta(A \,\&\, B) = \vartheta(A) \;\dot{\&}\; \vartheta(B) = ndv(\vartheta(A), \vartheta(B))$, if only one between $\vartheta(A)$ and $\vartheta(B)$ belongs to I;

$= Max(\vartheta(A), \vartheta(B))$, in the remaining cases.

For 11-15, use the matrix $M_2 = \langle N, I, \dot{\rightarrow}, \dot{\&} \rangle$, whose difference from M_1 is only in the definition of $\dot{\rightarrow}$, which is now the following:

$\vartheta(A \rightarrow B) = \vartheta(A) \;\dot{\rightarrow}\; \vartheta(B) = 2$ if $\vartheta(A) \in I$ and $\vartheta(B) \in N - I$, or $\vartheta(A) \in N - I$ and is less than $\vartheta(B) \in N - I$;

$= 1$ in the remaining cases. ■

DEFINITION 2.2. We say that a formal system is $\rightarrow$-*finitely trivializable* if there is a formula F such that $\vdash F \rightarrow A$ for any formula A; and the system is said to be $\vdash$-*finitely trivializable* if there is a formula F such that $F \vdash A$ for any formula A.

THEOREM 2.4. *$\mathcal{G}$ is not* $\vdash$-*finitely trivializable.*

PROOF: We proceed as in [1], by constructing a system $\mathcal{G}_{\vdash}$ (with a *formalized deduction*) whose axiom schemata are those of $\mathcal{G}$ preceded by the symbol $\vdash$, plus Gentzen's structural postulates; and the deduction rules are those of $\mathcal{G}$ with the symbol $\vdash$ preceding each premiss and conclusion. Let $\underline{\vdash}$ be the deduction symbol in $\mathcal{G}_{\vdash}$, then we easily prove that $\Gamma \vdash A$ in $\mathcal{G}$ if and only if $\underline{\vdash}\; \Gamma \vdash A$ in $\mathcal{G}_{\vdash}$. By the matrix $M_3 = \langle N, I, \dot{\rightarrow}, \dot{\&}, \dot{\vdash}, \dot{;} \rangle$, where N, I, $\dot{\rightarrow}$, and $\dot{\&}$ are defined as in M_1 and $\dot{\vdash}$ and $\dot{;}$ are defined, respectively, as $\dot{\rightarrow}$ and $\dot{\&}$, we prove that $F \vdash A$ (F being a fixed formula) is not valid in $\mathcal{G}$ because $\underline{\vdash}\; F \vdash A$ is not valid in $\mathcal{G}_{\vdash}$. ■

COROLLARY 2.2. *$\mathcal{G}$ is not* $\rightarrow$-*finitely trivializable.*

PROOF: If there is in $\mathcal{G}$ a fixed formula F such that $\vdash F \rightarrow A$, for any formula A, then $F \vdash A$ will be also a theorem of $\mathcal{G}$, which is not possible by the Theorem. A direct proof of this corollary may be obtained by the matrix M_3. ■

COROLLARY 2.3. *The deduction theorem is not valid in $\mathcal{G}$.*

PROOF: Similar to the proof of the Theorem, but defining the function $\dot{\vdash}$ of M_3 as $\vartheta(A) \dot{\vdash} \vartheta(B) = \vartheta(A \vdash B) = 2$ if $\vartheta(A) \in I$ and $\vartheta(B) \in N - I$, and $\vartheta(A) \dot{\vdash} \vartheta(B) = \vartheta(A \vdash B) = 1$ in the remaining cases. ■

To simplify the proof of the next theorem let us define $A \underset{p}{\rightarrow} (A \rightarrow B)$ as

$$A \underset{0}{\rightarrow} (A \rightarrow B) =_{\text{def}} A \rightarrow B,$$

$$A \underset{n+1}{\rightarrow} (A \rightarrow B) =_{\text{def}} A \rightarrow (A \underset{n}{\rightarrow} B).$$

Now, let us consider the matrix $M_4 = \langle N, I, \dot{\rightarrow}, \dot{\&} \rangle$, where N, I and $\dot{\&}$ are defined as in M_1, and $\dot{\rightarrow}$ as follows:

$\vartheta(A \rightarrow B) = \vartheta(A) \dot{\rightarrow} \vartheta(B) = \vartheta(B) - 2$ if $\vartheta(A) \in N - I$ and is less than $\vartheta(B) \in N - I$, or
$\vartheta(A) \in I$, and $\vartheta(B) \in N - I$ and greater than 2;

$= 2$ if $\vartheta(A) \in I$ and $\vartheta(B) = 2$;

$= 1$ in the remaining cases.

LEMMA 2.1. *The schema* $(A \underset{q}{\rightarrow} (A \rightarrow B)) \rightarrow (A \underset{p}{\rightarrow} (A \rightarrow B))$, $q > p$, *is not valid in $\mathcal{G}$.*

PROOF: By M_4, taking, for example, $\vartheta(A) \in I$ and $\vartheta(B) = 2n$ (i. e., $\vartheta(B) \in N - I$), with $2n > 2q$. To exemplify, we write in the first line bellow the formulas, and in the second the respective values:

$$A \rightarrow B, \quad A \underset{1}{\rightarrow} (A \rightarrow B), \quad A \underset{2}{\rightarrow} (A \rightarrow B), \quad \ldots, \quad A \underset{p}{\rightarrow} (A \rightarrow B), \quad \ldots, \quad A \underset{q}{\rightarrow} (A \rightarrow B)$$

$$2n - 2, \quad 2n - 4, \quad 2n - 6, \quad \ldots, \quad 2n - (2p + 2), \quad \ldots, \quad 2n - (2q + 2).$$

As we can see, the values of the formulas decrease when the number of $\rightarrow$ in them increases, and so the Lemma is proved. ■

THEOREM 2.5. *$\mathcal{G}$ is not decidable by finite matrices.*

PROOF: If $\mathcal{G}$ were decidable by a finite matrix M, then we could find m and n, $m > n$, such that, for each valuation ϑ of the atomic components we have:

$\vartheta(A \underset{m}{\rightarrow} (A \rightarrow B)) = \vartheta(A \underset{n}{\rightarrow} (A \rightarrow B))$. Then we would have $\vdash (A \underset{m}{\rightarrow} (A \rightarrow B)) \rightarrow (A \underset{n}{\rightarrow} (A \rightarrow B))$, which is impossible by the above Lemma. ■

3. GRISS' LOGIC OF SPECIES $\mathcal{E}$.

After constructing the system $\mathcal{G}$, Griss used it to construct a logic of species, which we name $\mathcal{E}$.

The main conceptions of Griss for the construction of such a logic of species are the following (quoted from [9]):

"We start from a species u (i. e., the set of natural numbers, the continuum, a space) all other species being subspecies of u. If a and b represent two such species, $a \cup b$ is their *union*, if a and b have an element in common, $a \cap b$ is their *intersection*; $a \subset b$ means that a is a *subspecies* of b."

"In negationless intuitionistic mathematics the notion of *distinguishability* is equally fundamental as the notion of *identity*. ...we shall suppose that u contains at least two distinguishable elements. Then we can define: a *proper* subspecies of u is a subspecies so that at least one element of u is distinguishable from all elements of a. If a is a proper subspecies of u, then the *complementary species* (*complement*) $\neg a$ is the species of all elements that are distinguishable from the elements of a. Each element of a is distinguishable from each element of $\neg a$, a and $\neg a$ are disjoint. $a \neq u$, in words, 'a is a proper subspecies of u' is the condition that is necessary to form the complement $\neg a$. There is also a condition for the existence of an intersection $a \cap b$, a so called *touch condition*, $a \times b$ expressing that a common element of a and b can be indicated.The appearence of these two conditions, $a \times b$ and $a \neq b$, is essential in negationless mathematics. It results from the *rejection of empty species*."

After these remarks, Griss gives the specific postulates of his logic of species, $\mathcal{E}$, without specifying (except for the postulate $\cap 4$ below) the conditions about the symbols $\times$ and $\neq$ that are necessary in the postulates where the symbols $\cap$ and $\neg$ occur. And in the same way he gives a list of theorems.

The language of $\mathcal{E}$ is composed of: 1) a constant u (the *universal species*); 2) variables for species (*subspecies of* u); 3) the operators: $\cap$ (*intersection*), $\cup$ (*union*), $\subset$ (*subspecies*), $\neq$ (*distinguishability*), χ (*existence of common elements*), and $\neg$ (*complement*); 4) the connectives: $\rightarrow$ (*implication*), and & (*conjunction*).

The notions of *term* and *formula* are defined as usual, but we need to observe that some terms and formulas are not realizable.

The notions of proof, and deduction are the usual ones. The fact that two species a and b are *disjoint* is expressed by one of the two equivalent formulas: $a \subset \neg b$, or $b \subset \neg a$. Consequently, for two species to be disjoint it is necessary that $a \neq u$ and $b \neq u$.

A substitution of $a \cap b$ or $\neg a$ for c in a term or formula, should be intuitively permitted if and only if we have proved that $a \chi b$ or $a \neq b$. But if we add such a rule of substitution to the axiomatics of $\mathcal{E}$ (as Vredenduin mentioned in [14], p. 260), it will be almost useless, because we cannot prove in general if $a \chi b$ or $a \neq b$ are theorems or not. Nevertheless, by the following formulation of the axiomatics of $\mathcal{E}$, we can use a substitution theorem for syntactical variables for species, if we add to the axioms (or theorems) the appropriate conditions for the symbols χ and $\neq$.

The postulates of $\mathcal{E}$ are those of $\mathcal{G}$ plus the following:

χ1. $a \chi a$.

χ2. $a \chi b \rightarrow b \chi a$.

χ3. $a \chi b \ \& \ b \subset c \rightarrow b \chi c$.

$\neq$1. $a \neq u \ \& \ b \subset a \rightarrow b \neq u$.

$\neq$2. $a \neq u \rightarrow \neg a \neq u$.

$\subset$1. $a \subset b \ \& \ b \subset c \rightarrow a \subset c$.

$\subset$2. $a \subset u$.

$\cap$1. $a \subset a \cap a$.

$\cap$2. $a \chi b \rightarrow a \cap b \subset b \cap a$.

$\cap$3. $a \chi b \rightarrow a \cap b \subset a$.

$\cap$4. $a \chi c \ \& \ a \subset b \rightarrow a \cap c \subset b \cap c$.

$\cup$1. $a \subset a \cup b$.

$\cup$2. $a \cup b \subset b \cup a$.

$\cup$3. $a \subset c \;\&\; b \subset c \rightarrow a \cup b \subset c$.

$\cup$4. $a \times b \;\&\; b \times c \rightarrow (a \cup b) \cap c \subset (a \cap c) \cup (b \cap c)$.

$\urcorner$1. $a \neq u \rightarrow (a \cup b) \cap (\urcorner a \cup b) \subset b$.

$\urcorner$2. $b \neq u \;\&\; c \neq u \;\&\; a \times b \;\&\; a \times c \;\&\; a \cap b \subset \urcorner c \rightarrow a \cap c \subset \urcorner b$.

DEFINITION 3.1. $a = b =_{\text{def}} (a \subset b) \;\&\; (b \subset a)$.

The only interpretation compatible with the above axioms seems to be the following: The theorems of $\mathcal{E}$ are the formulas which are realizable (true) in any interpretation, or those whose antecedent cannot be realizable.

The axiom schema $\neq 2$ does not occur in the original formulation of Griss for the axiomatics of $\mathcal{E}$, but we consider it not only because it is in accordance with Griss' ideas, but because it simplifies the formulation of many theorems.

Griss did not mention that it is necessary to take conditions on the symbols $\times$ and $\neq$ in the axiom schemata $\cap 3$, $\urcorner 1$, and $\urcorner 2$. Nonetheless, it is clear that these conditions are necessary, and their formulation is evident. In the axiom schema $\cap 4$ we have added only the condition $a \times c$ because $b \times c$ is a consequence of $a \times c$ and $b \subset c$ (see axiom schema $\times 3$).

Griss wanted his logic of species to be an "interpretation" of Heyting's propositional calculus. The only way to construe this "interpretation" is the following. Erase all axioms about $\times$ and $\neq$, and all the conditions about these symbols in the remaining axiom schemata; then, substitute $\rightarrow$, &, $\vee$, $\sim$ respectively for $\subset$, $\cap$, $\cup$ and $\urcorner$. In this way the axiom achemata of $\mathcal{E}$ are transformed into axiom schemata or theorems of $\mathcal{H}$. It is surprising that the transforms of the axiom schemata $\urcorner 1$ and $\urcorner 2$ of $\mathcal{E}$ are respectively equivalent to the axiom schemata 10 and 11 of $\mathcal{H}$ (i. e., the axiom schemata of negation in $\mathcal{H}$), as pointed out in the Introduction.

Now we list some of the theorems mentioned by Griss, specifying the conditions about the symbols $\times$ and $\neq$ (the theorems with a * are not mentioned by Griss).

1.1. $a \subset b \rightarrow a \times b$.

1.2. $a \subset b \ \& \ a \subset c \rightarrow b \times c$.

1.3. $a \times b \rightarrow a \cap b \times b$.

1.4. $b \times c \ \& \ a \times b \cap c \rightarrow a \times b \ \& \ c \times a \cap b$.

1.5. $a \times b \rightarrow a \times b \cup c$.

1.6 $a \times u$.

2.1. $a \cup b \neq u \rightarrow a \neq u \ \& \ b \neq u$.

2.2. $a \times b \ \& \ a \neq u \rightarrow a \cap b \neq u$.

2.3*. $a \cup b \neq u \rightarrow \neg a \times \neg b$.

3.1. $a \subset a$.

3.2. $a \cap a = a$.

3.3. $a \times c \ \& \ a \subset b \rightarrow c \cap a \subset c \cap b$.

3.4. $a \times c \ \& \ b \times d \ \& \ a \subset b \ \& \ c \subset d \rightarrow a \cap c \subset b \cap d$.

3.5. $a \times b \rightarrow a \cap b \subset b$.

3.6. $a \subset b \ \& \ a \subset c \rightarrow a \subset b \cap c$.

3.7. $b \times c \ \& \ a \subset b \cap c \rightarrow a \subset b \ \& \ a \subset c$.

3.8. $a \times b \ \& \ a \cap b \times c \rightarrow (a \cap b) \cap c \subset a \cap (b \cap c)$.

3.9. $a \times b \ \& \ a \subset c \rightarrow a \cap b \subset c$.

4.1. $(a \cup b) \cup c \subset a \cup (b \cup c)$.

4.2. $a \cup a = a$

4.3. $a \subset b \ \& \ c \subset d \rightarrow a \cup c \subset b \cup d$.

4.4. $a \times c \ \& \ a \subset b \rightarrow (a \cap c) \subset (b \cup d)$.

4.5. $a \subset b \leftrightarrow a \cup b = b$.

4.6. $a \times c \ \& \ b \times c \rightarrow (a \cap c) \cup (b \cap c) \subset (a \cup b) \cap c$.

4.7. $a \times b \rightarrow (a \cap b) \cup c \subset (a \cup c) \cap (b \cup c)$.

4.8. $a \subset b \cup c \ \& \ b \subset d \ \& \ c \subset e \rightarrow a \subset d \cup e$.

4.9. $a \cup u = u \ \& \ a \cap u = a$.

4.10*. $b = (a \cup b) \cap b$.

4.11*. $a \times b \to (a \cap b) \cup b = b.$

5.1. $a \neq u \;\&\; b \neq u \;\&\; a \subset \neg b \to b \subset \neg a.$

5.2. $a \neq u \to a \subset \neg\neg a.$

5.3. $a \neq u \;\&\; b \neq u \;\&\; a \subset b \to \neg b \subset \neg a.$

5.4. $a \neq u \;\&\; b \neq u \;\&\; a \subset b \to \neg\neg a \subset \neg\neg b.$

5.5. $a \neq u \to \neg a = \neg\neg\neg a.$

5.6. $b \neq u \;\&\; c \neq u \;\&\; a \times b \;\&\; a \times \neg c \;\&\; a \cap b \subset c \to a \cap \neg c \subset \neg b.$

5.7. $a \neq u \;\&\; a \times b \to (\neg a \cup b) \cap a \subset b \cap a.$

5.8. $a \neq u \;\&\; b \times \neg a \to (a \cup b) \cap \neg a \subset b.$

5.9. $c \neq u \;\&\; b \times c \;\&\; a \subset \neg c \to (a \cup b) \cap c \subset b \cap c.$

5.10. $b \neq u \;\&\; a \subset \neg b \to (a \cup c) \cap (b \cup c) \subset c.$

5.11. $a \neq u \;\&\; a \subset b \cup \neg a \to a \subset b.$

5.12. $a \cup b \neq u \to \neg(a \cup b) \subset \neg a \cap \neg b.$

5.13. $a \cup b \neq u \;\&\; \neg a \times b \to \neg a \cap \neg b \subset \neg(a \cup b).$

5.14. $a \neq u \;\&\; b \neq u \;\&\; a \times b \;\&\; b \cap a \subset \neg a \to b \subset \neg a.$

5.15*. $a \neq u \;\&\; a \times b \;\&\; b \cap a \subset \neg a \to a \times \neg a.$

5.16*. $a \neq u \;\&\; a \times b \;\&\; b \subset \neg a \to b \cap a \subset \neg a.$

6.1. $a = a.$

6.2. $a = b \to b = a.$

6.3. $a = b \;\&\; b = c \to a = c.$

6.4*. $a \neq u \;\&\; b \neq u \;\&\; a = b \to \neg a = \neg b.$

6.5*. $a \times c \;\&\; b \times c \;\&\; a = b \to a \cap c = b \cap c.$

6.6*. $a \times c \;\&\; b \times c \;\&\; a = b \to c \cap a = c \cap b.$

6.7*. $a = b \to a \cup c = b \cup c.$

6.8*. $a = b \to c \cup a = c \cup b.$

Now we prove some of the above theorems:

1.4. $b \chi c \ \& \ a \chi b \cap c \to b \cap c \subset b$ by $\cap 3$.
$\to b \cap c \chi a \ \& \ b \cap c \subset b$
$\to a \chi b$ by $\chi 3$. (I)

$b \chi c \ \& \ a \chi b \cap c \to a \cap (b \cap c) \subset c$ by $\cap 3$.
$\to a \cap (b \cap c) \chi c$ by 1.1. (II)

$b \chi c \ \& \ a \chi b \cap c \to a \chi b$ by (I).
$\to a \cap (b \cap c) \subset b$ by $\cap 3$.
$\to a \chi b \ \& \ a \cap (b \cap c) \subset b$
$\to a \cap (b \cap c) \subset a \cap b$ by $\cap 3$. (III)

$b \chi c \ \& \ a \chi b \cap c \to a \cap (b \cap c) \chi c \ \& \ a \cap (b \cap c) \subset a \cap b$ by $\cap 3$, 1.3.
$\to c \chi a \cap b$. (IV)

Now, the theorem follows from (I) and (IV). ■

3.9. $a \chi b \ \& \ a \subset c \to a \cap b \subset c \cap b$ by $\cap 4$.
$\to b \chi c$ by $\chi 3$.
$\to c \cap b \subset c$ by $\cap 3$.
$\to a \cap b \subset c$ by $\subset 1$. ■

5.1. $b \neq u \ \& \ a \subset \neg b \to a \cup b \chi a \ \& \ a \cup b \chi b$ by $\cup 1$ and $\chi 3$.
$\to \neg b \neq u \ \& \ a \subset \neg b$. by $\neg 2$.
$\to a \neq u$ by $\neq 1$.
$\to (a \cup b) \cap a \subset a$ by 3.5.
$\to (a \cup b) \cap a \subset \neg b$ by $\subset 1$.
$\to (a \cup b) \cap b \subset \neg a$ by $\neg 1$.
$\to b \subset (a \cup b) \cap b$ by 3.6.
$\to b \subset \neg a$ by $\subset 1$. ■

5.2. $a \neq u \to a \subset \neg\neg a$. By 4.10*, substituting $\neg a$ for b, and 5.1, and $\neg 1$. ■

5.3. $a \neq u \ \& \ b \neq u \ \& \ a \subset b \to \neg b \subset \neg a$. By $\subset 1$, 5.2, and 5.1. ■

2.3*. $a \cup b \neq u \to \neg(a \cup b) \neq u \ \& \ \neg a \neq u \ \& \ \neg b \neq u$ by $\neq 1$, and $\neq 2$.
$\to \neg(a \cup b) \subset \neg a$ by $\cup 1$, and 5.3.

$\rightarrow \neg(a \cup b) \chi \neg a$ by 1.1.
$\rightarrow \neg(a \cup b) \subset \neg b$ by $\cup 1$, $\cup 2$, and 5.3.
$\rightarrow \neg a \chi \neg b$ by $\chi 3$. ∎

5.14. $a \neq u$ & $b \neq u$ & $a \chi b$ & $b \cap a \subset \neg a \rightarrow a \cap b \subset \neg a$ by $\cap 2$.
$\rightarrow a \cap a \subset \neg b$ by $\neg 1$.
$\rightarrow a \subset a \cap a$ by $\cap 1$.
$\rightarrow a \subset \neg b$ by $\subset 1$.
$\rightarrow b \subset \neg a$ by 5.1. ∎

5.15*. $a \neq u$ & $a \chi b$ & $b \cap a \subset \neg a \rightarrow a \cap b \subset \neg a$ by $\cap 2$ and $\subset 1$.
$\rightarrow a \chi b$ & $\neg a \chi (a \cap b)$ by 1.1.
$\rightarrow \neg a \chi a$ by 1.4. ∎

5.16*. $a \neq u$ & $a \chi b$ & $b \subset \neg a \rightarrow b \cap a \subset \neg a$. By 3.9 substituting $\neg a$ for c. ∎

Analysing the Theorems 5.14 - 5.16*, we conclude that the antecedents of them are unrealizable formulas. What is really surprizing is that Theorems 5.15 and 5.16 are proved without using axioms $\neg 1$ and $\neg 2$. Theorem 5.14 is not so surprizing because it seems that the only way to prove it is through axiom $\neg 1$ whose transform is equivalent to axiom schema 10 of $\mathcal{H}$. By these theorems we may conclude that in $\mathcal{E}$ there are what we may call *weak forms of reasoning by reduction ad absurdum*. This fact is at first sight surprizing in the context of negationless intuitionistic mathematics, but it is actually a corollary of Griss' interpretation of the relation of implication (as quoted in the Introduction) and of the fact that on a theoretical level we cannot know in general whether, for any two species a and b, $a \neq u$, $a \chi b$, or $a \subset \neg b$ are true or not.

4. THE PROPOSITIONAL CALCULUS $\mathcal{G}_\vartheta$.

In [14] Vredenduin constructed his version of a logic for negationless intuitionistic mathematics (up to the level of a higher order predicate calculus). Vredenduin showed that the predicate calculus of Hilbert and

Ackermann, as well as Heyting's one, can be interpreted in his logic, but that only a partial interpretation of Griss' logic of species is possible. To obtain this interpretation, Vredenduin weakened the system $\mathcal{G}$, replacing the axiom schemata A4 and A5 by the corresponding rules:

$$\text{R4.}\quad \frac{A \to B}{A \,\&\, C \to B \,\&\, C} \qquad\qquad \text{R5.}\quad \frac{A \to B \qquad B \to C}{A \to C}$$

The system so obtained will be denoted by $\mathcal{G}_{\vartheta}$.

To Vredenduin, the modifications made in $\mathcal{G}$ are justified not only for the sake of a partial interpretation of Griss' logic of species in his own logic, but also for another reason as he himself explains: "Personally I prefer these rules (R4 and R5) because double implications between propositions are avoided by them. I have some doubts with respect to the fact that these double implications are in accordance with Griss' meaning of implication." (see [14], p. 259.)

Nevertheless, it is necessary to make some remarks on Vredenduin's criticism of Griss system $\mathcal{G}$. At first, the substitution of R4 and R5 for A4 and R5 does not avoid double implication between propositions, if R2 is maintained. For example, for each theorem of the form $A \to B$, we obtain by R2 another theorem of the form $C \to (A \to B)$. On the other hand, without R2 (or its equivalent R2') we cannot prove most of the theorems of Griss' logic of species. We conjecture that the only way to avoid double implication between propositions in Griss' system is the following. To proceed as Vredenduin did, but eliminating the rule R2, and introducing the rule $A\,,\ B / A \,\&\, B$ and also formulating the specific postulates of &, not as implications but as rules. If this conjecture is correct, we obtain a relevant Griss' logic of species.

Now we develop the system $\mathcal{G}_{\vartheta}$, maintaining the usual definitions of formula, theorem and deduction.

The postulates of $\mathcal{G}_{\vartheta}$ are the following.

A1. $A \to A \,\&\, A$.

A2. $A \,\&\, B \to B \,\&\, A$.

A3. $A \,\&\, B \to A$.

R1. $\dfrac{A \quad A \to B}{B}$ R2. $\dfrac{C \quad A \to B}{A \to B \,\&\, C}$

R4. $\dfrac{A \to B}{A \,\&\, C \to B \,\&\, C}$ R5. $\dfrac{A \to B \quad B \to C}{A \to C}$

With the appropriate substitution of $\vdash$ for $\to$, Theorems 2.1 and 2.2 are valid in $\mathcal{G}_\vartheta$, and Theorem 2.3 is valid without modifications, together with its Corollary 2.1. As $\mathcal{G}_\vartheta$ is a proper subsystem of $\mathcal{G}$, Theorem 2.4 is true for $\mathcal{G}_\vartheta$ as well as Theorem 2.5 and its Corollaries 2.2 and 2.3.

THEOREM 3.1. *In $\mathcal{G}_\vartheta$ the following deductions are not valid:*
$A \leftrightarrow B \vdash (C \to A) \leftrightarrow (C \to B)$,
$A \leftrightarrow B \vdash (A \to C) \leftrightarrow (B \to C)$.

PROOF: By the matrix $M_5 = \langle N, I, \dot{\to}, \dot{\&} \rangle$ where N, I, and $\dot{\&}$ are as in M_1, and $\dot{\to}$ is defined as follows (it is easy to prove that M_5 is addequate for $\mathcal{G}_\vartheta$):

$$\vartheta(A \to B) = \vartheta(A) \dot{\to} \vartheta(B) = \begin{cases} \vartheta(A) + \vartheta(B) + 1 & \text{if } \vartheta(A) \in I \text{ and } \vartheta(B) \in N - I; \\ \vartheta(A) + \vartheta(B) & \text{if } \vartheta(A), \vartheta(B) \in N - I, \text{ and } \vartheta(A) \neq 2 < \vartheta(B) \neq 4; \\ 1 & \text{in the remaining cases.} \end{cases}$$ ■

To simplify the proof of the following theorem let us define:

$$(A \to B) \underset{1}{\to} A =_{\text{def}} (A \to B) \to A,$$
$$(A \to B) \underset{p+1}{\to} A =_{\text{def}} ((A \to B) \underset{p}{\to} A) \to A.$$

Now, let us consider the following matrix $M_6 = \langle N, I, \dot{\to}, \dot{\&} \rangle$ where N, I and $\dot{\&}$ are as in M_1 and $\dot{\to}$ is defined as follows:

$\vartheta(A \to B) = \vartheta(A) \dot{\to} \vartheta(B) = \vartheta(B) - 2$ if $\vartheta(A) \in I$, and $\vartheta(B) \in N - I$ and is greater than 2, or
$\vartheta(A)$, $\vartheta(B) \in N - I$, and $\vartheta(A) \neq 2$ and less than $\vartheta(B)$;

$= 2$ if $\vartheta(A) \in I$, and $\vartheta(B) = 2$, or

$\vartheta(A) = 2$, and $\vartheta(B) \in N - I$ and is greater than 2;

$= 1$ in the remaining cases.

LEMMA 3.1. *The schema* $((A \to B) \underset{q}{\to} A) \to ((A \to B) \underset{p}{\to} A)$, $p<q$, *is not valid in* $\mathcal{G}_\vartheta$.

PROOF: By $\mathcal{M}_6$, taking, for example, $\vartheta(A) = 2n > 2q$, and $\vartheta(B) \in I$. To examplify, we write in the first line bellow the formulas, and in the second, the respective values according to the values assigned to A and B:

$$(A \to B) \underset{1}{\to} A,\quad (A \to B) \underset{2}{\to} A,\ \ldots,\ (A \to B) \underset{p}{\to} A,\ \ldots,\ (A \to B) \underset{q}{\to} A,\ \ldots$$

$$2n - 2,\quad 2n - 4,\quad \ldots,\quad 2n - 2p,\quad \ldots,\quad 2n - 2q,\quad \ldots$$

As we can see, the value of the formula decreases when the number of symbols $\to$ in the formula increases. Thus, the Lemma is proved. ■

THEOREM 3.2. $\mathcal{G}_\vartheta$ *is not decidable by finite matrices.*

PROOF: If $\mathcal{G}_\vartheta$ were decidable by a finite matrix $\mathcal{M}$, then we could find m and n, $m<n$, such that for each valuation ϑ of the atomic components we would have $\vartheta((A \to B) \underset{m}{\to} A) = \vartheta((A \to B) \underset{n}{\to} A)$. Consequently, we would have $\vdash ((A \to B) \underset{m}{\to} A) \to ((A \to B) \underset{n}{\to} A)$, which cannot be the case because in the above Lemma we proved that this formula is not a theorem of $\mathcal{G}_\vartheta$. ■

5. THE ALGEBRIZATION OF $\mathcal{G}$.

In this section we introduce first the Lindenbaum algebra associated to $\mathcal{G}$, and show that generalizing it we obtain the dual of the Griss' algebra as it was defined by Imai and Isēki in [12].

By Theorem 2.2 we see that the equivalence relation ($\equiv$) defined as follows: $A \equiv B =_{\text{def}} (A \to B)\ \&\ (B \to A)$ is also a congruence relation: hence we may consider $\mathbb{F}/\equiv$, where $\mathbb{F}$ is the class of the formulas of $\mathcal{G}$. But since in $\mathcal{G}$ we have: if $\vdash A$, then $\vdash B \to A$, therefore all theorems of $\mathcal{G}$ are in the same equivalence class, which we denote by 1.

The Lindenbaum algebra associated with $\mathcal{G}$ is a structure $G = \langle \mathbb{F}/_\equiv, 1,$

$\Rightarrow, \wedge \rangle$, where the operation $\wedge$ and $\Rightarrow$ are defined, respectively, as $|A| \wedge |B| = |A \,\&\, B|$ and $|A| \Rightarrow |B| = |A \to B|$. Therefore, we can define an order relation in G in the following way: $|A| \leqslant |B| =_{def} |A \to B| = 1$, i. e., $\vdash A \to B$. It is easy to prove in G:

1. $|A| \leqslant |A| \wedge |A|$.

2. $|A| \wedge |B| \leqslant |B| \wedge |A|$.

3. $|A| \wedge |B| \leqslant |A|$.

4. $|A| \Rightarrow |B| \leqslant (|A| \wedge |C| \Rightarrow |B| \wedge |C|)$.

5. $(|A| \Rightarrow |B| \wedge |B| \Rightarrow |C|) \leqslant |A| \Rightarrow |C|$.

6. $|A| \leqslant 1$.

7. If $|A| \leqslant |B|$ and $|B| \leqslant |A|$, then $|A| = |B|$.

Hence, G is a semilattice with last element, in which is defined an operation $\Rightarrow$ satisfying conditions 4 and 5.

Generalizing, we introduce an algebra $G' = \langle S, 1, \Rightarrow, \wedge \rangle$, where $a \leqslant b =_{def} a \Rightarrow b = 1$, and the operations $\Rightarrow$ and $\wedge$ satisfy the following conditions:

1. $a \leqslant a \wedge a$.

2. $a \wedge b \leqslant b \wedge a$.

3. $a \wedge b \leqslant a$.

4. $a \Rightarrow b \leqslant a \wedge c \Rightarrow b \wedge c$.

5. $a \Rightarrow b \wedge b \Rightarrow c \leqslant a \Rightarrow c$.

6. $a \leqslant 1$.

7. If $a \leqslant b$ and $b \leqslant a$, then $a = b$.

This algebra is the "dual" of Imai and Isēki's algebra defined in the Section 1 of [12].

It is easy to prove that the algebra G' equivalent to the algebra $G = \langle S, 1, \Rightarrow, \wedge \rangle$ such that:

1. $\langle S, 1, \wedge \rangle$ is a semilattice with last element;

2. $a \Rightarrow a = 1$;

3. If $a \leqslant b$ ($a \leqslant b =_{def} a \wedge b = a$), then $a \Rightarrow c \leqslant b \Rightarrow c$ for all ele-

ments a , b, and c of S.

4. $c \Rightarrow a \wedge b = (c \Rightarrow a) \wedge (c \Rightarrow b)$ for all elements a , b , c of S.

This last structure G is the "dual" of the algebra defined in Section 2 of Imai and Isēki's [12], which is named by them *Griss' algebra*. The notions of filter, of homomorphism, etc., are obtainable in an analogous manner.

Since in $\mathcal{G}_\vartheta$ the connective $\rightarrow$ is not compatible with the equivalence relation $\equiv$ defined above (even if it is defined as: $A \vdash B$ *and* $B \vdash A$), it is not possible to obtain a natural Lindenbaum algebra corresponding to this calculus. Nonetheless, employing the methods of da Costa [3], it is possible to present an "algebrization" of $\mathcal{G}_\vartheta$ without help of the Lindenbaum algebra. This question and the details of the algebraic study of $\mathcal{G}$ and $\mathcal{G}_\vartheta$ will be left for a future paper.

6. Open problems.

Concluding our paper, we list some problems about the logic of negationless intuitionistic mathematics which we think interesting for the study of the subject.

1- To obtain semantics (even classical) and decision methods for $\mathcal{G}$ and $\mathcal{G}_\vartheta$. (As $\mathcal{G}$ and $\mathcal{G}_\vartheta$ are similar to the system $\mathbb{P}$ (see [2]) it seems possible to solve this problem through the methods of A. Loparic̄ (cf. [13]).)

2- To study the similarities of $\mathcal{G}$ and specially $\mathcal{G}_\vartheta$ with certain relevant logics (for this kind of logic, see A. R. Anderson and N. D. Belnap Entailment, Princeton University Press, 1975).

3- To develop a sort of relevant logic of species for negationless intuitionistic mathematics, as mentioned in Section 4.

4- To develop in detail the algebraic versions of $\mathcal{G}$ and $\mathcal{G}_\vartheta$.

REFERENCES.

[1] A. I. Arruda, *Sur certaines hiérarchies de calculs propositionnels*, C. R. Acad. Sc. Paris, 266 A (1968), 37-39.

[2] A. I. Arruda and N. C. A. da Costa, *O paradoxo de Curry - Moh Shaw-Kwei*, Boletin da Sociedade Matemática de São Paulo, vol. 18, 1º - 2º (1965), 83-89.

[3] N. C. A. da Costa, *Opérations non monotones dans les treillis*, C. R. Acad. Sc. Paris, 263 A (1966), 429-432.

[4] G. F. C. Griss, *Negatieloze intuitionïstische Wiskunde*, Versl. Ned. Akcd. v. Wetensch., 53 (1944), 261-268.

[5] G. F. C. Griss, *Negationless intuitionistic mathematics*, Indagationes Mathematicae, 8 (1946), 675-681.

[6] G. F. C. Griss, *Sur la négation (dans les mathématiques et la logique)*, Synthese, 7 (1948/9), 71-74.

[7] G. F. C. Griss, *Negationless intuitionistic mathematics, II*, Indagationes Mathematicae, 12 (1950), 108-115.

[8] G. F. C. Griss, *Logique des mathématiques intuitionistes sans négation*, C. R. Acad. Sc. Paris, 227 (1949), 946-948.

[9] G. F. C. Griss, *Logic of negationless intuitionistic mathematics*, Indagationes Mathematicae, 13 (1951), 41-49.

[10] G. F. C. Griss, *Negationless intuitionistic mathematics, III - IV*, Indagationes Mathematicae, 13 (1951), 193-199, 452-462, and 463-471.

[11] G. F. C. Griss, *La mathématique intuitioniste sans négation*, Nieuw Archief voor Wiskunde, vol. 3, nº 3 (1955), 134-142.

[12] Y. Imai and K. Iséki, *On Griss' algebra, I*, Proceedings of the Japan Academy of Sciences, 42 (1966), 213-216.

[13] R. Routley and A. Loparić, *Semantical investigations on Arruda-da Costa P systems and of adjacent systems*, to appear.

[14] P. G. J. Vredenduin, *The logic of negationless mathematics*, Compositio Mathematica , 11 (1953), 204-270.

Departamento de Matemática
Universidade Estadual de Campinas
13.100 Campinas, SP., Brazil.

Bernays' Class Theory.

by *ROLANDO CHUAQUI.*

ABSTRACT. This is a (mainly) expository paper about a system of axiomatic set theory introduced by P. Bernays in *Zur Frage der Unendlichkeits schemata in der Axiomatischen Mengenlehre,* Essays on the Foundations of Mathematics, Magnes Press, Jerusalem, 1961. This system is a theory of classes with two types of objects: classes and sets. However, we use only one type of variables and express 'x is a set' by '$\exists u(x \in u)$'. The axioms use only $\in$ as non-logical symbol and are expressed in first-order logic with identity. These axioms are: the impredicative axiom of class specification, extensionality, axiom of subsets (i. e. subclass of a set is a set), and a reflection principle.

In section 1 the axioms are given and it is shown that they are intuitively justified. It is also proved, that the theory contains MT, the impredicative theory of classes (see Kelley, General Topology, Appendix, for this theory).

In section 4, results of Tharp, *On a set theory of Bernays*, J. of Symb. Logic 32 (1967), 319-321, are given which indicate that the Π_n^1-indescribable cardinals are exactly those which can be obtained in the theory.

Another end of this paper is to show, by giving an example, how the methods developped in Kreisel and Lévy, *Reflection principles*, Zeitschr. f. math. Logik und Grundl. der Math. 14 (1968), 97-112, are used to study the complexity of axiom systems. This is done in Section 3, where a detailed proof is given of the impossibility of axiomatizing the theory by adding a set of sentences of bounded quantifier depth to MT. This is, probably, the only new result of the paper.

In his paper [1], Bernays introduced an elegant set theory based on a reflection principle. The main purpose of this (mainly) expository paper is to present a simple axiomatization for this theory and prove some metamathematical results about it.

In Section 1 I give the axioms and show (or try to show) that they are intuitively justified. I also prove, that the theory contains MT, the impredicative theory of classes (for MT see Kelley [5], or Chuaqui [2]). In Section 4, results of Tharp (see [13]) are given which indicate the large cardinals whose existence is implied by Bernays' theory.

Another end of this paper is to show, by giving an example, how the methods developped in Kreisel-Lēvy [6] are used to study the complexity of axiom systems. This is done in Section 3 (after some metamathematical preliminaries in Section 2), where a detailed proof is given of the impossibility of axiomatizing the theory by adding a set of sentences of bounded quantifier depth to MT. I believe this is the only new result of this paper. It is, however, an easy consequence of methods of Kreisel-Lēvy [6].

These methods combine two types of reflection principles, set-theoretical and semantical. Both type of principles, can be traced back to Montague (see [8] and [9]) ; and both were studied and its uses developped by Lēvy (see [6] and [7]).

1. THE THEORY.

The point of view adopted in this paper with respect to axiomatic set theory is realistic: we start from objective notions of set and class and choose some properties of these notions as axioms. From these axioms , we try to derive as many properties of the notions as possible.

The intuitive concepts of set and class are not clearly determined; at least in principle, there are several possible notions of set or class. Thus, before giving the axioms, I shall delimit the basic notions that I intend to formalize and explain them sufficiently so as to be able to see that the axioms that will be selected are true for these concepts.

A *class* is an arbitrary collection of objects, which may be numbers, functions, physical objects, sets, etc. . Since there are no restrictions with respect to the nature of these objects, we might think of a class as specified when, for each object, it is possible to determine whether it belongs to the class or not. In particular, this means that to each property, defined in any way, corresponds the collection of the objects which have this property. It is well known, that this way of considering classes leads to Russell's paradox. Thus, we cannot consider classes as extensions of arbitrary properties.

The course we shall follow here is to limit the class extensions to a given universe or domain V. All elements of this universe are also classes, the so-called *sets*. The main intuitive concepts about sets that we shall formalize are those of "*element of* ..." and "*set of elements of* ...". Instead of ... we put a given arbitrary collection C.

There is in V, the class of all sets, an initial collection of elements, u, that may be called urelements. This initial collection may also be empty. V is closed under the two notions we want to study. Whenever a collection (a set) a is in V, all elements and all sets of elements (subsets) of a are also in V.

Since we limit the extension of properties to V, all classes are subclasses of V; therefore every element of a class is in V. This means, that an object is in V if and only if it belongs to some class. That is, X is a set if and only if X belongs to a class. Classes that are not sets will be called *proper classes*.

All the objects of the theory are classes, in particular, sets are also classes, namely extensions of properties restricted to V. In our definition, we left the inicial class u completely undetermined. In particular, once we have a domain V closed under the notions specified above, we can take this V as a set of urelements in another universe V'.

I shall give, now, the theory. This theory, $\mathbb{B}$, is given in first-order logic with identity with only one non logical constant $\in$. As variables we use capital or lower-case italic letters.

The first two axioms formalize the idea that classes are extensions of properties limited to sets. As we saw above, we can say that x is a set by the formula '$\exists u(x \in u)$'. Following Skolem and Fraenkel, since we want a first-order theory, we take for properties formulas. Thus, we have:

(I) AXIOM SCHEMA OF CLASS SPECIFICATION:

$$\exists A \forall x(x \in A \leftrightarrow \phi \wedge \exists u(x \in u))$$

where ϕ is any formula in which A is not free.

(II) AXIOM OF EXTENSIONALITY:

$$\forall x(x \in A \leftrightarrow x \in B) \rightarrow A = B \ .$$

The third axiom formalizes the fact that our universe is closed under sets of elements of b, for any set b.

(III) AXIOM OF SUBSETS:

$(\exists u(b \in u) \wedge a \subseteq b) \rightarrow \exists u(a \in u)$

here $\subseteq$ has its usual meaning, i. e.

$A \subseteq B \leftrightarrow \forall x(x \in A \rightarrow x \in B)$

The fourth axiom embodies the principle that we can take any universe already given as set of urelements in another universe. In order to formulate this principle, we need some definitions.

We introduce the class abstractor, an operator which forms the term $\{x: \phi\}$, for each variable x and formula ϕ. The definition of this term is, thus,

$x \in \{x: \phi\} \leftrightarrow \phi \wedge \exists u(x \in u)$

Since, by axioms (I) and (II) there is a unique A such that

$\forall x(x \in A \leftrightarrow \phi \wedge \exists u(x \in u))$;

this terms can be eliminated in the usual way (and, hence, the theory with $\{ : \}$ is a conservative extension of the theory without $\{ : \}$).

$\cap$ has its usual meaning, i. e. $A \cap B = \{x: x \in A \wedge x \in B\}$.

For each formula ϕ, let ϕ^u be the formula obtained from ϕ by replacing each part of the form $\exists x \psi$ by $\exists x(x \subseteq u \wedge \psi^u)$ and $\forall x \psi$ by $\forall x(x \subseteq u \rightarrow \psi^u)$. The term $\{x: \phi\}^u$ is $\{x: x \in u \wedge \phi^u\}$.

I shall write $\phi(X_0 \dots X_{n-1})$ whenever I want to indicate that all free variables of ϕ are among $X_0, \dots, X_{n-1}$; in this case, $\phi(Y_0 \dots Y_{n-1})$ is the formula obtained from $\phi(X_0 \dots X_{n-1})$ by replacing (avoiding collisions with quantifiers) the variables $X_0, \dots, X_{n-1}$, by $Y_0, \dots, Y_{n-1}$.

The fourth axiom is:

(IV) AXIOM SCHEMA OF REFLECTION:

$\phi(A) \rightarrow \exists u(\exists y(u \in y) \wedge \forall x \forall y(x \in y \in u \rightarrow x \in u) \wedge \phi^u(A \cap u))$

where $\phi(A)$ is any formula which does not contain u.

It is easy to see, that even with axiom (IV), $\{ : \}$ is eliminable, and hence the theory with $\{ : \}$ is a conservative extension of the theory without this operator. Thus, in some metamathematical contexts we shall consider $\mathbb{B}$ without $\{ : \}$.

I shall now prove the axioms for the impredicative theory of classes which appear in Chuaqui [2]. This will prove that $\mathbb{B}$ is an extension of this theory (MT). MT is essentially the same theory (without the axiom of choice) which appears in Kelley [5]. We shall abbreviate "u is transitive" by Tran(u), thus:

$$\text{Tran}(u) \leftrightarrow \forall x \forall y (x \in y \in u \rightarrow x \in u)$$

"u is supertransitive" is abbreviated by Stran(u), thus:

$$\text{Stran}(u) \leftrightarrow \text{Tran}(u) \wedge \forall x \forall y (x \subseteq y \in u \rightarrow x \in u)$$

V stands for the universal class $\{x : x = x\}$. We have in $\mathbb{B}$:

$$x \in V \leftrightarrow \exists u (x \in u) .$$

In order to avoid axiom of regularity, we shall use a modification of the axiom of infinity that appears in Chuaqui [2]. The axioms for MT (without regularity) are the following:

(I) Same as for $\mathbb{B}$.

(II) Same as for $\mathbb{B}$.

(III') AXIOMS OF UNIONS.

$$b \in V \wedge \forall x (x \in a \leftrightarrow \exists y (x \in y \in b)) \rightarrow a \in V .$$

(IV') AXIOM OF THE POWER SET.

$$b \in V \wedge \forall x (x \in a \leftrightarrow x \subseteq b) \rightarrow a \in V .$$

(V') AXIOM OF REPLACEMENT.

$$b \in V \wedge \forall x (x \in a \rightarrow \exists y (y \in b \wedge \forall z (z \in a \rightarrow (y \in z \leftrightarrow z = x)))) \rightarrow a \in V.$$

(VI') AXIOM OF INFINITY.

$$\exists x \exists a (x \in a \in V \wedge \forall x (x \in a \rightarrow (\exists y (x \in y \in a) \wedge \forall z (z \subseteq x \rightarrow z \in a)))).$$

It is easy to show that these axioms imply that the empty class, 0, is a set, and that there is an infinite set.

We pass now to show that $\mathbb{B} \vdash \mathrm{MT}$.

LEMMA 1.1.

$$\mathbb{B} \vdash a, b \in V \to \{a,b\} \in V .$$

PROOF: Let $c = \{a, b\}$. Suppose that $a \neq b$. We have,

$$\exists x \exists y (x \neq y \wedge x, y \in c).$$

Applying (IV), we obtain:

$$\exists u (\mathrm{Tran}(u) \wedge u \in V \wedge \exists x \exists y (x,y \subseteq u \wedge x \neq y \wedge x,y \in c \cap u).$$

But then,

$$\exists x \exists y (x \neq y \wedge x, y \in c \cap u)$$

Hence,

$$c \cap u = c .$$

By (III), $c \in V$.

When $a = b$, the proof is similar. ■

LEMMA 1.2. $\mathbb{B} \vdash 0 \in V$.

PROOF: From (IV), we obtain,

$$X = X \to \exists u (u \in V \wedge \mathrm{Tran}(u) \wedge X \cap u = X \cap u).$$

Hence, $\exists u (u \in V)$. But, $0 \subseteq u$. Therefore, by (III), $0 \in V$. ■

LEMMA 1.3. *For each formula* $\phi(A)$ *we have,*

$$\mathbb{B} \vdash \phi(A) \to \exists u (\mathrm{Tran}(u) \wedge \forall x \forall y (x, y \in u \to \{x,y\} \in u) \wedge 0 \in u \wedge u \in V \wedge \phi^u (A \cap u)).$$

PROOF: Suppose $\phi(A)$. Then, by 1.1,

$$\phi(A) \wedge \forall x \forall y (x, y \in V \to \{x, y\} \in V) \wedge 0 \in V.$$

Applying, now, (IV) to this formula, we obtain the theorem. ■

LEMMA 1.4. *For each formula* $\phi(A,B)$, *we have*

$$\mathbb{B} \vdash \phi(A, B) \to \exists u (\mathrm{Tran}(u) \wedge u \in V \wedge \phi^u(A \cap u, B \cap u)).$$

PROOF: Let $\psi(X)$ be the formula

$$\exists A \exists B (\phi(A,B) \wedge \forall x (x \in A \leftrightarrow \langle x,0\rangle \in X) \wedge \forall x (x \in b \leftrightarrow \langle x,1\rangle \in X)).$$

Let $\phi(A, B)$. Taking $X = A \times \{0\} \cup B \times \{1\}$, we have, $\psi(X)$. Applying, now, 1.3 to this formula, we obtain the desired conclusion. ■

LEMMA 1.5. *For each formula* $\phi(A,b)$ *we have,*

$$\mathbb{B} \vdash b \in V \wedge \phi(A,b) \to \exists u(\mathrm{Tran}(u) \wedge b \in u \in V \wedge \phi^u(A \cap u, b))$$

PROOF: Let $b \in V$. Then, by 1.1, $\{b\} \in V$. Let $a = \{b\}$. Let $\phi(A,b)$. Then we have

$$\exists b(\phi(A,b) \wedge \forall x(x \in a \leftrightarrow x = b)).$$

Let us call this formula $\psi(A,a)$. By 1.4, we have,

$$\exists u(\mathrm{Tran}(u) \wedge u \in V \wedge \psi^u(A \cap u, a \cap u))$$

Then,

$$\exists b(b \subseteq u \wedge \phi^u(A \cap u, b) \wedge \forall x(x \in a \cap u \leftrightarrow x = b)).$$

Hence,

$$a \cap u = \{b\} = a.$$

Thus, $b \in u$. ■

LEMMA 1.6. *For each formula* $\phi(A,b)$, *we have:*

$$\mathbb{B} \vdash \phi(A,b) \wedge b \in V \to \exists u(\mathrm{Stran}(u) \wedge b \in u \in V \wedge \phi^u(A \cap u, b)).$$

PROOF: Suppose $\phi(A,b) \wedge b \in V$. Then, we have,

$$\phi(A,b) \wedge b \in V \wedge \forall x \forall y(x \subseteq y \in V \to x \in V).$$

Applying, now, 1.5, we obtain the theorem. ■

THEOREM 1.7. (Axiom of Unions.)

$$\mathbb{B} \vdash b \in V \wedge \forall x(x \in a \leftrightarrow \exists y(x \in y \in b)) \to a \in V.$$

PROOF: Let $\cup b = \{x : \exists y(x \in y \in b)\}$. By (I), we have,

$$\forall x(x \in \cup b \leftrightarrow \exists y(x \in y \in b)).$$

Hence,

$$b \in V \wedge \exists a \forall x(x \in a \leftrightarrow \exists y(x \in y \in b)).$$

Thus, by 1.6,

$$\exists u(\mathrm{Stran}(u) \wedge b \in u \in V \wedge \exists a(a \subseteq u \to \forall x(x \subseteq u \to (x \in a \leftrightarrow$$
$$\exists y(y \subseteq u \wedge x \in y \in b))))).$$

Since, $\cup u \subseteq u$ (i. e., u is transitive), we obtain,

$$\exists u(\mathrm{Stran}(u) \wedge b \in u \in V \wedge \exists a(a \subseteq u \to \forall x(x \in a \leftrightarrow \exists y(x \in y \in b)))).$$

Hence,

$$\cup b = a \subseteq u \in V.$$

By (III), we get

$$a \in V. \ ■$$

THEOREM 1.8. (Axiom of Power Set.)

$\mathbb{B} \vdash b \in V \wedge \forall x (x \in a \leftrightarrow x \subseteq b) \to a \in V.$

PROOF: Let $\mathbf{S}b = \{x: x \subseteq b\}$. By (I) and (III),

$x \in \mathbf{S}b \leftrightarrow x \subseteq b.$

Hence,

$\exists a \forall x (x \in a \leftrightarrow x \subseteq b).$

Using 1.6, we get

$\exists u (\text{Stran}(u) \wedge b \in u \in V \wedge \exists a (a \subseteq u \wedge \forall x (x \subseteq u \to (x \in a \leftrightarrow x \subseteq b)))).$

Since Stran(u), we obtain,

$\exists u (\text{Stran}(u) \wedge b \in u \in V \wedge \exists a (a \subseteq u \wedge \forall x (x \in a \leftrightarrow x \subseteq b))).$

Then, $\mathbf{S}b = a \subseteq u \in V.$

Thus, by (III), $\mathbf{S}b \in V.$ ■

THEOREM 1.9. (Axiom of Replacement.)

$\mathbb{B} \vdash b \in V \wedge \forall x (x \in a \to \exists y (y \in b \wedge \forall z (z \in a \to (y \in z \leftrightarrow z = x))))$
$\to a \in V.$

PROOF: Let the hypothesis of the axiom be satisfied and let $b' = b \cap \cup a$. By (III), $b'' \in V$. Also,

(*) $\forall x (x \in a \to \exists y (y \in b' \wedge \forall z (z \in a \to (y \in z \leftrightarrow z = x)))).$

Since $b' \subseteq \cup a$, we have,

$\forall y (y \in b' \to \exists x (y \in x \in a)).$

Using 1.6, we obtain,

$\exists u (\text{Stran}(u) \wedge b'' \in u \in V \wedge \forall y (y \subseteq u \wedge y \in b' \to \exists x (x \subseteq u \wedge$
$y \in x \in a \cap u))).$

Since Stran(u), we simplify this to:

(**) $\exists u (b' \in u \in V \wedge \forall y (y \in b' \to \exists x (y \in x \in a \cap u))).$

We shall prove that $a \cap u = a$. Let $x \in a$; then by (*) there is a $y \in b'$ such that

$\forall z (z \in a \to (y \in z \leftrightarrow z = x)).$

But, from (**) we obtain that there is a $z \in a \cap u$ such that $y \in z$. Hence $z = x$, and $x \in a \cap u$.

Therefore,

$a = a \cap u \subseteq u \in V.$

Thus, by (III), $a \in V$. ■

THEOREM 1.10. (Axiom of Infinity.)

$$\mathbb{B} \vdash \exists a \exists x (x \in a \in V \wedge \forall x (x \in a \rightarrow (\exists y (x \in y \in a) \wedge \forall z (z \subseteq x \rightarrow z \in a)))).$$

PROOF: By 1.2, $0 \in V$. Hence,

$$\exists x (x \in V).$$

Also, if $x \in V$, then $x \in \mathbf{S}x \in V$, by 1.4; and by (III), if $z \subseteq x \in V$, then $z \in V$. Thus,

$$\exists a \exists x (x \in a \wedge \forall x (x \in a \rightarrow \exists y (x \in y \in a) \wedge \forall z (z \subseteq x \rightarrow z \in a))).$$

Using 1.6, we get,

$$\exists u (u \in V \wedge \mathrm{Stran}(u) \wedge \exists a \exists x (a \subseteq u \wedge x \subseteq u \wedge x \in a \wedge \forall x (x \subseteq u \wedge x \in a \rightarrow \exists y (y \subseteq u \wedge x \in y \in a) \wedge \forall z (z \subseteq u \wedge z \subseteq x \rightarrow z \in a)))).$$

Since Stran(u), we simplify this to,

$$\exists u (u \in V \wedge \exists a \exists x (a \subseteq u \wedge x \in a \rightarrow \forall x (x \in a \rightarrow \exists y (x \in y \in a) \wedge \forall z (z \subseteq x \rightarrow z \in a)))).$$

From (III), we obtain the axiom of infinity. ■

It is usual to add to the axioms of $\mathbb{B}$ the axiom of regularity and some form of the axiom of choice. Actually, Bernays in his paper, considers the theory with these axioms. We shall not use these axioms until Section 4. Thus, we shall introduce them there.

It is interesting to notice that Axiom Schema (I) can be replaced in $\mathbb{B}$ by a finite set of axioms, namely, the axioms of groups A and B of Gödel [4]. In [4], it is proved that these axioms imply the predicative axiom of class specification, i. e., (I) with the bound variables in ϕ restricted to sets. In Bernays [1] (this result is attributed to Specker), it is proved that this axiom and (IV) imply our (I). This modification may be of interest since ϕ in (I) has an arbitrary number of free variables, while in (IV) it has only one.

2. METAMATHEMATICAL NOTIONS.

In this section, I define the main metamathematical notions which will be used in the rest of the paper. We work in a weak class theory GC (General Class Theory) which was introduced in Chuaqui [2]. Most of the contents of this section are sketched in that paper. The version of the axioms for GC given here are a trivial modification of those presented there.

GC has the following axioms. (I) and (II) are the same as for $\mathbb{B}$.

(I) CLASS SPECIFICATION.

$$\exists A\, \forall x\, (x \in A \leftrightarrow \phi \wedge \exists u\, (x \in u)),$$

where ϕ is any formula which does not contain A free.

(II) EXTENSIONALITY.

$$\forall x\, (x \in A \leftrightarrow x \in B) \rightarrow A = B.$$

(III'') PAIRING.

$$\forall x\, (x \in A \rightarrow (x = b \vee x = c)) \rightarrow \exists u\, (A \in u).$$

We have also the class abstractor $\{x\colon \phi\}$ with the same meaning as in $\mathbb{B}$, i. e.,

$$x \in \{x : \phi\} \leftrightarrow \phi \wedge \exists u\, (x \in u).$$

In GC it is possible to develop Peano's arithmetic. We first introduce the class of Zermelo's natural numbers Z,

$$Z = \cap \{X : 0 \in X \wedge \forall x\, (x \in X \rightarrow \{x\} \in X)\}$$

It is not possible to use in GC the usual (von Neumann) natural numbers: all axioms of GC are satisfied in the class of all subsets of the set of all sets that are hereditarily of cardinality less than or equal to two.

It is easy to prove all Peano's axioms for Z with the operation of successor given by

$$x' = \{x\}.$$

Functions with domain included in Z are also available. Thus, we can introduce functions defined by recursion on Z.

I shall give a sketch of the arithmetization of metamathematics that we need for our purposes. For details see Tarski, Mostowski, Robinson [12].

We shall use k, l, m, n, p, q as variables for elements of Z. That is,

$\exists k \phi$ will stand for $\exists k(k \in Z \wedge \phi)$ and $\forall k \phi$ for $\forall k(k \in Z \rightarrow \phi)$.

Let $\mathbf{N}$ be the set of natural numbers in the metalanguage. For each $n \in \mathbf{N}$, we define the term Δ_n of $\mathcal{L}_{GC}$ (the language of GC) by:

$$\Delta_0 = 0$$

$$\Delta_{n+1} = \{\Delta_n\}.$$

A formula ϕ (or term τ) of $\mathcal{L}_{GC}$ is an n-formula (or n-term) if its free variables are among $v_0, \ldots, v_{n-1}$.

Let R be an n-relation in $\mathbf{N}$ (i. e., $R \subseteq {}^n\mathbf{N}$). We say that the n-formula ϕ of $\mathcal{L}_{GC}$ *represents* (or *locally defines*) R in GC iff for every m_0, $\ldots$, $m_{n-1} \in \mathbf{N}$,

$$\langle m_0, \ldots, m_{n-1}\rangle \in R \text{ implies } GC \vdash \phi(\Delta_{m_0}, \ldots, \Delta_{m_{n-1}}),$$

and,

$$\langle m_0, \ldots, m_{n-1}\rangle \notin R \text{ implies } GC \vdash \neg\phi(\Delta_{m_0}, \ldots, \Delta_{m_{n-1}})$$

ϕ *weakly represents* R *in* GC, iff for every $m_0, \ldots, m_{n-1} \in \mathbf{N}$,

$$\langle m_0, \ldots, m_{n-1}\rangle \in R \text{ iff } GC \vdash \phi(\Delta_{m_0}, \ldots, \Delta_{m_{n-1}}).$$

If F is a function from ${}^n\mathbf{N}$ to $\mathbf{N}$ ($F: {}^n\mathbf{N} \rightarrow \mathbf{N}$) an n-term τ *strongly represents* F *in* GC iff for every $m_0, \ldots, m_{n-1} \in \mathbf{N}$,

$$GC \vdash \tau(\Delta_{m_0}, \ldots, \Delta_{m_{n-1}}) = \Delta_{F(m_0, \ldots, m_{n-1})}.$$

In the usual way, we prove that all recursive relations are representable, all recursive functions are strongly representable, and all recursively enumerable relations are weakly representable.

We can associate a Gödel number to each expression in such a way that the following classes, relations, and functions are recursive. We shall use the same abbreviation beginning with the corresponding lower-case letter for the formula or the term that represents them in GC.

Vl : the class of variables.

At form : the class of atomic formulas.

Fm : the class of formulas.

$Fmwa$: the class of formulas that do not contain the abstractor operator $\{:\}$.

Tm : the class of terms.

$Fv(k,l)$: k is a free variable of l ; $Fv(k)$ is the class of free variables of k.

$E(k,l) = k \in l.$

$Id(k,l) = k = l.$

$Ng(k) = \neg k.$

$I(k,l) = k \to l.$

$Ds(k,l) = k \vee l.$

$Cn(k,l) = k \wedge l.$

$Qu(k,l) = \exists k\, l.$

$Qu^n(k,l) = \exists k_0 \ldots \exists k_{n-1}\, l.$

$Rqu(k,l) = \{\exists k(\exists m(k \in m) \wedge l) : m \in V\ell \text{ and } m \neq k\}.$

$S(k,l) = k(\Delta_l)$ (i. e., Δ_l substituted for v_0 in k).

$Dk = k(\Delta_k)$ (i. e., $S(k,k)$).

If T is any recursively enumerable theory, let $Prov_T(k,l)$ be the recursive relation that holds between k and l iff l is a sentence and k is a proof of l. We need the following properties of $prov_T$:

THEOREM 2.1.

(i) (Schema). *For every sentence* ϕ of $\mathcal{L}_T$ (*the language of* T), $T \vdash \phi$ *iff for some number* m, $GC \vdash prov_T(\Delta_m, \Delta_\phi)$.

(ii) $GC \vdash \exists p\, prov_T(p, i(k,l)) \wedge \exists q\, prov_T(q, i(l,m)) \to \exists n\, prov_T(n, i(k,m)).$

The notion of unrestricted quantifier alternation depth, which is introduced in the following definition, appears in Kreisel-Lévy [6]. It was also defined in a different way in Chuaqui [2]. Definition 2.2 stresses the fact that the notion is recursive.

DEFINITION 2.2.

(i) *If* $k \notin Fmwa$, *then* $Dp(k) = 0$.

(ii) *If* $k \in Fmwa$, *then*,

(a) *if* $k \in Atform$, *then* $Dp(k) = 1$,

(b) *if* $k = Ng(l)$, *then* $Dp(k) = Dp(l)$,

(c) *if* $k = Ds(l, m)$, *then* $Dp(k) = max(Dp(l), Dp(m))$,

(d) *if* $k \in Rqu(l, m)$, *then* $Dp(k) = Dp(m)$,

(e) *if* $k = Qu(l, m)$ *and* $m = Qu(p, n)$, *and for all* $q \leqslant m$, $m \notin Rqu(p, q)$, *then* $Dp(k) = Dp(m)$,

(f) *if* $k = Qu(l, m)$ *and* ($m = Ds(p, n)$ *or* $m = Ng(p)$ *or* $m = Rqu(p, n)$), *then* $Dp(k) = Dp(m) + 1$.

The general principle of definition of operations by recursion over well-founded relations, which was proved in Chuaqui [2] for MT, is also valid for GC, and the proof is the same. Therefore, the following definition is justified.

Because of the fact mentioned above, i. e., that there is a model of GC consisting of classes whose members are hereditarily of cardinality less than or equal to two, it is not true, in general, that

$$GC \vdash a, b \in V \rightarrow a \times b \in V.$$

However, we have,

$$GC \vdash a, b \in V \rightarrow \{\langle a\ b\rangle\} \in V.$$

In order to work with relations inside of classes, we introduce the corresponding function, and define, for r a function,

$$\bar{r} = \cup\{r(u) \times \{u\} : u \in dom(r)\}.$$

It is clear that, $r(u) = \bar{r}^*\{u\}$, where for any relation R, $R^*A = \{y : \exists x \in A\ (\langle x, y\rangle \in R)\}$, i. e. R^*A is the image of A under R.

We introduce now, the notions of satisfaction and truth for formulas of bounded unrestricted alternation of quantifiers depth (also in Chuaqui [2] ; similar notions appear in Montague [8]).

DEFINITION SCHEMA 2.4. (GC)

(i) *Define the binary operation* K_1, *by:*

(a) *If* $dp(k) \neq \Delta_1$ *or* $dom\ R \not\subseteq fv\, k$, *then* $K_1(k, R) = 0$.

(b) *If* $dp(k) = \Delta_1$ *and* $dom R \not\subseteq fv\, k$, *then:*

(b1) $k = e(l, m) \wedge l, m \in v\ell \rightarrow K_1(k, R) = \{r : dom(r) \cap dom\, R = 0 \wedge r$ *is a function* $\wedge\ dom\, r \subseteq fv\, k \wedge (R \cup \bar{r})^* \{l\} \in (R \cup \bar{r})^* \{m\}\}$.

(b2) $k = id(l, m) \wedge l, m \in v\ell \rightarrow K_1(k, R) = \{r : dom\, r \cap dom\, R = 0 \wedge r$ *is a function* $\wedge\ dom\, r \subseteq fv\, k \wedge (R \cup \bar{r})^* \{l\} = (R \cup \bar{r})^* \{m\}\}$.

(b3) $k = ng(l) \rightarrow K_1(k, R) = \{r : dom\, r \cap dom\, R = 0 \wedge r$ *is a function* $\wedge$ $dom\, r \subseteq fv\, k \wedge r \notin K_1(l, R)\}$.

(b4) $k = ds(l, m) \rightarrow K_1(k, R) = \{r : dom\, r \cap dom\, R = 0 \wedge r$ *is a function* $\wedge\ dom\, r \subseteq fv\, k \wedge (r \restriction Fv\, l \in K_1(l, R \restriction Fv\, l) \vee r \restriction fv\, m \in K_1(m, R \restriction fv\, m))\}$.

(b5) $k \in rqu(l, m) \wedge l \notin fv\, m \rightarrow K_1(k, R) = K_1(m, R)$.

(b6) $k \in rqu(l, m) \wedge l \in fv\, m \rightarrow K_1(k, R) = \{r : dom\, r \cap dom\, R = 0 \wedge$

$dom\, r \subseteq fv\, k \wedge \exists s (dom\, s \subseteq \{l\} \wedge r \cup s$ *is a function* $\wedge$ $r \cup s \in K_1(m, R))\}$.

(ii) *Define the binary operation* K_{n+1} *for* $n \geqslant 1$ *by:*

(a) *If* $dp(k) > \Delta_{n+1}$, $dp(k) = \Delta_0$ *or* $dom\, R \not\subseteq fv\, k$, *then* $K_{n+1}(k, R) = 0$.

(b) *If* $\Delta_1 \leqslant dp(k) \leqslant \Delta_{n+1}$ *and* $dom\, R \subseteq fv\, k$, *then:*

(b1) $dp(k) \leqslant \Delta_n \rightarrow K_{n+1}(k, R) = K_n(k, R)$,

(b2) $k = Qu^m(l, p) \wedge dp(p) = \Delta_n \rightarrow K_{n+1}(k, R) =$

$\cup \{K_n(p, S) : dom\, S \subseteq fv p \wedge S \restriction fv\, k = R\}$,

(b3)-(b6) *as in* (i).

(iii) $Sat_n(k, R) \leftrightarrow \Delta_1 \leqslant dp(k) \leqslant \Delta_n \wedge 0 \in K_n(k, R \restriction fvk)$.

(iv) $Tr_n(k) \leftrightarrow fvk = 0 \wedge Sat_n(k, 0)$.

It is easy to prove the following theorem:

THEOREM SCHEMA 2.5. *For each formula* ϕ *with* $1 \leqslant Dp(\phi) \leqslant n$, *we have in* GC:

(i) $\Delta_\phi = \Delta_{v_i \in v_j} \rightarrow (Sat_n(\Delta_\phi, R) \leftrightarrow R^*\{\Delta_{v_i}\} \in R^*\{\Delta_{v_j}\})$.

(ii) $\Delta_\phi = \Delta_{v_i = v_j} \rightarrow (Sat_n(\Delta_\phi, R) \leftrightarrow R^*\{\Delta_{v_i}\} = R^*\{\Delta_{v_j}\})$.

(iii) $\Delta_\phi = \Delta_{\neg\psi} \rightarrow (Sat_n(\Delta_\phi, R) \leftrightarrow \neg Sat_n(\Delta_\psi, R))$.

(iv) $\Delta_\phi = \Delta_{\psi_1 \vee \psi_2} \rightarrow (Sat_n(\Delta_\phi, R) \leftrightarrow (Sat_n(\Delta_{\psi_1}, R) \vee Sat_n(\Delta_{\psi_2}, R)))$.

(v) $\Delta_\phi = \Delta_{\exists v_i (v_i \in V \wedge \psi)} \rightarrow (Sat_n(\Delta_\phi, R) \leftrightarrow \exists x (x \in V \wedge Sat_n(\Delta_\psi, R - R \restriction \{\Delta_{v_i}\} \cup x \times \{\Delta_{v_i}\})))$.

From 2.5 we can derive, by induction:

THEOREM SCHEMA 2.6.

(i) *Let* ϕ *be a formula with* $1 \leqslant Dp(\phi) \leqslant n$ *and free variables* $v_{i_0}, \ldots, v_{i_m}$. *Then*

GC $\vdash Sat_n(\Delta_\phi, R) \leftrightarrow \phi(R^*\{\Delta_{v_{i_0}}\}, \ldots, R^*\{\Delta_{v_{i_m}}\})$.

(ii) *Let* ϕ *be a sentence with* $1 \leqslant Dp(\phi) \leqslant n$. *Then*

GC $\vdash Tr_n(\Delta_\phi) \leftrightarrow \phi$.

Since the proof of 2.6 can be done in GC (or any extension of GC), we have:

THEOREM SCHEMA 2.7. *Let* T *be any extension of* GC. *Then*

GC $\vdash \exists q\, prov_{\mathrm{T}}(q, i(s(\Delta_{Tr_n(v_0)}, k), k))$

(*i. e., there is a proof in* T *of* $\forall k(Tr_n(\Delta_k) \rightarrow k)$).

We also need the notions of satisfaction of a formula in a set. For this, we work in MT or any extension of MT. This definition appears in Mostowski [10] and Chuaqui [2].

DEFINITION 2.8. (MT).

(i) *Define by recursion the operation* H, *such that:*

(a) *If* $k \notin fm$, *then* $H(k) = 0$.

(b) *If* $k \in fm$, *then:*

(b1) $k = e(l, m) \to H(k) = \{\langle u, x \rangle : x \in {}^{\{l, m\}}u \wedge x(l) \in x(m)\}$;

(b2) $k = id(l, m) \to H(k) = \{\langle u, x \rangle : x \in {}^{\{l, m\}}u \wedge x(l) = x(m)\}$;

(b3) $k = ng(l) \to H(k) = \{\langle u, x \rangle\} : x \in {}^{fv\,k}u \wedge \langle u, x \rangle \notin H(l)\}$;

(b4) $k = ds(l, m) \to H(k) = \{\langle u, x \rangle : \langle u, x \upharpoonright fv\, l \rangle \in H(l) \vee \langle u, x \upharpoonright fv\, m \rangle \in H(m)\}$;

(b5) $k = qu(l, m) \wedge l \notin fv\, m \to H(k) = H(m)$;

(b6) $k = qu(l, m) \wedge l \in fv\, m \to H(k) = \{\langle m, x \rangle : \exists a (a \in u \wedge \langle u, x \cup \{\langle a, l \rangle\}\rangle \in H(m)\}$.

(ii) $St = \{\langle u, k, x \rangle : \langle u, x \rangle \in H(k)\}$,

(*we write also,* $St(u, k, y_0, \ldots, y_m)$ *instead of* $\langle u, k, x \rangle \in St$, *where* $x(\Delta_{v_{i_l}}) = y_l$ *for every free variable of* k, v_{i_l}).

We have the following:

THEOREM SCHEMA 2.9. (MT). *For any formula* $\phi(v_{i_0}, \ldots, v_{i_m})$ *and* $x \in {}^{Fv\,\phi}\mathbf{S}u$, *we have*

$$MT \vdash \phi^u(x(\Delta_{v_{i_0}}), \ldots, x(\Delta_{v_{i_m}})) \leftrightarrow \langle \mathbf{S}u, \Delta_\phi, x \rangle \in St.$$

Hence we can formulate 1.6 (which is a strengthening of (IV), by

$$b \in V \wedge \phi(A,b) \rightarrow \exists u(\mathit{Stran}(u) \wedge b \in u \in V \wedge \mathit{St}(\mathbf{S}u, \Delta_\phi, A \cap u, b).$$

We can strengthen this schema by noticing that if ψ is a sentence provable in $\mathbb{B}$, we can obtain,

$$b \in V \wedge \phi(A,b) \rightarrow \exists u(\mathit{Stran}(u) \wedge b \in u \in V \wedge \psi^u \wedge \phi^u(A \cap u, b)).$$

It is clear that for each instance of (I), ψ, if $\mathit{Stran}(u)$, we have,

$$\mathit{St}(\mathbf{S}u, \Delta_\psi, 0).$$

Also, we can require that $\mathbf{S}u$ satisfy Axioms (III') - (IV'). If $\mathbf{S}u$ satisfies all the axioms of a theory T, we say that u is a supertransitive (or supercomplete) model of T, in symbols $STM_T(u)$. Thus, (IV) can be strengthened to

$$b \in V \wedge \phi(A, b) \rightarrow \exists u(STM_{\mathrm{MT}}(u) \wedge b \in u \in V \wedge \mathit{St}(\mathbf{S}u, \Delta_\phi, A \cap u, b).$$

From 2.6 we can obtain:

THEOREM 2.10. (MT).

$$\Delta_1 \leqslant dp(k) \leqslant \Delta_n \wedge STM_{\mathrm{GC}}(u) \rightarrow (\mathit{St}(\mathbf{S}, \Delta_{\mathit{Satn}(v_0, v_1)}, k, \bar{r})$$
$$\leftrightarrow \langle \mathbf{S}u, k, r \rangle \in \mathit{St}).$$

Finally, we state the relation between proof and satisfaction:

THEOREM SCHEMA 2.11. (MT). *Let T be a theory, then for each formula $\phi(n)$ we have*

$$\mathrm{MT} \vdash \mathbf{S}u \textit{ is a model of } T \wedge \exists q\, \mathit{prov}_T(q, s(\Delta_\phi, n)) \rightarrow \mathit{St}(\mathbf{S}u, \Delta_\phi, n).$$

3. Complexity of the Axiom System.

Since $\mathbb{B}$ is an extension of GC with the same symbols, we obtain, from of Chuaqui [2] that $\mathbb{B}$ is not axiomatizable by sentences of bounded alternation of unrestricted quantifier depth. In this section we shall give a detailed proof of: $\mathbb{B}$ is not axiomatizable by $\mathrm{MT} \cup \Sigma$ where Σ is a set of sentences ϕ, with $1 \leqslant \mathcal{D}p(\phi) \leqslant n$ for some n.

Let T, T' be theories such that $\mathrm{GC} \subseteq T \subseteq T'$. We say that T' is an extension of bounded depth of T if T' is obtained from axioms $T \cup \Sigma$ where Σ is a set of sentences ϕ with $1 \leqslant \mathcal{D}p(\phi) \leqslant n$ (for some n).

The main theorem of this section is the following:

THEOREM 3.1. *Let* $\mathrm{MT} \subseteq T \subseteq T$, T, T' *theories. Suppose that*

$$T' \vdash n \in Z \wedge \phi(n) \rightarrow \exists u(\mathbf{S}u \text{ is a model of } T \wedge n \in u \wedge St(\mathbf{S}u, \Delta_\phi, n)).$$

Then no consistent extension of T' *is an extension of bounded depth of* T.

Thus theorem clearly implies:

COROLLARY 3.2. *No consistent extension of* $\mathbb{B}$ *with the same symbols is an extension of bounded depth of* MT.

The proof of 3.1 uses results of Kreisel-Lévy [6]. Since this is an expository paper I shall give the proofs of the main results used.

LEMMA 3.3. (The standard diagonalization lemma.) *For every formula* $\psi(x)$ *of* GC *there exists a sentence* χ *of* GC *such that,*

$$\mathrm{GC} \vdash \chi \leftrightarrow \psi(\Delta_\chi).$$

PROOF: We have that the recursive function $\mathcal{D}$ defined by $\mathcal{D}k = k(\Delta_k)$ is strongly representable in GC by the term $d(v_0)$. Let

$$\phi(v_0) = \psi(d(v_0)).$$

Then,

$$\chi = \phi(\Delta_\phi).$$

We have:

$$\begin{aligned}\phi(\Delta_\phi) &\leftrightarrow \psi(d(\Delta_\phi))\\ &\leftrightarrow \psi(\Delta_{\mathcal{D}\phi})\\ &\leftrightarrow \psi(\Delta_{\phi(\Delta_\phi)})\end{aligned}$$

Thus, χ satisfies the conclusion of the lemma. ∎

From now on, we assume that T, T', T'' satisfy the hypothesis of Theorem 3.1.

LEMMA 3.4. *For every formula* $\phi(n)$ *we have:*

$$T' \vdash \exists q\, prov_T(q, s(\Delta_\phi, n)) \to \phi(n).$$

(This is called the *uniform reflection principle*. To distinguish it from our set-theoretical reflection principle, we shall call it the *syntactical reflection principle for* T).

PROOF: Suppose $\neg\phi(n)$. Then we have,

$$\exists u(\mathbf{S}u \text{ is a model of } T \wedge n \in u \wedge St(\mathbf{S}u, \Delta_{\neg\phi}, n)).$$

Suppose also that $\exists q\, prov_T(q(s, \Delta_\phi, n))$.

Since $\mathbf{S}u$ is a model of T, we have,

$$St(\mathbf{S}u, \Delta_\phi, n) \quad \text{(by 2.11)}.$$

But we cannot have $St(\mathbf{S}u, \Delta_\phi, n) \wedge St(\mathbf{S}u, \Delta_{\neg\phi}, n)$. Hence the lemma is proved. ∎

LEMMA 3.5. (Kreisel-Lēvy [6], 8.) *For every formula* ψ *and every* n, *we have*

$$T' \vdash \exists k(\exists q\, prov_T(q, i(k, \Delta_\psi)) \to (Tr_n(k) \to \psi)).$$

PROOF: Assume $\exists q\, prov_T(q, i(k, \Delta_\psi))$.

By 2.7,

$$\exists p\, prov_T(p, i(s(\Delta_{Tr_n(v_0)}, k), k)).$$

Hence, by 2.1 (ii),

$$\exists r\, prov_T(r, i(s(\Delta_{Tr_n(v_0)}, k), \Delta_\psi)).$$

But, we have,

$$i(s(\Delta_{Tr_n(v_0)}, k), \Delta_\psi) = s(i(\Delta_{Tr_n(v_0)}, \Delta_\psi), k).$$

Thus,

$$\exists r\, prov_T(r, s(i(\Delta_{Tr_n(v_0)}, \Delta_\psi), k)).$$

Using the uniform syntactic reflection principle (3.4), we obtain,

$$Tr_n(k) \to \psi. \ ∎$$

PROOF OF THEOREM 3.1. (Kreisel-Lévy [6], 1, 2, 4.) Let T'' be an extension of T' which is an extension of T of bounded depth. That is, let $T \cup \Sigma$ be axioms for T'' where $\phi \in \Sigma$, implies $1 \leqslant \mathcal{D}p(\phi) \leqslant n$.

Let $\psi(v_0)$ be the formula,

$$\forall k\,(\mathcal{T}r_n(k) \rightarrow \neg \exists p\; \mathit{prov}_T(p, i(k, v_0))).$$

By Lemma 3.3, there is a sentence χ such that:

$$T' \vdash \chi \leftrightarrow \psi(\Delta_\chi).$$

We shall prove:

(1) $\quad T' \vdash \chi.$

By 3.4, we have,

$$T' \vdash \exists p\; \mathit{prov}_T(p, i(k, \Delta_\chi)) \rightarrow (\mathcal{T}r_n(k) \rightarrow \chi).$$

Hence,

$$T' \vdash \neg\chi \rightarrow \forall k(\mathcal{T}r_n(k) \rightarrow \neg \exists p\; \mathit{prov}_T(p, i(k, \Delta_\chi)).$$

From the definition of χ, we obtain,

$$T' \vdash \neg\chi \rightarrow \chi;$$

hence,

$T' \vdash \chi$ and (1) is proved.

Since T'' is an extension of T', we have,

$$T \cup \Sigma \vdash \chi.$$

Hence, for some $\phi_0, \ldots, \phi_{k-1} \in \Sigma$,

$$T \vdash \Lambda_{i<k}\, \phi_i \rightarrow \chi .$$

Let $\phi = \Lambda_{i<k}\, \phi_i$. Since $T \vdash \phi \rightarrow \chi$, there is a number q such that,

$$T' \vdash \mathit{prov}_T(\Delta_q, \Delta_{\phi \rightarrow \chi});$$

i. e.,

$$T' \vdash \mathit{prov}_T(\Delta_q, i(\Delta_\phi, \Delta_\chi)).$$

Thus,

$$T' \vdash \exists p\; \mathit{prov}_T(p, i(\Delta_\phi, \Delta_\chi)).$$

From the definition of χ, we obtain,

$$T' \vdash \chi \to (Tr_n(\Delta_\phi) \to \neg \exists p\ prov_T(p, i(\Delta_\phi, \Delta_\chi))).$$

Hence,

$$T' \vdash \chi \to \neg Tr_n(\Delta_\phi).$$

By 2.6 (ii), since $1 \leqslant \mathcal{D}p(\phi) \leqslant n$,

$$T' \vdash Tr_n(\Delta_\phi) \leftrightarrow \phi.$$

Hence,

$$T' \vdash \chi \to \neg\phi.$$

But, by (1), we get,

$$T' \vdash \neg\phi.$$

Since T'' is an extension of T' and $T'' \vdash \phi$, we get that T'' is inconsistent. ■

Similarly as in Chuaqui [2], if we define for any extension of GC, T:

$\mathrm{T}^* = \{\phi : \phi$ *is in the language of* $\mathrm{ZF} \wedge \mathrm{T} \vdash \phi^V\}$, we can prove that $\mathbb{B}^*$ is not an extension of bounded depth (now in the sense of Chuaqui [2]) of MT^*. In fact, for any theory $T \supseteq \mathrm{GC}$ in which it is possible to prove that there is a model of MT, this result holds, i. e., T^* is not an extension of bounded depth of MT^*. This is obtained immediately from Theorem 10 of Kreisel-Lēvy [6].

4. Large cardinals.

In this section I shall present results of Tharp [13] that show the large cardinals whose existence is provable in $\mathbb{B}$. We assume the theory $\mathbb{B}$ with the axioms of foundation (or regularity) and choice, call it $\mathbb{B}\mathbb{C}$:

(V) FOUNDATION.

$\exists x (x \in A) \to \exists x (x \in A \wedge A \cap x = 0).$

(VI) CHOICE.

$a \in V \wedge 0 \notin a \to \exists b \forall x (x \in a \to \exists y (b \cap x = \{y\})).$

With the axiom of choice we identify cardinals with initial ordinals in the usual way. As is custumary we introduce $R(\alpha)$, the set of sets of rank less than α. Greek lower-case letters stand for ordinals. **OR** is the class of ordinals. From (V), we obtain:

$$V = \cup\{R(\alpha) : \alpha \in \mathbf{OR}\}.$$

Shepherdson ([11], II, p. 326) proved that $Stran(u)$ (i.e., u is a supertransitive) iff $u = R(\alpha)$ for some ordinal α. Thus, (IV) is equivalent in $\mathbb{BC}$ to the following schema:

$$(\mathrm{IV}_1) \quad \phi(X) \to \exists\alpha(St(R(\alpha+1), \Delta_\phi, X \cap R(\alpha))).$$

We also have that $SMT_{\mathrm{MT}}(u)$ iff $u = R(\theta)$ for some θ inaccessible.

We introduce, now, the definitions of the main cardinals needed. For details see Drake [3], Chapter 9.

DEFINITION 4.1.

(i) *α is described by a formula $\phi(v_0)$ with parameter U, iff $St(R(\alpha+1), \Delta_\phi, U)$ but for all $\beta<\alpha$, $\neg St(R(\beta+1), \Delta_\phi, U\cap R(\beta))$.*

(ii) *α is described by ϕ iff for some $U \subseteq R(\alpha)$, α is described by ϕ with parameter U.*

(iii) *Let $\Gamma \subseteq Fm$; then α is Γ-indescribable iff α is not described by any formula of Γ. $In(\Gamma) = \{\alpha : \alpha$ is Γ-indescribable$\}$.*

We shall consider some special sets of formulas:

DEFINITION 4.2.

(i) $Fm_n = \{\phi : \phi \in Fm \wedge 1 \leqslant Dp(\phi) \leqslant n\}$.

(ii) $\Pi^1_n = \{\phi : \phi \in Fm \wedge \phi = \forall X_1 \ldots \forall X_{n_1} \exists X_{n_1+1} \ldots \exists X_{n_2} \forall X_{n_2+1} \ldots \mathcal{Q}\psi \wedge \psi \in Fm_1\}$.

(iii) $\Sigma^1_n = \{\phi : \phi \in Fm \wedge \phi = \exists X_1 \ldots \exists X_{n_1} \forall X_{n_1+1} \ldots \forall X_{n_2} \exists X_{n_2+1} \ldots \mathcal{Q}\psi \wedge \psi \in Fm_1\}$.

Π^1_n-indescribable cardinals are very large. Thus, the Π^1_0-indescribable are the inaccessibles cardinals, and the Π^1_1-indescribable are the weakly compact cardinals. Uncountable weakly compact cardinals are fixed points in the enumeration of inaccessibles (see Drake [3], Chapter 9).

THEOREM 4.3. (Tharp.) *For each* n, $\mathbb{BC} \vdash In(Fm_n) \notin V$.

PROOF: By (IV_1), we have,

$$Sat_n(k, X) \rightarrow \exists\alpha(St(R(\alpha+1), \Delta_{Sat_n(v_0, v_1)}, X \cap R(\alpha), k)).$$

Applying again (IV_1), we obtain that this sentence, call it ϕ, is satisfied in some $R(\theta+1)$. It is easy to see that given any $\beta \geqslant \omega$, we can take $\theta > \beta$.

We show, now, that θ is Fm_n-indescribable. Suppose that $St(R(\theta+1)$, $\Delta_\psi, U)$ for some $U \subseteq R(\theta)$ and $\psi \in Fm_n$.

Then, by 2.10,

$$St(R(\theta+1), \Delta_{Sat_n}, U, \Delta_\psi).$$

Since ϕ is satisfied in $R(\theta+1)$, there is an $\alpha < \theta$, such that in $R(\theta+1)$ the following is valid,

$$(+) \quad St(R(\alpha+1), \Delta_{Sat_n}, U \cap R(\alpha), \Delta_\psi).$$

But satisfaction is absolute for supertransitive sets (see Chuaqui [2], Ch. IV, for a proof). Hence, (+) is valid in the universe. Applying, again, 2.10, we obtain,

$$St(R(\alpha+1), \Delta_\psi, U).$$

Hence θ is ψ-indescribable, and thus, Fm_n-indescribable. ■

Let $Con(BC) = \neg\exists q\ prov_{BC}(q, \Delta_{\exists x(x \neq x)})$. That is, $Con(BC)$ express the consistency of $\mathbb{BC}$. We have,

THEOREM 4.4. (Tharp.)

$$\mathbb{BC} \vdash \forall n\, \exists\theta(\theta \in In(\Pi^1_n)) \rightarrow Con(BC).$$

PROOF: We shall prove that for each axiom ϕ of $\mathbb{BC}$, there is an n such that if $\theta \in In(\Pi^1_n)$, then $R(\theta+1)$ satisfies ϕ. For axioms (I)-(III), (V), (VI), this is obvious. Let ϕ be an instance of (IV_1):

$$\psi(X) \rightarrow \exists\alpha\, St(R(\alpha+1), \Delta_\psi, X \cap R(\alpha)).$$

Since the axiom of choice (VI) holds we have, for any θ inaccessible,

$$St(R(\theta+1), \Delta_{\psi \leftrightarrow \psi'}, Y)$$

where $\psi' \in \Pi^1_n$ for some n (Drake [3], Chapter 5, 7.2).

Then, if $\theta \in In(\Pi^1_n)$ we have, $St(R(\theta+1), \Delta_\phi, X)$ because satisfaction and rank are absolute (for a proof see, for instance, Chuaqui [2], Chapter IV).

Therefore if we have a finite set Σ of axioms of $\mathbb{BC}$, there is an n, such that if $\theta \in In(\Pi^1_n)$, $R(\theta+1)$ satisfies Σ. Thus,

$\forall n\, \exists\theta\, (\theta \in \Pi'_n)$ implies $Con(BC)$. ■

To conclude the paper, I want to mention two results about $\mathbb{B}$ (or $\mathbb{BC}$).

1) It is easy to prove that if (IV) holds in the universe, then it holds in the constructible universe. Hence, we obtain the relative consistence of $V \simeq L$ with $\mathbb{B}$ (see Chuaqui [2], Chapter V, for $V \simeq L$).

2) The existence of a measurable cardinal implies the consistency of $\forall n\, \exists\theta\, (\theta \in In(\Pi^1_n))$ (and much more), and thus, $Con(BC)$ (Tharp [13]).

REFERENCES.

[1] P. Bernays, *Zur Frage der Unendlichkeitsschemata in der Axiomatischen Mengenlehre*, Essays on the Foundation of Mathematics, Magnes Press, Jerusalem , 1961, 3-49.

[2] R. Chuaqui, *Internal and forcing models for the impredicative theory of classes*, to appear in Dissertationes Mathematicae.

[3] F. R. Drake, Set Theory, Noth-Holland Pub. Co., Amsterdam, 1974.

[4] K. Gödel, The Consistency of the Continnun Hypothesis, Annals of Mathematical Studies, nº 3, Princeton U. Press, Princeton, N.J., 1940.

[5] J. L. Kelley, General Topology, Van Nostrand Pub. Co., Princeton, N. J., 1955.

[6] G. Kreisel and A. Lēvy, *Reflection principles and their use for establishing the complexity of axiomatic systems*, Zeitschr. f. Math. Logik und Grundlagen d. Math., vol. 14 (1968), 97-112.

[7] A. Lēvy, *Axiom schemata of strong infinity in axiomatic set theory*, Pacific J. of Math., vol. 10 (1960), 223-238.

[8] R. Montague, *Semantic closure and non-finite axiomatizability I*, Infinitistic Methods, Warsaw, 1961, 45-69.

[9] R. Montague, *Fraenkel's addition to the axioms of Zermelo*, Essays in the Foundations of Mathematics, Magnes Press, Jerusalem, 1961, 91-114.

[10] A. Mostowski, Constructible Sets, North-Holland Publ. Co., 1969.

[11] J. C. Shepherdson, *Inner models for set theory I, II, III*, The Journal of Symbolic Logic, vol. 16 (1951), 161-190; vol. 17 (1952), 225-237; vol. 18 (1953), 145-167.

[12] A. Tarski, A. Mowstowski and R. Robinson, Undecidable Theories, North-Holland Pub. Co., Amsterdam, 1953.

[13] L. Tharp, *On a set theory of Bernays*, The Journal of Symbolic Logic, vol. 32 (1967), 319-321.

Instituto de Matemática
Universidad Católica de Chile
Santiago 8, Chile
and
Departamento de Matemática
Universidade Estadual de Campinas
13.100 Campinas, S.P., Brazil.

Deductive Systems and Congruence Relations in Ortholattices

by *ROBERTO CIGNOLI.*

ABSTRACT. Let **L** be an ortholattice and let $'$ denote the orthocomplement. The weak implication is the binary operation $x \to y = x' \vee y$ and the strong implication is defined (cp. G. Kalmbach, *Orthomodular Logic*, Proc. Univ. Houston Lattice Theory Conf., Houston 1973, 498-503) as $x \twoheadrightarrow y = (y \to x) \to ((y' \to x) \to x \wedge (x \to y))$. A (strong) deductive system is a subset $\mathbf{D} \subseteq \mathbf{L}$ such that $1 \in \mathbf{D}$ and if x and $x \to y$ are in **D** (x and $x \twoheadrightarrow y$ are in **D**) then $y \in \mathbf{D}$. A Finch filter is a filter **F** of **L** such that if $z \in \mathbf{F}$, then $x \to (x \wedge y) \in \mathbf{F}$ for all x in **L**. It is shown that the notions of deductive system, strong deductive system and Finch filter are mutually equivalent in every ortholattice.

A deductive system **D** is said to be Boolean (orthomodular) if the relation $(x \to y) \wedge (y \to x) \in \mathbf{D}$ $((x \twoheadrightarrow y) \wedge (y \twoheadrightarrow x) \in \mathbf{D})$ is a congruence on **L**, and it is shown that the Boolean (orthomodular) homomorphic images of **L** are in one-to-one correspondence with the Boolean (orthomodular) deductive systems. Moreover, **D** is a Boolean deductive system if and only if it is an intersection of prime filters of **L**.

Finally, by using results of P. D. Finch (J. Austral. Math. Soc., 6 (1966), 46-54) it is shown that if **L** is an orthomodular lattice, then there is a one-to-one correspondence between deductive systems and congruence relations on **L**. In particular it follows that if **L** is an orthomodular lattice satisfying the chain condition, then the congruence lattice of **L** is anti-isomorphic to the center of **L**, and therefore, it is a Boolean algebra.

Recently the theory of orthomodular lattices has received considerable attention, due mainly to its connections with the algebras of operators on Hilbert spaces [7] and with the foundations of quantum mechanics [3], [8], [11].

In 1966 P. D. Finch [5] characterized a class of filters, that we shall call Finch filters, that are in one-to-one correspondence with congruence relations on orthomodular lattices.

In 1973 G. Kalmbach [9] defined an orthomodular logic as a propositional calculus having as models the orthocomplemented lattices, and he proved that a certain binary operation, that we shall call strong implication, leads to the completeness theorem when orthomodular logic is axiomatized by a set of tautologies and a rule of detachment respect to this implication. The classical material implication, that we shall call weak implication, can also be defined in orthomodular logic, and it was observed by Kalmbach that a set of formulas is closed under the rule of detachment respect to the strong implication if and only if it is closed under the rule of detachment respect to the weak implication.

The algebraic counterparts of the notion of set of formulas closed under a rule of detachment is that of deductive system, and in Boolean algebras, deductive systems and filters coincide, and they are in one-to-one correspondence with the congruence relations.

The above mentioned facts motivated the present paper. In § 1 we recall the definitions of ortholattices and orthomodular lattices and some of the properties that are needed in this paper. In § 2 we introduce deductive systems (with respect to weak implication) in general ortholattices and we show that they coincide with Finch filters, and in § 3 we show that they also coincide with the strong deductive systems.

§ 4 is the main section of this paper, and in it we consider congruences in ortholattices. We show that the shell of any congruence is a deductive system, and we single out some deductive systems that are shells of congruences. We show that equivalence with respect to weak and strong implication respectively are in one-to-one correspondence with the Boolean and orthomodular images of an ortholattice. Finally, in § 5, we restate the above mentioned results of Finch in terms of deductive systems, and we show in particular that when an orthomodular lattice **L** satisfies the chain condition (for instance, when **L** is finite) the congruence lattice of **L** is anti-isomorphic with the center of **L** and, therefore, a Boolean algebra.

1. Ortholattices and orthomodular lattices.

An *ortholattice* $\langle \mathbf{L}, \wedge, \vee, ', 0, 1 \rangle$ or, simply, **L**, is an algebra of type (2, 2, 1, 0, 0) such that $\langle \mathbf{L}, \wedge, \vee, 0, 1 \rangle$ is a bounded lattice with zero 0 and unit 1 and $'$, which is called the orthocomplement, has the following properties:

O1) $(x')' = x$.

O2) $(x \wedge y)' = x' \vee y'$

O3) $x \wedge x' = 0$

An *orthomodular lattice* is an ortholattice in which the following equality, called the orthomodular property, holds:

OM) $x \vee ((x \vee y) \wedge x') = x \vee y$.

Observe that the distributive ortholattices are just the Boolean algebras.

Ortholattices and orthomodular lattices are considered in [1], [2] and [10]. An excellent survey article by S. S. Holland [7] is devoted to orthomodular lattices and their connection with operator algebras. Orthomodular lattices are studied in [3], [8] and [11] from the point of view of the foundations of quantum mechanics.

In what follows, we shall denote by $\mathcal{O}$, $\mathcal{M}$ and $\mathcal{B}$ the (equational) classes of ortholattices, orthomodular lattices and Boolean algebras, respectively. We have that $\mathcal{B} \subset \mathcal{M} \subset \mathcal{O}$.

Note that the orthomodular property (OM) is equivalent to:

OM') If $x \leqslant y$, then $y = x \vee (y \wedge x')$.

Following [10], if a, b and c are three elements of a lattice $\mathbf{L}$ we write $(a, b, c)\mathrm{D}$ if $(a \vee b) \wedge c = (a \wedge c) \vee (b \wedge c)$, and $(a, b, c)\mathrm{D}^*$ if $(a \wedge b) \vee c = (a \wedge c) \vee (b \wedge c)$. If $(a, b, c)\mathrm{D}$ and $(a, b, c)\mathrm{D}^*$ hold for all permutations of a, b and c we say that $\{a, b, c\}$ is a *distributive triple* and write $(a, b, c)\mathrm{T}$.

Let $\mathbf{L}$ be a bounded lattice with zero 0 and unit 1. An element $z \in \mathbf{L}$ is is said to be *central* if z has a complement and $(x, y, z)\mathrm{T}$ for each pair x, y of elements of $\mathbf{L}$. The set of central elements of $\mathbf{L}$ forms a Boolean sublattice of $\mathbf{L}$, called the *center* of $\mathbf{L}$ that we shall denote by $\mathbf{B}(\mathbf{L})$.

Let $\mathbf{L} \in \mathcal{O}$ we say that x *commutes with* y, and we write $x\mathbf{K}y$ if $x = (x \wedge y') \vee (x \wedge y)$. Then $\mathbf{L} \in \mathcal{M}$ if and only if $\mathbf{K}$ is a symmetric relation (i.e. $x\mathbf{K}y$ implies $y\mathbf{K}x$) and in this case $\mathbf{B}(\mathbf{L}) = \{ z \in \mathbf{L} : z\mathbf{K}x \text{ for each } x \text{ in } \mathbf{L}\}$.

All these properties can be found in any of the books [1], [2] or [10].

We are going to say that a subset $\mathbf{F}$ of a bounded lattice $\mathbf{L}$ is a filter if the following properties hold:

F1) $1 \in \mathbf{F}$,

F2) If x, y are in $\mathbf{F}$, then $x \wedge y \in \mathbf{F}$,

F3) If $x \in \mathbf{F}$ and $x \leqslant y$, then $y \in \mathbf{F}$.

A filter $\mathbf{F}$ is *proper* if $\mathbf{F} \neq \mathbf{L}$. A filter $\mathbf{F}$ is said to be *prime* if it is proper, and $x \vee y \in \mathbf{F}$ implies that $x \in \mathbf{F}$ or $y \in \mathbf{F}$.

Finally we shall use freely the language of Universal Algebra, as it is given, for instance, in [1], Chapter VI.

2. DEDUCTIVE SYSTEMS.

We shall call *weak implication* the binary operation defined in each $\mathbf{L} \in \mathcal{O}$ by the formula:

(W) $x \to y = x' \vee y$.

We note the following properties of weak implication, that follows at once from (W):

(W1) If $x \leqslant y$, then $x \to y = 1$,

(W2) $1 \to x = x$,

(W3) $x \to (y \to (x \wedge y)) = 1$.

Regarding the converse of (W1), we have that:

2.1. LEMMA. *The following are equivalent conditions for each* $\mathbf{L} \in \mathcal{O}$:

(i) *If* $x \to y = 1$, *then* $x \leqslant y$.

(ii) $\mathbf{L} \in \mathcal{B}$.

PROOF: It is well known that (ii) implies (i). In order to prove that (i) implies (ii), note than since $x \to y = 1$ is equivalent to $x \wedge y' = 0$, it follows that $x \to y = 1$ implies $x \leqslant y$ for each y in $\mathbf{L}$ if and only if x' is the pseudocomplement of x. Therefore (i) implies that $'$ coincide with the pseudocomplement, and it follows from Glivenko's Theorem ([6], p. 58) that $\mathbf{L} \in \mathcal{B}$. ■

2.2. REMARK. It was proved in [3], p. 22, that if $\mathbf{L} \in \mathcal{M}$ and the orthocomplement coincides with the pseudocomplement, then $\mathbf{L} \in B$.

2.3. DEFINITIONS. *Let* $\mathbf{L} \in O$. *A set* $\mathbf{D} \subseteq \mathbf{L}$ *is said to be a deductive system if the following two conditions hold:*

D1) $1 \in \mathbf{D}$, *and*

D2) *if* x *and* $x \to y$ *belong to* $\mathbf{D}$, *then* $y \in \mathbf{D}$.

A deductive system $\mathbf{D}$ *is said to be proper if* $\mathbf{D} \neq \mathbf{L}$.

It follows at once from properties W1, W2 and W3 that:

2.4. PROPOSITION. *Each deductive system of* $\mathbf{L} \in O$ *is a filter.*

2.5. PROPOSITION. *The following are equivalent conditions for each element* z *of* $\mathbf{L} \in O$:

(i) *The principal filter* $[z)$ *is a deductive system of* $\mathbf{L}$.

(ii) $z \to x = 1$ *implies* $z \leqslant x$ *for each* x *in* $\mathbf{L}$.

(iii) z' *is the pseudocomplement of* z.

(iv) $z \leqslant x \to (z \wedge x)$ *for each* x *in* $\mathbf{L}$.

(v) $x \to z = x \to (z \wedge x)$ *for each* x *in* $\mathbf{L}$.

PROOF: It follows at once from the definition of deductive system that (i) implies (ii). The equivalence between (ii) and (iii) was shown in the proof of Lemma 2.1. Since by W3 $x \to (x \to (z \wedge x)) = 1$, it follows that (ii) implies (iv). If (iv) is true, then $x \to z = x' \vee z \leqslant x' \vee (z \wedge x) \leqslant x' \vee z$ and (v) holds. Finally, suppose (v) and let x, y be elements in $\mathbf{L}$ such that $z \leqslant x \wedge (x \to y)$. Then: $y \vee z = y' \to z = y' \to (z \wedge y') = y \vee (z \wedge y') \leqslant y \vee (x \wedge (x \to y) \wedge y') = y \vee ((x \to y)' \wedge (x \to y)) = y$, and it follows that $[z)$ is a deductive system.■

From Lemma 2.1 and the above Proposition we get that:

2.6. COROLLARY. *The following are equivalent conditions for each* $\mathbf{L} \in O$:

(i) *Each filter of* $\mathbf{L}$ *is a deductive system.*

(ii) *Each pricipal filter of* $\mathbf{L}$ *is a deductive system.*

(iii) $\mathbf{L} \in B$.

In order to characterize the deductive systems among the filters we start with the following:

2.7. DEFINITION. *A filter* $\mathbf{F}$ *of* $\mathbf{L} \in O$ *is said to be a Finch filter if* $z \in \mathbf{F}$ *implies that* $x \to (x \wedge z) \in \mathbf{F}$ *for each* x *in* $\mathbf{L}$.

2.8. PROPOSITION. *A subset of* $\mathbf{L} \in O$ *is a deductive system if and only if it is a Finch filter.*

PROOF: Let $\mathbf{F}$ be a Finch filter and suppose that x and $x \to y$ belong to $\mathbf{F}$. Then $x \wedge (x \to y) \in \mathbf{F}$ and $y' \to (y' \wedge (x \wedge (x \to y))) \in \mathbf{F}$, i. e., $y \vee (y' \wedge x \wedge (x' \vee y)) = y \in \mathbf{F}$. Conversely, suppose that $\mathbf{D}$ is a deductive system and that $z \in \mathbf{F}$. Then it follows at once from (W3) and D2) that $x \to (z \wedge x) \in \mathbf{D}$.■

3. STRONG IMPLICATION.

We shall call *strong implication* the binary operation $\leftrightarrow$ defined in each $\mathbf{L} \in O$ by the formula:

(S) $x \leftrightarrow y = (y \to x) \to ((y' \to x) \to x \wedge (x \to y))$.

Strong implication plays an important role in the theory of orthomodular logics, as was shown by G. Kalmbach [9]. We shall prove that the deductive systems with respect to strong implication are the same as the deductive system with respect to weak implication, a fact that clarifies some of Kalmbach's results.

3.1.LEMMA. *The following properties hold in each* $\mathbf{L} \in O$:

S1) *If* $x \leqslant y$, *then* $x \leftrightarrow y = 1$.

S2) *If* $x \leqslant y$, *then* $y \leftrightarrow (x \vee (y \wedge x')) = 1$.

S3) $x \leqslant x \leftrightarrow (y \leftrightarrow (x \wedge y))$.

S4) $x \leftrightarrow y \leqslant x \to y$.

PROOF: S1) If $x \leqslant y$, then $x \to y = 1$ and $y' \to x = y$. So we have: $x \leftrightarrow y = (y \to x) \to (y \to x) = 1$.

S2) Set $\alpha = x \vee (y \wedge x')$. Since $y \to \alpha = 1$, we have that $y \leftrightarrow \alpha = (\alpha \leftrightarrow y) \to ((\alpha' \to y) \to y)$. But $(\alpha' \to y) \to y = (x' \wedge (y' \vee x) \wedge y') \vee y = (x' \wedge y') \vee y = 1$ because $y' \leqslant x'$. Therefore (W1) implies that $y \leftrightarrow \alpha = (y \to \alpha) \to 1 = 1$

S3) Since $x \wedge y \leqslant y$, it follows that $x \wedge y \to y = 1$. Then by (W2) we have that: $y \leftrightarrow (x \wedge y) = ((x \wedge y)' \to y) \to (y \wedge (y \to (x \wedge y)) = y' \vee (y \wedge (y' \vee (x \wedge y))) \geqslant y' \vee (y \wedge (x \wedge y)) = y \to (x \wedge y)$.

Set $\alpha = y \leftrightarrow (x \wedge y)$ and $\beta = y \to (x \wedge y)$. We have just seen that $\alpha \geqslant \beta$. Therefore:

$$x \leftrightarrow (y \leftrightarrow (x \wedge y)) = x \leftrightarrow \alpha = (\alpha \to x)' \vee (\alpha' \to x)' \vee (x \wedge (x \to \alpha)) \geqslant$$
$$\geqslant x \wedge (x \to \alpha) = x \wedge (x' \vee \alpha) \geqslant x \wedge (x' \vee \beta) =$$
$$= x \wedge (x \to (y \to (x \wedge y))) = x \text{ by (W3).}$$

S4) $x \leftrightarrow y = (y \to x)' \vee (y' \to x)' \vee (x \wedge (x \to y)) =$
$(y \wedge x') \vee (y' \wedge x') \vee (x \wedge (x \to y)) \leqslant x' \vee (x \to y) = x \to y$. ■

3.2. PROPOSITION.(G. Kalmbach [9]). *The following are equivalent conditions for each* $\mathbf{L} \in \mathcal{O}$:

(i) $x \leftrightarrow y = 1$ *implies* $x \leqslant y$.

(ii) $\mathbf{L} \in \mathcal{M}$.

PROOF: (i) *implies* (ii). Let $x \leqslant y$. By (S2) $y \leftrightarrow (x \vee (x \wedge y')) = 1$, then (i) implies that $y \leqslant x \vee (x \wedge y') \leqslant y$ and the orthomodular law holds in $\mathbf{L}$.
(ii) *implies* (i). Observe first that in $\mathbf{L} \in \mathcal{M}$, if $x \to y = 1$ and $x \mathbf{K} y$, then $x \leqslant y$. Suppose now that $x \leftrightarrow y = 1$. Since by (S4) $x \to y = 1$, in order to complete the proof we need to show that $x \mathbf{K} y$. But now, $x \leftrightarrow y = (y \to x) \to ((y' \to x) \to x) = ((y \to x) \wedge (y' \to x)) \to x = 1$. Set $\alpha = (y \to x) \wedge (y' \to x)$. Since $x \leqslant \alpha$, $x \mathbf{K} \alpha$ and $\alpha \to x = 1$ implies then that $\alpha \leqslant x$. But this last inequality is equivalent to $x \mathbf{K} y$. ■

3.3. DEFINITION. *Let* $\mathbf{L} \in \mathcal{O}$: *A set* $\mathbf{D} \subseteq \mathbf{L}$ *is called a strong deductive system if the following conditions hold:*

SD1) $1 \in \mathbf{D}$, *and*

SD2) *if* x *and* $x \leftrightarrow y$ *bolong to* $\mathbf{D}$, *then* $y \in \mathbf{D}$.

3.4 LEMMA. *Each strong deductive system of* $\mathbf{L} \in \mathcal{O}$ *is a filter.*

PROOF: It follows at once from properties (S1) and (S3). ■

3.5. THEOREM. *Let* $\mathbf{L} \in O$. *A set* $\mathbf{D} \subseteq \mathbf{L}$ *is a strong deductive system if and only if it is a deductive system.*

PROOF: Assume first that $\mathbf{D}$ is a strong deductive system and that x and $x \to y$ belong to $\mathbf{D}$. Since $\mathbf{D}$ is a filter and $x \leftrightarrow y \geqslant x \wedge (x \to y)$, it follows that $x \leftrightarrow y \in \mathbf{D}$, and (SD2) implies that $y \in \mathbf{D}$. Conversely, suppose that $\mathbf{D}$ is a deductive system and that x and $x \leftrightarrow y$ belong to $\mathbf{D}$. It follows from (S4) that $x \to y \in \mathbf{D}$ and (D2) implies that $y \in \mathbf{D}$. ∎

4. Boolean and orthomodular images of ortholattices.

In this section we shall study some connections between deductive systems and congruence relations in ortholattices. We start with the following:

4.1. DEFINITION. *If* $h: \mathbf{L} \to \mathbf{L}'$ *is an homomorphism* ($\mathbf{L}, \mathbf{L}' \in O$), *the shell of* h *is the set* $\mathbf{S}(h) = h^{-1}\{1\} = \{x \in \mathbf{L}: h(x) = 1\}$.

The proof of the following lemma is immediate:

4.2. LEMMA. *Let* $\mathbf{L}, \mathbf{L}' \in O$ *and* $\mathbf{D}'$ *be a deductive system of* $\mathbf{L}'$. *For each homomorphism* $h: \mathbf{L} \to \mathbf{L}'$, $h^{-1}(\mathbf{D}')$ *is a deductive system of* $\mathbf{L}$. *In particular,* $\mathbf{S}(h)$ *is always a deductive system.*

4.3. DEFINITION. *Let* $\mathbf{F}$ *be a filter of* $\mathbf{L} \in O$. *Define a relation* $\Theta(\mathbf{F})$ *by* $\Theta(\mathbf{F}) = \{(x,y) \in \mathbf{L}^2: x \wedge u = y \wedge u$ *for some* $u \in \mathbf{F}\}$: *The filter* $\mathbf{F}$ *is said to be regular if* $\Theta(\mathbf{F})$ *is a congruence relation on* $\mathbf{L}$.

4.4. REMARKS. Since $\mathbf{F}$ is a filter, it is plain that $(x,y) \in \Theta(\mathbf{F})$ if and only if there are u, v in $\mathbf{F}$ such that $x \wedge u = y \wedge v$. Moreover, it is well known that $\Theta(\mathbf{F})$ is an equivalence relation on $\mathbf{L}$ that is compatible with $\wedge$. So $\mathbf{F}$ is a regular filter if and only if $(x,y) \in \Theta(\mathbf{F})$ implies $(x',y') \in \Theta(\mathbf{F})$.

Since a regular filter $\mathbf{F}$ is the shell of the natural epimorphism $\mathbf{L} \to \mathbf{L}/\Theta(\mathbf{F})$, it follows at once from Lemma 4.2 that:

4.5. LEMMA. *Each regular filter of* $\mathbf{L} \in \mathcal{O}$ *is a deductive system.*

4.6. PROPOSITION. *For each element* z *of* $\mathbf{L} \in \mathcal{O}$ *the following are equivalent conditions:*

(i) *The principal filter* $[z)$ *is regular.*

(ii) *For each* x *in* $\mathbf{L}$, $(z' \vee x) \wedge z = x \wedge z$.

NOTE. In the nomenclature of [10], condition (ii) says that $z' \nabla z$.

PROOF: (i) *implies* (ii): Since $(z \wedge x') \wedge z = x' \wedge z$, it follows from (i) that $(z' \vee x) \wedge z = (z \wedge x')' \wedge z = x \wedge z$.

(ii) *implies* (i): suppose $x \wedge z = y \wedge z$. Then $x' \vee z' = y' \vee z'$ and it follows from (ii) that $z \wedge x' = (z' \vee x') \wedge z = (z' \vee y') \wedge z = y' \wedge z$. ■

An important class of regular filters, and therefore, of deductive systems, is singled out by the following:

4.7. DEFINITION. *A filter* $\mathbf{F}$ *of a bounded lattice* $\mathbf{L}$ *is said to be a Stone filter* (cf. [4]) *if it is the filter of* $\mathbf{L}$ *generated by a filter of* $\mathbf{B}(\mathbf{L})$. *In other words,* $\mathbf{F}$ *is a Stone filter if for each* x *in* $\mathbf{L}$, $x \in \mathbf{F}$ *if and only if there is* $z \in \mathbf{F} \cap \mathbf{B}(\mathbf{L})$ *such that* $z \leqslant x$.

4.8. PROPOSITION. *Each Stone filter of* $\mathbf{L} \in \mathcal{O}$ *is a regular filter.*

PROOF: According to Remark 4.4 we need to prove that if $x \wedge u = y \wedge u$ for some $u \in \mathbf{F}$, then there is v in $\mathbf{F}$ so that $x' \wedge v = y' \wedge v$. But, since $\mathbf{F}$ is a Stone filter, we can find $z \in \mathbf{F} \cap \mathbf{B}(\mathbf{L})$ such that $z \leqslant u$, and it is plain that $x \wedge z = y \wedge z$. This equality implies that $(x' \vee z') \wedge z = (y' \vee z') \wedge z$, and since z is the center of $\mathbf{L}$, we get that $x' \wedge z = y' \wedge z$. ■

Prime filters form another important class of regular filters.

Let $\mathbf{P}$ be a prime filter of $\mathbf{L} \in \mathcal{O}$ and suppose that $(x, y) \in \Theta(\mathbf{P})$. From this relation it follows that $x \in \mathbf{P}$ if and only if $y \in \mathbf{P}$, and since $\mathbf{P}$ is prime and $x \vee x' = 1$, it follows that $x \in \mathbf{P}$ or $x' \in \mathbf{P}$. So we have that one (and only one) of the following relations holds: $x \wedge y \in \mathbf{P}$ or $x' \wedge y' \in \mathbf{P}$. In the first case we have that $x' \wedge (x \wedge y) = 0 = y' \wedge (x \wedge y)$ and in the second one, $x' \wedge (x' \wedge y') = x' \wedge y' = y' \wedge (x' \wedge y')$, thus in both cases $(x', y') \in \Theta(\mathbf{P})$, and it follows from Remark 4.4 that $\mathbf{P}$ is a regular filter. In particular, $\mathbf{P}$ is a deductive system, and since each deductive system is a filter and a prime filter in $\mathbf{L} \in \mathcal{O}$ is maximal, it follows that $\mathbf{P}$ is a maximal deductive system, in the sense that it is a proper deductive system that is properly contained in no proper deductive system.

So we have that:

4.9. PROPOSITION. *Each prime filter of* $\mathbf{L} \in \mathcal{O}$ *is a regular filter that, moreover, is a maximal deductive system.*

4.10. REMARK. Let $\mathbf{M}_6$ be the orthomodular lattice represented in Fig.1:

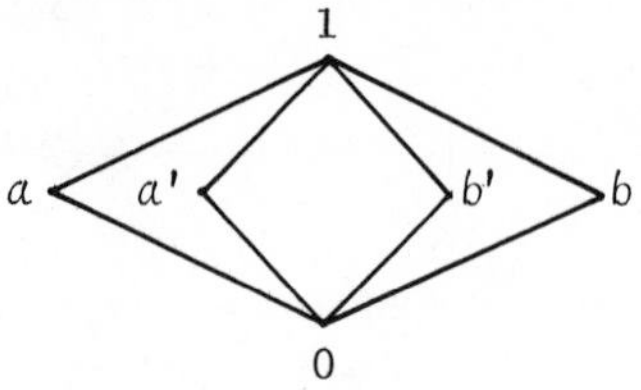

Fig. 1.

Then $[a)$ is a maximal filter that is not a deductive system. On the other hand $[1)$ is a maximal deductive system (because it is the only proper deductive system of $\mathbf{M}_6$) that is not a prime filter.

Consider now the (non-orthomodular) ortholattice $\mathbf{L}_6$ represented in Fig. 2:

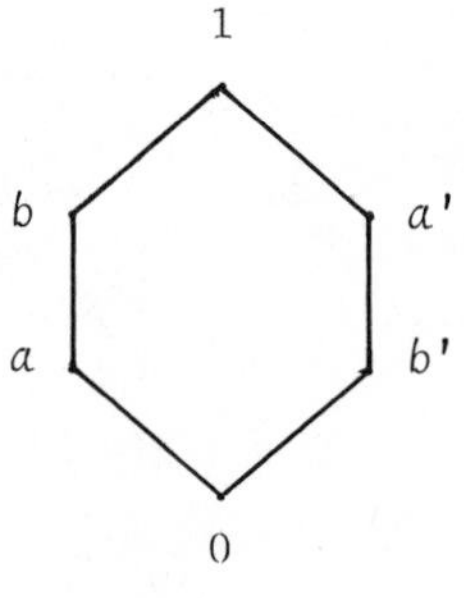

Fig. 2.

The proper deductive systems of $\mathbf{L}_6$ are $[a)$, $[b')$ and $[1)$ and they are all regular filters. $[a)$ and $[b')$ are maximal deductive systems (because they are prime filters) but they are not Stone filters. $[1)$ is a Stone filter.

Note that if we define the relation β on $\mathbf{L}_6$ by the prescription that $(x,y) \in \beta$ if and only if $x \to y = 1$ and $y \to x = 1$, then β is a congruence relation on $\mathbf{L}_6$ and $\mathbf{L}_6/\beta$ is the four-element Boolean algebra. Thus [1) is the shell of the identity and also it is the shell of the natural epimorphism $\mathbf{L}_6 \to \mathbf{L}_6/\beta$.

This remark motivates the following definitions:

4.11. DEFINITION. *Let* $\mathbf{D}$ *be a deductive system of* $\mathbf{L} \in O$. *Define a relation* $\beta(\mathbf{D})$ *by* $\beta(\mathbf{D}) = \{(x,y) \in \mathbf{L}^2: (x \to y) \wedge (y \to x) \in \mathbf{D}\}$. $\mathbf{D}$ *is said to be a Boolean deductive system if* $\beta(\mathbf{D})$ *is a congruence relation on* $\mathbf{L}$.

4.12. THEOREM. *the following are equivalent conditions for each proper filter* $\mathbf{F}$ *of* $\mathbf{L} \in O$:

(i) $\mathbf{F}$ *is a proper Boolean deductive system.*

(ii) $\mathbf{F}$ *is the shell of an epimorphism from* $\mathbf{L}$ *onto an* $\mathbf{L}^* \in B$.

(iii) $\mathbf{F}$ *is an intersection of prime filters of* $\mathbf{L}$.

PROOF: (i) *implies* (ii): Since $\beta(\mathbf{F})$ in a congruence relation on $\mathbf{L}$ it is plain that $\mathbf{L}/\beta(\mathbf{F}) \in O$. It is also clear that if $h: \mathbf{L} \to \mathbf{L}/\beta(\Theta)$ is the natural epimorphism, them $\mathbf{S}(h) = \mathbf{F}$. Suppose, now, that $h(x) \vee h(y) = 1$ and $h(x) \wedge h(y) = 0$. Then, $x' \to y = x \vee y \in \mathbf{F}$ and $y \to x' = y' \vee x' \in \mathbf{F}$, i. e. $(x',y) \in \beta(\mathbf{F})$ and hence $h(x)' = h(y)$. Since h is an onto map, we have seen that $\mathbf{L}/\beta(\mathbf{F})$ is a uniquely complemented ortholattice, and it is well known (see, for instance, [2] p. 18) that this implies that $\mathbf{L}/\beta(\mathbf{F}) \in B$.

(ii) *implies* (iii): Let $h: \mathbf{L} \to \mathbf{L}^* \in B$ be an epimorphism. If $\{\mathbf{P}_j\}_{j \in \mathbf{J}}$ is the set of prime filters of $\mathbf{L}^*$, it is easy to see that $\mathbf{S}(h) = \bigcap_{j \in \mathbf{J}} h^{-1}(\mathbf{P}_j)$

(iii) *implies* (i): If $\mathbf{P}$ is a prime filter of $\mathbf{L} \in O$, it is easy to see that $\beta(\mathbf{P})$ is a congruence. The proof is completed by observing that the class of Boolean deductive systems is closed under intersections. ∎

4.13. REMARK. Assume that $h_i: \mathbf{L} \to \mathbf{L}_i$ are epimorphisms, where $\mathbf{L} \in O$ and $\mathbf{L}_i \in B$, $i = 1, 2$. If $\mathbf{S}(h_1) \subseteq \mathbf{S}(h_2)$, then there is a unique epimorphism $h: \mathbf{L}_1 \to \mathbf{L}_2$ such that $hh_1 = h_2$. Therefore, the Boolean homomorphic images of $\mathbf{L} \in O$ are in one-to-one correspondence with the proper Boolean deductive systems of $\mathbf{L}$.

From Proposition 4.9 and the previous theorem it follows that a prime filter **P** of $\mathbf{L} \in O$ is simultaneously a regular filter and a Boolean deductive system and that $\Theta(\mathbf{P}) = \beta(\mathbf{P})$: The example preceding Definition 4.11 shows that in general, $\Theta(\mathbf{D}) \neq \beta(\mathbf{D})$. Moreover, {1} is a regular filter of $\mathbf{M}_6$ (see Fig. 1) that it is not a Boolean deductive system.

In the ortholattice $\mathbf{L}_{10}$ represented in Fig. 3, the deductive systems are $[a)$, $[c)$, $[d)$, $[b')$, $[c')$, $[d')$, and $[1)$. The prime filters are $[a)$, $[c)$ and $[d)$. Thus $\mathbf{L}_{10}$ is an example of an ortholattice in which all deductive systems are Boolean. Note that $[c)$ is not a regular filter, because $(c' \vee a) \wedge c = b \neq c \wedge a$ (see Proposition 4.6):

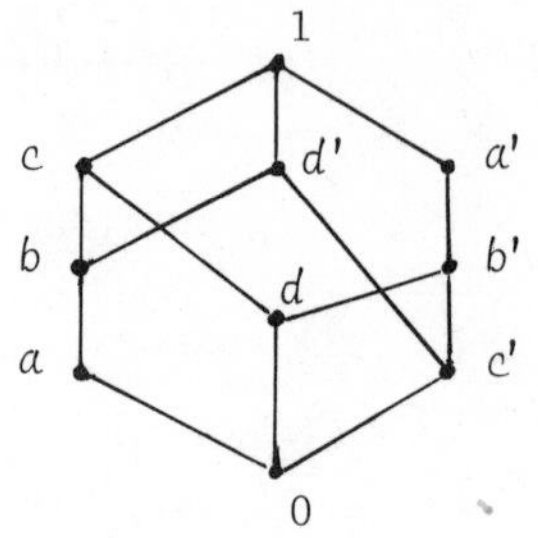

Fig. 3.

Having studied the Boolean images of ortholattices, we shall consider now the orthomodular images. We start with the following:

4.14. DEFINITION. *Let* **D** *be a deductive system of* $\mathbf{L} \in O$. *Define a relation* $\upsilon(\mathbf{D})$ *by* $\upsilon(\mathbf{D}) = \{(x,y) \in \mathbf{L}^2 : (x \leftrightarrow y) \wedge (y \leftrightarrow x) \in \mathbf{D}\}$. **D** *is said to be an orthomodular deductive system if* $\upsilon(\mathbf{D})$ *is a congruence relation on* **L**.

4.15. THEOREM. *The following are equivalent conditions for each* $\mathbf{D} \subseteq \mathbf{L} \in O$:

(i) **D** *is a proper orthomodular deductive system of* **L**.

(ii) **D** *is the shell of an epimorphism from* **L** *onto an* $\mathbf{L}^* \in M$.

PROOF: (i) *implies* (ii). Set $\mathbf{L}^* = \mathbf{L}/_{\upsilon(\mathbf{D})}$, and $h: \mathbf{L} \to \mathbf{L}^*$ be the natural epimorphism. Since $\mathbf{D} = \mathbf{S}(h)$, it is enough to prove that $\mathbf{L}^* \in M$. Suppose that $h(x) \leq h(y)$. Then it follows from (S2) that $h(y) \leftrightarrow (h(x) \vee (h(y)$

$\wedge\ h(x)')) = 1$. So $y \leftrightarrow (x \vee (y \wedge x')) \in \mathbf{D}$. On the other hand, since $h(x) \vee (h(y) \wedge h(x)') \leqslant h(y)$, we also have $x \vee (y \wedge x') \leftrightarrow y \in \mathbf{D}$. Therefore $(y, x \vee (y \wedge x')) \in \upsilon(\mathbf{D})$ and $h(y) = h(x) \vee (h(y) \wedge h(x)')$. Sine h is an onto map, this shows that $\mathbf{L}^* \in \boldsymbol{M}$.

(ii) *implies* (i). Let $\mathbf{L}^* \in \boldsymbol{M}$ and $h: \mathbf{L} \to \mathbf{L}^*$ be an epimorphism. It follows at once from Proposition 3.2 that $h(x) \leqslant h(y)$ if and only if $x \leftrightarrow y \in \mathbf{S}(h)$. So $h(x) = h(y)$ if and only if $(x,y) \in \upsilon(\mathbf{S}(h))$, and this shows that $\mathbf{S}(h)$ is an orthomodular deductive system of $\mathbf{L}$. ■

4.16. REMARK. As in Remark 4.13, we can easily see that there is a one-to-one correspondence between the orthomodular images of an $\mathbf{L} \in \mathcal{O}$ and the proper orthomodular deductive systems of $\mathbf{L}$.

Since $B \subset \boldsymbol{M}$, it follows from Remark 4.13 and 4.16 that:

4.17. PROPOSITION. *Each Boolean deductive system* $\mathbf{D}$ *of* $\mathbf{L} \in \mathcal{O}$ *is an orthomodular deductive system and* $\beta(\mathbf{D}) = \upsilon(\mathbf{D})$.

Consider the ortholattice $\mathbf{L}_8$ represented in Figure 4:

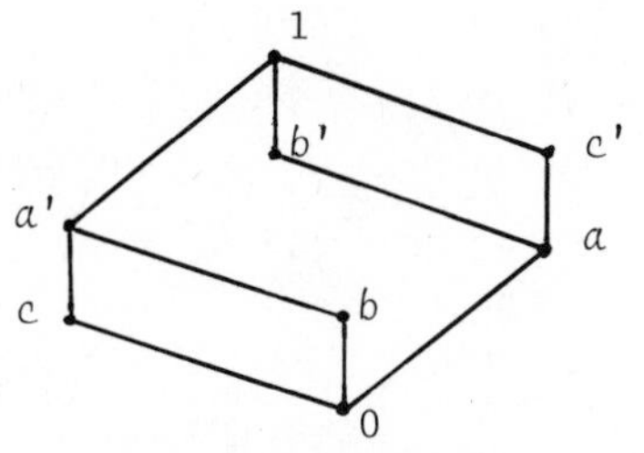

Fig. 4.

Then $\{1\}$ is a regular filter of $\mathbf{L}_8$ that it is not an orthomodular deductive system, because (a,b') and (a,c') belong to $\upsilon(\{1\})$ but $b' \leftrightarrow c' = b' \notin \{1\}$.

5. CONGRUENCES IN ORTHOCOMPLEMENTED LATTICES.

The following theorem has been proved by P. D. Finch [5]. For completeness, we sketch the proof.

5.1. THEOREM (Finch). *Each Finch filter of an* $\mathbf{L} \in \boldsymbol{M}$ *is regular.*

PROOF: Let $\mathbf{F}$ be a Finch filter of $\mathbf{L}$ and suppose that $(x,y) \in \Theta(\mathbf{F})$. This relation means that there are elements u, v in $\mathbf{F}$ such that $x \wedge u = y \wedge v$. Set $\alpha = x \to (x \wedge u)$ and $\beta = y \to (y \wedge v)$. Since $\mathbf{F}$ is a Finch filter, α and β belong to $\mathbf{F}$, and the orthomodular law (OM') implies that $x \wedge \alpha = y \wedge \beta$. Moreover, since $\alpha' \leqslant x$ and $\beta' \leqslant y$, a new application of (OM') yields $x' \wedge \beta = \alpha \wedge \beta \wedge (x \wedge \alpha)' = \alpha \wedge \beta \wedge (y \wedge \beta)' = \alpha \wedge y'$. Therefore, $(x',y') \in \Theta(\mathbf{F})$ and by Remark 4.4, it follows that $\mathbf{F}$ is a regular filter. ∎

Finch has also proved that the correspondence $\mathbf{D} \to \Theta(\mathbf{D})$ establishes a bijection between the set of Finch filters of $\mathbf{L} \in \boldsymbol{M}$ and the set of all congruence on $\mathbf{L}$. We shall obtain this result from the following remarks.

Suppose that $\mathbf{D}$ is a deductive system of $\mathbf{L} \in \boldsymbol{M}$. According to Proposition 2.8 and Theorem 5.1, $\Theta(\mathbf{D})$ is a congruence relation, and since $\boldsymbol{M}$ is an equational class, $\mathbf{L}/_{\Theta(\mathbf{D})} \in \boldsymbol{M}$. If h: $\mathbf{L} \to \mathbf{L}/_{\Theta(\mathbf{D})}$ is the natural epimorphism, Theorem 4.15 allows us to conclude that $\mathbf{D}$ is an orthomodular deductive system and that $\Theta(\mathbf{D}) = \upsilon(\mathbf{D})$. So we have:

5.2. COROLLARY. *The following are equivalent conditions for each subset* $\mathbf{D}$ *of an* $\mathbf{L} \in \boldsymbol{M}$:

(i) $\mathbf{D}$ *is a deductive system of* $\mathbf{L}$.

(ii) $\mathbf{D}$ *is an orthomodular deductive system of* $\mathbf{L}$.

(iii) $\mathbf{D}$ *is a regular filter of* $\mathbf{L}$.

Moreover, if $\mathbf{D}$ *satisfies these conditions,* $\Theta(\mathbf{D}) = \upsilon(\mathbf{D})$ *and the map* $\mathbf{D} \to \Theta(\mathbf{D})$ *establishes an order anti-isomorphism between the set* $D(\mathbf{L})$ *of deductive systems of* $\mathbf{L}$ *ordered by inclusion and* $C(\mathbf{L})$, *the congruence lattice of* $\mathbf{L}$.

5.3. REMARK. It is worthwhile to note that $\mathbf{L}_8$ (see fig. 3) is an example of a non-orthocomplemented ortholattice such that the map $\mathbf{D} \to \Theta(\mathbf{D})$ establishes a bijection from $D(\mathbf{L}_8)$ onto $C(\mathbf{L}_8)$ For, the proper deductive systems of $\mathbf{L}_8$ are $[a)$ and $[1)$. Since $[a)$ is a prime filter, $[a)$ determines only one congruence. By examining the possible partitions of $\mathbf{L}_8$, we see that $\Theta([1))$ is the only congruence having $[1)$ as an equivalence class. Thus $C(\mathbf{L}_8)$ is the three-element chain $\Theta([1)) < \Theta([a)) < \Theta([0))$ = improper congruence, corresponding to the improper filter $[0)$.

We have seen in Proposition 4.8 that each Stone filter is regular. We shall see now that Stone filters are the only principal regular filters of orthomodular lattices.

5.4. PROPOSITION. *A principal filter* [z) *of an* $\mathbf{L} \in \boldsymbol{M}$ *is regular in and only if* $z \in \mathbf{B}(\mathbf{L})$.

PROOF: If $z \in \mathbf{B}(\mathbf{L})$, then [z) is a Stone filter, and Proposition 4.8 implies that [z) is a regular filter. Conversely, if [z) is regular, it is a deductive system, and, by Proposition 2.5 (iii), z' is the pseudocomplement of z; therefore z' is the only complement of z and this implies that $z \in \mathbf{B}(\mathbf{L})$ (see [10], Theorem 36.9). Another proof of this proposition can be derived from Proposition 4.6. ■

As a consequence of Corollary 5.2 and Proposition 5.4 we have that:

5.5. COROLLARY. *If* $\mathbf{L} \in \boldsymbol{M}$ *satisfies the chain condition (i. e., every filter of* $\mathbf{L}$ *is principal),* $C(\mathbf{L})$ *is anti-isomorphic to* $\mathbf{B}(\mathbf{L})$, *and therefore,* $C(\mathbf{L}) \in B$.

A simple consequence of this result is:

5.6. COROLLARY. *The following are equivalent conditions for an orthomodular lattice* $\mathbf{L}$ *satisfying the chain condition:*

(i) $\mathbf{B}(\mathbf{L}) = \{0,1\}$.

(ii) $\mathbf{L}$ *is irreducible.*

(iii) $\mathbf{L}$ *is subdirectly irreducible.*

(iv) $\mathbf{L}$ *is simple.*

5.7. REMARK. Let L be the orthomodular lattice formed by the closed subspaces of the infinite dimensional separable Hilbert space $\mathbf{H}$. It is well known that $\mathrm{B}(L) = \{(\mathbf{0}), \mathbf{H}\}$, so L is irreducible, and the only Stone filter is $[\mathbf{H})$. It was proved by Finch [5] that the set of all closed subspaces of finite codimension in $\mathbf{H}$ forms a Finch filter. So we have an example of an infinite irreducible orthomodular lattice that it is not simple and, moreover, of a deductive system that it is not a Stone filter.

REFERENCES.

[1] G. Birkhoff, Lattice Theory, Amer. Math. Soc. Colloq. Pub. 25, 3nd. ed., Providence, R. I., 1967.

[2] T. S. Blyth and M. F. Janowitz, Residuation Theory, Pergamon Press, Oxford - N. York, 1972.

[3] G. Bodiou, Théorie Dialectique des Probabilités Englobant leurs Calculs Classique et Quantique, Gauthier-Villars, Paris, 1964.

[4] R. Cignoli, *Stone filters and ideals in distributive lattices*, Bull. Math. Soc. Scie. Math. R. S. Roumainie, 15 (16) (1971), 131-137.

[5] P. D. Finch, *Congruence relations on orthomodular lattices*, J. Austral. Math. Soc. 6 (1966), 46-54.

[6] C. G. Gratzer, Lattice Theory: First concepts and distributive lattices, W. H. Freeman, San Francisco, Ca. 1971.

[7] S. S. Holland, *The current interest in orthomodular lattices*, Trends in Lattice Theory (J. C. Abott Editor), Van Nostrand Reinhold Math. Studies 31, N. York, 1970.

[8] J. M. Jauch, Foundations of Quantum Mechanics, Addison-Wesley, Reading, Ma., 1968.

[9] G. Kalmbach, *Orthomodular logic*, Proc. Univ. Houston Lattice Theory Conf., Houston, 1973, 498-503.

[10] F. Maeda and S. Maeda, Theory of Symmetric Lattices, Springer-Verlag, Berlin, Heidelberg, N. York, 1970.

[11] V. S. Varadarajan, Geometry of Quantum Theory, vol. I, Van Nostrand, Princeton, N. J., 1968.

Departamento de Matemática
Universidade Estadual de Campinas
13.100 Campinas, S.P., Brazil.

INTERNAL MODELS FOR ANY FINITE SUBSET OF THE AXIOMS FOR THE IMPREDICATIVE THEORY OF CLASSES.*

by *M. CORRADA* and *R. CHUAQUI*.

ABSTRACT. In *Global and local choice functions* (Israel J. Math., 22 (1975)), Gaifman proved that ZFGC (i.e. Zermelo-Fraenkel with a global choice function) is a conservative extension of ZFC (ZF with the axiom of choice). In order to carry out this proof, he constructed internal models for finite subsets of the axioms of ZFGC. His methods do not use forcing, as was the case with the previous proof by Felgner (cf. *Comparison of the axioms of local and universal choice*, Fund. Math. 49 (1960)).

The purpose of this paper is to adapt Gaifman's method to the impredicative theory of classes. This is the theory, due in essence to Morse, which uses an impredicative form of the axiom of class specification (for this theory see the Appendix of Kelley, General Topology; Mostowski, Constructible Sets with Applications or; Chuaqui, *Internal and forcing models for the impredicative theory of classes*, to appear in Diss. Math.).

INTRODUCTION.

In [8] Gaifman proved that ZFGC (i. e., Zermelo-Fraenkel with a global choice function) is a conservative extension of ZFC (ZF with the axiom of choice). In order to carry out this proof, he constructed internal models for finite subsets of the axioms of ZFGC. His methods do not use forcing, as was the case with the previous proof by Felgner [5].

---

* This article was partially written when the first author was at the State University of Campinas, São Paulo, Brazil with a fellowship from the Brazilian Government.

The purpose of this paper is to adapt Gaifman's method to the impredicative theory of classes. This is the theory, due in essence to Morse, which uses an impredicative form of the axiom of class specification (for this theory see [11] Appendix, [13] or [2]).

The main problem in giving a proof for this theory is to include in the models a class structure sufficient to satisfy a finite number of the axioms of class specification, and, at the same time, the Axiom of Replacement. The chief tool for this purpose is, what we call the Reduction Theorem (Theorem 2.2.4). Intuitively this theorem corresponds to the fact that classes are specified by formulae and, in a sense to be made precise later, if formulae substitute the respective classes the basic relation of satisfaction is preserved.

This paper is divided into three sections. The first section contains the main set-theoretical notions and notations used throughout and gives the axioms of the impredicative class theory in a two-sorted first-order language without equality. Section two deals with some metamathematical aspects of the theory. In particular, it contains a satisfaction relation for a two-sorted first-order language. In order to do this, it is necessary to arithmetize metamathematics. Semantics is developed in § 2, and finally, the Reduction Theorem is proved (Theorem 2.2.4). The final section contains a modified version of Gaifman's proof.

1. A SURVEY OF TWO-SORTED IMPREDICATIVE THEORY OF CLASSES.

We shall formulate the impredicative theory of classes in a two-sorted language without equality. This theory can be found in [11], [13], [1], [2] or [3] in a one-sorted language with equality. Although it may appear simpler to formulate the theory in this last language, it is advisable for the purpose of this paper to do it in a language without equality and with two types of variables, one for classes and the other for sets.

Equality is introduced as defined relation by means of extensionality. This possibility was first pointed out by Fraenkel in [7].

1.1. PRELIMINARIES.

The theory will be formulated within two-sorted predicate claculus without equality whose unique non-logical constant is $\in$ (*membership rela-*

tion). As the underlying language we take

$$\mathcal{L} = \{ \vee, \neg, \exists, \in \}$$

where the logical constants have their usual sense and the non-logical constant $\in$ is a two place predicate. We assume that the variables are of two types: the first type of variables - called *set variables* are arranged in an infinite sequence without repetitions like this

$$\nu = \langle \nu_0, \ldots, \nu_k, \ldots \rangle_{k \in N};$$

and the second type of variables - called *class variables* - are arranged in an infinite sequence without repetitions like this

$$\mathbf{X} = \langle X_0, \ldots, X_k, \ldots \rangle_{k \in N}.$$

We adopt the following convention: a, b, c ... will denote sets; x, y, z the firts three terms of the sequence ν. A, B, C will denote classes, and X, Y, Z the first three terms of the sequence $\mathbf{X}$.

Note that the symbols $\wedge$ (*conjunction*), $\rightarrow$ (*implication*), $\leftrightarrow$ (*equivalence*), $\forall$ (*universal quantifier*) are introduced by definition by means of $\neg$, $\vee$, and $\exists$ as usual, and will be considered as abbreviations.

Equality is introduced as a defined relation by means of extensionality, that is to say it is defined by

$$y_i = y_j \leftrightarrow \forall \nu_k (\nu_k \in y_i \leftrightarrow \nu_k \in y_j)$$

where y_i, y_j are variables of any type and in the case y_i or y_j are set variables then ν_k is the next following variable after y_i and y_j. Note that $=$ is a defined predicate and not a symbol of the theory.

The *atomic formulae* of our language are the expressions of the form $x \in y$ and $x \in Y$. The class of *$\mathcal{L}$-formulae* is defined as usual. The expressions $X \in Y$ will be considered as an abbreviation for the formula:

$$\exists u (X = u \wedge u \in Y).$$

Lower case Greek letters φ, ψ, θ with or without subscripts and superscripts will denote $\mathcal{L}$-formulae. Capital Greek letters Σ, Λ, Π will denote arbitrary sets of $\mathcal{L}$-formulae.

In the following, we shall adopt usual logical conventions, definitions and theorems. $\vdash$ will denote logical inference.

For an $\mathcal{L}$-formula φ, and $P(x)$ an $\mathcal{L}$-formula, the *relativization* φ^P of φ is obtained by replacing all the quantifiers $\forall x$ and $\exists x$ that occur in

φ by $\forall x(P(x) \rightarrow \ldots)$ and $\exists x(P(x) \wedge \ldots)$ respectively. When u is a set, the relativization φ^u of φ means the relativization φ^P of φ where $P(x) = x \in u$. If $P(x)$ is an $\mathcal{L}$-formula with a free set variable and $Q(X)$ is an $\mathcal{L}$-formula with a free class variable the relativization $\varphi^{\langle P,Q \rangle}$ of φ is obtained by replacing all the quantifiers $\forall x, \forall X$, $\exists x$ and $\exists X$ that occur in φ by $\forall x(P(x) \rightarrow \ldots)$, $\forall X(Q(X) \rightarrow \ldots)$, $\exists x(P(x) \wedge \ldots)$, and $\exists X(Q(X) \wedge \ldots)$ respectively. In the case u, v are sets the relativization $\varphi^{\langle u,v \rangle}$ of φ stands for the relativization $\varphi^{\langle P,Q \rangle}$ of φ where $P(x) = x \in u$ and $Q(X) = X \in v$.

An $\mathcal{L}$-formula φ is said to be *predicative* if it is logically equivalent to an $\mathcal{L}$-formula ψ where all the quantifiers are restricted to sets.

1.2. AXIOMS.

We give an axiomatization for the impredicative theory of classes which is equivalent to the usual one (for instance, the axiomatizations that appears in [2] and [13]).

The *axioms* of the theory are the universal closure of the following formulae:

(I) AXIOM OF $\mathcal{L}$-EQUALITY.

$$a = b \rightarrow (a \in X \rightarrow b \in X)$$

(II) AXIOM SCHEMA OF CLASS SPECIFICATION.

for every formula φ in which A *is not free the following is an axiom* (φ *may contain any free variable, except* A):

$$\exists A \, \forall x \, (x \in A \leftrightarrow \varphi)$$

(III) AXIOM ASSERTING THAT EVERY SET IS A CLASS.

$$\forall x \, \exists X (x = X)$$

(IV) AXIOM OF THE EMPTY SET.

$$\exists a \, \forall x (x \notin a)$$

(V) AXIOM OF UNIONS.

$$\exists a \, \forall x (x \in a \leftrightarrow \exists y (x \in y \in b))$$

(VI) AXIOM OF THE POWER SET.

$\exists a\ \forall x(x \in a \leftrightarrow \forall y(y \in x \rightarrow y \in b))$

(VII) AXIOM OF INFINITY.

$\exists x\ \exists a(x \in a \wedge \forall x(x \in a \rightarrow \exists y(x \in y \in a)))$

(VIII) AXIOM OF REGULARITY OR FOUNDATION.

$\exists x(x \in A) \rightarrow \exists x(x \in A \wedge \forall y(y \in x \rightarrow y \notin A))$

In order to introduce the pairing and replacement axioms we need some auxialiary definitions. The unordered pair of the sets a and b is denoted by $\{a,b\}$. $\mathrm{Func}(X)$ means that X is a function. If f is a function $f^{*}x$ is the image of the set x under f, $f^{\bullet}x$ is the value of f at the point x, and $\mathbb{D}f$ is the domain of f.

(IX) AXIOM OF PAIRING.

$\forall x\ \forall y\ \exists z(z = \{x,y\})$

(X) REPLACEMENT AXIOM.

$\forall X(\mathrm{Func}(X) \rightarrow \forall z\ \exists y(X^{*}z = y))$

We call the theory obtained with axioms I - X, M2.

For our development we need to introduce some definitions and notations. We shall write below those that are non-standard. Apart from these exceptional cases, our notations and definitions will be standard. The unique class specified by Axiom II is denoted by $\{x\colon \varphi\}$. The class $\{x\colon x \in A \wedge \varphi(x)\}$ is also written by $\{x \in A\colon \varphi\}$. The rank function will be denoted by ρ. The notions of *transitive* and *supertransitive set* are defined as follows:

$\mathrm{Trans}(u) \leftrightarrow \forall a\ \forall b(a \in b \wedge b \in u \rightarrow a \in u)$.

$\mathrm{Strans}(u) \leftrightarrow \mathrm{Trans}(u) \wedge \forall a\ \forall b(b \in u \wedge a \subseteq b \rightarrow a \in u)$.

OR denotes the class of ordinal numbers. $\mathrm{Lim}(\alpha)$ means that α a limit ordinal. Capital or lower case letters f, g, h with or without subscripts and superscripts are used to represent functions. Greek letter $\alpha, \beta, \gamma, \delta, \xi$ with or without subscripts and superscripts are used to represent ordinals.

Capital Greek letters Γ, Δ with or without subscripts will represent classes of ordinals. i, j, l, m, p, q with or without subscripts will represent natural numbers, i. e., elements of ω (the set of finite ordinals). $R(\alpha)$ is defined as usual as the set of all sets of rank less than α. pr_i will be the ith projection function.

1.3. AXIOMS OF CHOICE.

We introduce the *local form of the axiom of choice:*

AC: $\forall x(0 \notin x \rightarrow \exists g(\mathrm{Func}(g) \wedge \mathcal{D}g = x \wedge \forall y(y \in x \rightarrow g^{\cdot}y \in y)))$

(the function g considered in AC is called a *local choice function*, or a *choice function for* x).

We call the theory obtained from M2 + AC, M2C.

In M2C we can develop the theory of cardinal numbers in the usual way. A cardinal is an initial ordinal. If x is a set, the cardinality of x is written $\bar{\bar{x}}$. If κ is a cardinal we define $S_\kappa(x) = \{y: y \subseteq x \wedge \bar{\bar{y}} < \kappa\}$.

Now extend the language $\mathcal{L}$ by adding a one-place operation symbol F, and we call this new language $\mathcal{L}(F)$.

We define as usual the notion of *set-term* and *class-term* : class variables are all the class-terms; a set variable is a set-term, and if τ is a set-term, then $F^{\cdot}\tau$ is also a set-term. The set of $\mathcal{L}(F)$-*formulae* is defined as usual from the atomic formulae $\sigma \in \tau$ where σ is a set-term and τ is a set or class-term. In practice,usage of metalinguistic (syntactical) variables that refer to $\mathcal{L}(F)$-formulae will be the same as those which refer to $\mathcal{L}$-formulae. Because $\mathcal{L} \subseteq \mathcal{L}(F)$ we can forlumate **M2C** in the language $\mathcal{L}(F)$. In the class of $\mathcal{L}(F)$-formulae can be defined an *extended substitution operation* in which the operation symbol is substituted everywhere by a function h; we denote by $\varphi\binom{F}{h}$ the result of performing the above operation to the $\mathcal{L}(F)$-formula φ.

Now we are able to introduce the *global form of the axiom of choice* (Axiom E of [10]):

GLOBAL AXIOM OF CHOICE

E: $\forall x(x \neq 0 \rightarrow F^{\cdot}x \in x)$.

AXIOM OF $\mathcal{L}(F)$ EQUALITY.

$$Eq(\mathcal{L}(F)): \qquad x = y \rightarrow F{}^{\bullet}x = F{}^{\bullet}y \ .$$

The theory obtained by M2 + E + Eq($\mathcal{L}(F)$) will be called M2E.

For other equivalent forms of the axiom of choice see [14].

2. METAMATHEMATICAL NOTIONS CONCERNING THE THEORY M2.

In this section the two metamathematical aspects of the theory **M2**: its syntax and its semantics, will be developed. Through the well known method of arithmetization due to Gödel [9] it is possible to express within the theory certain metalinguistic notion. A sketch of this method of arithmetization will be done in what follows. For further details the reader is referred to the paper by Feferman [4].

2.1. SATISFACTION.

In **M2** Peano's arithmetic is developed with the sucessor operation, ' , defined by $x' = x \cup \{x\}$.

Let N be the set of natural numbers in the metatheory. $n, n_0, n_1, \ldots$ will be elements of the set N. For each n we will define the *numeral* Δ_n of $\mathcal{L}(F)$ by the following recursive definition:

$$\Delta_0 = 0$$

$$\Delta_{n+1} = \Delta'_n$$

It is convenient to identify expressions with natural numbers - called *Gödel numbers* - and we assume that such identification has been made. The representation of relations by formulae of **M2C** is defined as usual; similarly for strong representability of a function (cf. [3], [4]).

In each row of the table below the expression in the left columm is understood as denoting that formula of **M2** which expresses in the "natural" way the class, relation or function indicated in the right columm. We shall use the same abbreviation beginning with the corresponding capital letter for the notions of the metalanguage indicated in the right columm. This convention will be used for all the metatheoretical notions defined inside **M2**.

vl_1	the class of set variables.
vl_2	the class of class variables.
vl	the class of variables.
sett	the class of set-terms.
sel(k)	the concatenation of the Gödel number corresponding to the operation symbol F and k.
$e_1(k,l)$	k belongs to l, $k, l \in$ sett.
$e_2(k,l)$	k belongs to l, $k \in$ sett, $l \in vl_2$.
ng(k)	the negation of k.
ds(k,l)	the disjunction of k and l.
$qu_1(k,l)$	there exists k such that l, $k \in vl_1$.
$qu_2(k,l)$	there exists k such that l, $k \in vl_2$.
qu(k,l)	there exists k such that l, $k \in vl$.
$qu_1\mathbf{k}$	the class of all existential set quantifications of k.
$qu_2\mathbf{k}$	the class of all existential class quantifications of k.
qu**k**	the class of all existential quantifications of k.
$fv_1\mathbf{k}$	the class of free set variables that occur in **k**.
$fv_2\mathbf{k}$	the class of free class variables that occur in k.
fv**k**	the class of free variables that occur in **k**.
$s(k,l) = k(\Delta_l)$	Δ_l substituted for v_0 in k.
fm	the class of $\mathcal{L}(F)$-formulae.
fmw.o.	the class of $\mathcal{L}$-formulae.
pred	the class of predicative $\mathcal{L}(F)$-formulae.
ucl(k)	the class of universal closures of **k**.
qn(k)	the number of quantifiers of **k**.
$for_n(\mathcal{L}(F))$	the class of formulae of $\mathcal{L}(F)$ with at most n quantifiers.

φ_m will demote the $\mathcal{L}(F)$-formula whose Gödel number is m, and Δ_ψ the numeral Δ_m where m is the Gödel number of ψ.

Because the general principle of definitions by recursion over well founded relations is valid in M2 (see [2], Theorem 4.2) it is possible to do the following definition. The recursion is done over the relation of subterm or subformula which are well founded.

DEFINITION 2.1.1.

(i) Let J be the ternary operation defined by:

(a) $k \notin \text{sett} \vee \neg \text{Func}(H) \vee \mathcal{D}f \neq \text{fv}_1\mathbf{k} \rightarrow J(k, f, H) = V$,

(b) $\mathbf{k} \in \text{vl}_1 \wedge \text{Func}(H) \wedge \mathcal{D}f = \text{fv}_1 k \rightarrow J(k, f, H) = f\cdot\mathbf{k}$,

(c) $k \in \text{sett} \wedge \text{Func}(H) \wedge \mathcal{D}f = \text{fv}_1\mathbf{k} \rightarrow J(\text{sel}(\mathbf{k}), f, H) = H\cdot J(\mathbf{k}, f, H)$;

(ii) Let K be the binary operation defined by:

(a) $\mathbf{k} \notin \text{fm} \vee \neg \text{Func}(H) \rightarrow K(\mathbf{k}, H) = 0$,

(b) $k \in \text{fm} \wedge \text{Func}(H) \rightarrow$

(b.1) $k = e_1(l, m) \rightarrow K(\mathbf{k}, H) = \{\langle f, 0\rangle : \mathcal{D}f = \text{fv}_1 k \wedge J(l, f, H) \in J(m, f, H)\}$,

(b.2) $k = e_2(l, m) \rightarrow K(k, H) = \{\langle f, g\rangle : \mathcal{D}f = \text{fv}_1 k \wedge \mathcal{D}g = \text{fv}_2 k \wedge J(l, f, H) \in g\cdot m\}$,

(b.3) $k = \text{ng}(l) \rightarrow K(k, H) = \{\langle f, g\rangle : \mathcal{D}f = \text{fv}_1 k \wedge \mathcal{D}g = \text{fv}_2 k \wedge \langle f, g\rangle \notin K(l, H)\}$,

(b.4) $k = \text{ds}(l, m) \rightarrow K(k, H) = \{\langle f, g\rangle : \mathcal{D}f = \text{fv}_1 k \wedge \mathcal{D}g = \text{fv}_2\,\mathbf{k} \wedge (\langle f \upharpoonright \text{fv}_1 l, g \upharpoonright \text{fv}_2 l\rangle \in K(l, H) \vee \langle f \upharpoonright \text{fv}_1 m, g \upharpoonright \text{fv}_2 m\rangle \in K(m, H))\}$,

(b.5) $\mathbf{k} = \text{qu}(l, p) \wedge l \notin \text{fvp} \rightarrow K(k, H) = K(p, H)$,

(b.6) $k = \text{qu}_1(l, p) \wedge l \in \text{fv}_1 p \rightarrow K(k, H) = \{\langle f, g\rangle : \mathcal{D}f = \text{fv}_1 k \wedge \mathcal{D}g = \text{fv}_2 k \wedge \exists x(\langle f \cup \{\langle x, l\rangle\}, g\rangle \in K(p, H))\}$,

(b.7) $k = \text{qu}_2(l, p) \wedge l \in \text{fv}_2 p \rightarrow K(k, H) = \{\langle f, g\rangle : \mathcal{D}f = \text{fv}_1 k \wedge \mathcal{D}g = \text{fv}_2 k \wedge \exists x(\langle f, g \cup \{\langle x, l\rangle\}\rangle \in K(p, H))\}$;

(iii) $\mathrm{Sat}(k,f,g,H) \leftrightarrow \langle f \restriction fv_1 k, g \restriction fv_2 k \rangle \in K(k,H)$.

The above formula $\mathrm{Sat}(k,f,g,H)$ denotes satisfaction of a formula whose Gödel number is k; f, g are the corresponding assignments to set and class variables. We make use of the fact that formula $\mathrm{Sat}(k,f,g,H)$ is absolute with respect to supertransitive sets.

We have the following metatheorem, whose proof consists in trivial verifications according to Definition 2.1.1.

THEOREM SCHEMA 2.1.2. (M2C) *For every* $\varphi \in \mathrm{Fm}$ *the following is a theorem:*

(i) $\Delta_\varphi = \Delta_{v_i \in v_j} \to (\mathrm{Sat}(\Delta_\varphi, f, g, H) \leftrightarrow f \cdot \Delta_{v_i} \in f \cdot \Delta_{v_j}$.

(ii) $\Delta_\varphi = \Delta_{v_i \in x_j} \to (\mathrm{Sat}(\Delta_\varphi, f, g, H) \leftrightarrow f \cdot \Delta_{v_i} \in g \cdot \Delta_{x_j}$.

(iii) $\Delta_\varphi = \Delta_{v_i \in F \cdot v_j} \to (\mathrm{Sat}(\Delta_\varphi, f, g, H) \leftrightarrow f \cdot \Delta_{v_i} \in H \cdot f \cdot \Delta_{v_j}$.

(iv) $\Delta_\varphi = \Delta_{F \cdot v_i \in v_j} \to (\mathrm{Sat}(\Delta_\varphi, f, g, H) \leftrightarrow H \cdot f \cdot \Delta_{v_i} \in f \cdot \Delta_{v_j}$.

(v) $\Delta_\varphi = \Delta_{F \cdot v_i \in F \cdot v_j} \to (\mathrm{Sat}(\Delta_\varphi, f, g, H) \leftrightarrow H \cdot f \cdot \Delta_{v_i} \in H \cdot f \cdot \Delta_{v_j}$.

(vi) $\Delta_\varphi = \Delta_{F \cdot v_i \in x_j} \to (\mathrm{Sat}(\Delta_\varphi, f, g, H) \leftrightarrow H \cdot f \cdot \Delta_{v_i} \in g \cdot \Delta_{x_j}$

(vii) $\Delta_\varphi = \Delta_{\neg\psi} \to (\mathrm{Sat}(\Delta_\varphi, f, g, H) \leftrightarrow \neg \mathrm{Sat}(\Delta_\psi, f, g, H))$;

(viii) $\Delta_\varphi = \Delta_{\psi \vee \theta} \to (\mathrm{Sat}(\Delta_\varphi, f, g, H) \leftrightarrow (\mathrm{Sat}(\Delta_\psi, f, g, H) \vee$
$\mathrm{Sat}(\Delta_\theta, f, g, H)))$;

(ix) $\Delta_\varphi = \exists v_i \psi \to (\mathrm{Sat}(\Delta_\varphi, f, g, H) \leftrightarrow$
$\exists x (\mathrm{Sat}(\Delta_\psi, f \sim f \restriction \{\Delta_{v_i}\} \cup \{\langle x, \Delta_{v_i} \rangle\}, g, H)))$;

(x) $\Delta_\varphi = \exists X_j \psi \to (\mathrm{Sat}(\Delta_\varphi, f, g, H) \leftrightarrow \exists x (\mathrm{Sat}(\Delta_\psi, f,$
$g \sim g \restriction \{\Delta_{x_j}\} \cup \{\langle x, \Delta_{x_j} \rangle\}, H)))$.

Due to the use of a two-sorted predicate calculus to define a predicate, whose intended interpretation correspond to the satisfaction notion for a fixed assignment, is rather special. It is defined a 6-ary relation between two sets, a natural number and three functions; the first two functions corresponds respectively to the assignments for set and class variables that occur in the formula whose Gödel number is in the relation, assigning to set variables elements of one of the sets and to class variables elements of the other. The third function interprets the operation symbol F.

The formal definition is as follows:

DEFINITION 2.1.3.

(i) Let L be the quaternary operation defined by:

(a) $k \notin \text{sett} \vee \mathbb{D}f \neq \text{fv}_1 k \vee \neg \text{Func}(h) \vee \mathbb{D}h \neq u \rightarrow L(k,f,h,u) = V$,

(b) $k \in \text{vl}_1 \wedge \mathbb{D}f = \text{fv}_1 k \wedge \text{Func}(h) \wedge \mathbb{D}h = u \rightarrow L(k,f,h,u) = f \cdot k$,

(c) $k \in \text{sett} \wedge \mathbb{D}f = \text{fv}_1 k \rightarrow L(\text{sel}(k),f,h,u) = h^{\cdot}(L(k,f,h,u))$.

(ii) Define by recursion the operation H such that:

(a) $k \notin \text{fm} \rightarrow H(k) = 0$,

(b) $k \in \text{fm} \rightarrow$

(b1) $k = e_1(l,m) \rightarrow H(k) = \{\langle u,v,x,0,h\rangle : x \in {}^{\{l,m\}}u \wedge \mathbb{D}h = u \wedge L(l,x,h,u) \in L(m,x,h,u)\}$,

(b2) $k = e_2(l,m) \rightarrow H(k) = \{\langle u,v,x,y,h\rangle : x \in {}^{\{l\}}u \wedge y \in {}^{\{m\}}v \wedge \mathbb{D}h = u \wedge L(l,x,h,u) \in y \cdot m\}$,

(b3) $k = \text{ng}(l) \rightarrow H(k) = \{\langle u,v,x,y,h\rangle : x \in {}^{\text{fv}_1 k}u \wedge y \in {}^{\text{fv}_2 k}v \wedge \mathbb{D}h = u \wedge \langle u,v,x,y,h\rangle \notin H(l)\}$,

(b4) $k = \text{ds}(l,m) \rightarrow H(k) = \{\langle u,v,x,y,h\rangle : \langle u,v,x \restriction \text{fv}_1 l, y \restriction \text{fv}_2 l, h\rangle \in H(l) \vee \langle u,v,x \restriction \text{fv}_1 m, y \restriction \text{fv}_2 m, h\rangle \in H(m)\}$,

(b5) $k = \text{qu}(l,m) \wedge l \notin \text{fv}m \rightarrow H(k) = H(m)$,

(b6) $k = \text{qu}_1(l,m) \wedge l \in \text{fv}_1 m \rightarrow H(k) = \{\langle u,v,x,y,h\rangle : \exists a(a \in u \wedge \langle u,v,x \cup \{\langle a,l\rangle\},y,h\rangle \in H(m))\}$,

(b7) $k = \text{qu}_2(l,m) \wedge l \in \text{fv}_2 m \rightarrow H(k) = \{\langle u,v,x,y,h\rangle : \exists a(a \in v \wedge \langle u,v,x,y \cup \{\langle a,l\rangle\},h\rangle \in H(m))\}$.

(iii) $St = \{\langle u,v,k,x,y,h\rangle : \langle u,v,x,y,h\rangle \in H(k)\}$.

We shall abbreviate $\langle u,v,k,x,y,h\rangle \in St$ by $St(u,v,k,x,y,h)$.

DEFINITION 2.1.4. Let u, v be sets, $\varphi \in Fm$. Then:

(i) $\langle u \cup v, v, h\rangle \models \varphi\,[f,g] \leftrightarrow \langle u,v,\Delta_\varphi, f,g,h\rangle \in St$.
if $u = v$ we simply write $\langle u,h\rangle \models \varphi[f,g]$;

(ii) $\langle u \cup v, v, h\rangle \models \varphi$ iff for every function f, $f \in {}^{fv_1\Delta\varphi}u$ and for every function g, $g \in {}^{fv_2\Delta\varphi}v$ we have $\langle u,v,\Delta_\varphi, f,g,h\rangle \in St$;

(iii) Let $k \in Fm$, $fv_1 k = \{i_0,\ldots,i_{m_1}\}$, $fv_2 = \{j_0,\ldots,j_{m_2}\}$.

Then:

$$St(u,v,k,a_0,\ldots,a_{m_1},b_0,\ldots,b_{m_2},h) \leftrightarrow \exists f \exists g\,(f{\cdot}i_0 = a_0 \wedge \ldots$$
$$\wedge f{\cdot}i_{m_1} = a_{m_1} \wedge g{\cdot}j_0 = b_0 \wedge \ldots \wedge g{\cdot}j_{m_2} = b_{m_2} \wedge \langle u,v,k,f,g,h\rangle \in St).$$

We also write $\langle u \cup v, v, h\rangle \models \varphi[a_0,\ldots,a_{m_1},b_0,\ldots,b_{m_2}]$ instead of $St(u,v,\Delta_\varphi, a_0,\ldots,a_{m_1},b_0,\ldots,b_{m_2},h)$.

(iv) Let $\mathcal{K}$ be a class of structures of the form $\langle u \cup v, v, h\rangle$.
$\mathcal{K} \models \varphi$ iff for every structure $\langle u \cup v, v, h\rangle \in \mathcal{K}$ we have $\langle u \cup v, v, h\rangle \models \varphi$.

If Σ is a set of $\mathcal{L}(F)$-formulae we write $\mathcal{K} \models \Sigma$ if $\mathcal{K} \models \varphi$ for every $\varphi \in \Sigma$.

The usual properties of the satisfaction relation, $\models$, are easily obtained from the previous definitions. In particular the satisfaction relation $\langle u,h\rangle \models \varphi$ coincides with the well known satisfaction relation for Zermelo-Fraenkel set theory in the case that the formula φ is predicative.

The following theorem is analogous to Theorem 2.9 in [3].

THEOREM 2.1.5. *For any* $\varphi(v_{i_0},\ldots,v_{i_m},X_{i_0},\ldots,X_{i_p}) \in Fm$, $f \in {}^{fv_1\Delta\varphi}u$ and $g \in {}^{fv_2\Delta\varphi}v$,

$$M2C \vdash \varphi\binom{F}{h}^{\langle u,v\rangle}(f{\cdot}\Delta_{v_{i_0}},\ldots,f{\cdot}\Delta_{v_{i_m}},\ldots,g{\cdot}\Delta_{X_{i_0}},\ldots,g{\cdot}\Delta_{X_{i_p}}) \leftrightarrow$$

$\langle u,v,\Delta_\varphi,f,g,h\rangle \in St$.

2.2 THE REDUCTION THEOREM.

Our aim is to obtain in M2C models for a finite number of instances of the axiom for class specification plus the other axioms of M2E including the axiom of replacement. The method used consists in starting with a model of Zermelo's set theory with the axiom of infinite. Then to add to this model a class structure that contains enough classes so that the model satisfy the axiom of class specification for formulae with at most n quantifiers. We then use Gaifman's method to insure that the model satisfies the axiom of replacement. This section is concerned with the first of these problems: to add a sufficiently rich class structure. The main problem here is that the parameters and quantifiers of the formulae which specify classes may refer to these same classes. Our Reduction Theorem insures that for each formula φ we can find an equivalent formula φ^* (with respect to the model) in which the parameters and class variables refer to the specifying formula instead of the corresponding class.

DEFINITION 2.2.1. Let u be a set and h be a function. For each $n \in N$, define:

$$C_n^{\langle u,h\rangle} = \{y\colon \exists m(m \in for_n(\mathcal{L}(F)) \wedge \exists u_1 \ldots u_q \in u \wedge$$
$$y = \{x\colon x \in u \wedge St(u,m,x,u_1,\ldots,u_q,h)\}\}.$$

DEFINITION SCHEMA 2.2.2. For each $n \in N$ and for every $\varphi \in For_n(\mathcal{L}(F))$ define $\varphi^*_{n_0} \in For_{n+k}(\mathcal{L}(F))$ $(n_0 = n+k)$, where $k \in N$ is a fixed number that depends of n, by the following recursive definition:

(i) $Ucl(\varphi) \subseteq Pred \rightarrow \varphi^*_{n_0} = \varphi$.

(ii) $\varphi = v_i \in X_j \rightarrow \varphi^*_{n_0} = Sat(X_j(\Delta_{v_i}), \{\langle v_i,\Delta_{v_i}\rangle\} \cup f, g, F)$.

(iii) $\varphi = F\cdot v_i \in X_j \rightarrow \varphi^*_{n_0} = Sat(X_j(\Delta_{v_i}), \{\langle F\cdot v_i, \Delta_{v_i}\rangle\} \cup f, g, F)$.

(iv) $\varphi = \neg\psi \rightarrow \varphi^*_{n_0} = \neg\psi^*_{n_0}$.

(v) $\varphi = \varphi \vee \theta \rightarrow \varphi^{\star}_{n_0} = \psi^{\star}_{n_0} \vee \theta^{\star}_{n_0}$.

(vi) $\varphi = \exists X \psi \rightarrow \varphi^{\star}_{n_0} = \exists X \in \mathrm{For}_{n_0}(\mathcal{L}(F))\, \exists f \exists g\, \psi^{\star}_{n_0}(X)$.

It is unnecessary to consider set-terms of a structure more complicated than $F \cdot v_i$, since they can be eliminated. We shall frequently omit the subscript n_0 in the formula $\varphi^{\star}_{n_0}$, and will depend upon the context.

DEFINITION 2.2.3. For each n, if $g \in^{t} C_n^{\langle u,h \rangle}$, where $t \subseteq vl_2$, let $As(g) \in^{t} \mathrm{For}_n(\mathcal{L}(F)) \times {}^{S_\omega(vl_1)}u \times {}^{S_\omega(vl_2)}u$ be the function defined by:

$As(g) \cdot s = \langle m, f, g \rangle \leftrightarrow g \cdot s = \{x\colon x \in u \wedge St(u, m, f, g, h) \wedge f \cdot \Delta_{v_0} = x\}$;

and for every set u and for every $\varphi \in \mathrm{For}_n(\mathcal{L}(F))$

$g_{\langle k, \varphi, u \rangle} \in {}^{S_\omega vl_k}u$ (k = 1,2) is the assignment for $\varphi^{\star}$ obtained from the assignments $pr_{k+1}\, As(g) \cdot s$.

THEOREM SCHEMA 2.2.4. (M2C) (The Reduction Theorem) *For every* $n_0 > n$ *and for every* $\varphi \in \mathrm{For}_n(\mathcal{L}(F))$ *the following is a theorem:*

$\forall h \forall u (\mathrm{Strans}(u) \wedge \omega \in u \wedge \mathrm{Lim}(\rho \cdot u) \rightarrow (\langle u \cup C_{n_0}^{\langle u,h \rangle}, C_{n_0}^{\langle u,h \rangle}, h \rangle$

$\models \varphi[f, g] \leftrightarrow \langle u, h \rangle \models \varphi^{\star}_{n_0}[f \cup g_{\langle 1, \varphi, u \rangle}, pr_1\, As(g) \cup g_{\langle 2, \varphi, u \rangle}]))$.

In words: if the $\mathcal{L}(F)$-formula φ is realized in the model $\langle u \cup C_{n_0}^{\langle u,h \rangle}, C_{n_0}^{\langle u,h \rangle}, h \rangle$ with the given assignments then the $\mathcal{L}(F)$-formula $\varphi^{\star}_{n_0}$ is realized in the model $\langle u, h \rangle$ assigning to class variables $\mathcal{L}(F)$-formulae and extending the assignments in such a way so as to include the assignments to the new variables that occur in the formulae that are assigned to classes; the converse also holds.

PROOF: By recursion over the complexity of the formula.

(a) $Ucl(\varphi) \subseteq \mathrm{Pred}$, the proof is trivial.

(b) Suppose $\varphi = v_r \in X_s$. Then by 2.1.3 (ii)

$$\langle u \cup C_{n_0}^{\langle u,h\rangle}, C_{n_0}^{\langle u,h\rangle}, h\rangle \models \varphi[f,g] \leftrightarrow f \in {}^{\{\Delta_{v_r}\}}u \wedge$$

$$g \in {}^{\{\Delta_{x_s}\}}C_{n_0}^{\langle u,h\rangle} \wedge f\cdot\Delta_{v_r} \in g\cdot\Delta_{x_s} .$$

By 2.2.1 this is equivalent to

$$\exists k\, \exists u_1, \ldots, \exists u_q (f \in {}^{\{\Delta_{v_r}\}}u \wedge k \in \mathrm{for}_{n_0}(\mathcal{L}(F)) \wedge u_1 \in u \wedge \ldots \wedge u_q \in u$$

$$\wedge\; g\cdot\Delta_{x_s} = \{x:\ x \in u \wedge \mathrm{St}(u,k,u_1,\ldots,u_q,h)\} \wedge f\cdot\Delta_{v_r} \in g\cdot\Delta_{x_s} \quad ;$$

this is :

$f \in {}^{\{\Delta_{v_r}\}}u \wedge k \in \mathrm{for}_{n_0}(\mathcal{L}(F)) \wedge u_1 \in u \wedge \ldots \wedge u_q \in u$ and for every $G_0 \in {}^{\mathrm{fv}_1 k(f\cdot\Delta_{v_r})}u$ and for every $G_1 \in {}^{\mathrm{fv}_2 k(f\cdot\Delta_{v_r})}u$

$\mathrm{St}(u,k(f\cdot\Delta_{v_r}), G_0, G_1, h)$, where $G_0 \cdot \Delta_{v_{i_k}} = u_k \quad (1 \leqslant k \leqslant q)$.

By the absoluteness of Sat(, , ,) with respect to the supertransitive sets we have

$f \in {}^{\{\Delta_{v_r}\}}u \wedge k \in \mathrm{for}_{n_0}(\mathcal{L}(F)) \wedge u_1 \in u \wedge \ldots \wedge u_q \in u$ and for every $G_0 \in {}^{\mathrm{fv}_1 k(f\cdot\Delta_{v_r})}u$ and for every $G_1 \in {}^{\mathrm{fv}_2 k(f\cdot\Delta_{v_r})}u$

$$\langle u,h\rangle \models \mathrm{Sat}(k(f\cdot\Delta_{v_r}), G_0, G_1, F);$$

this is :

$f \in {}^{\{\Delta_{v_r}\}}u$ and $\langle u,h\rangle \models \mathrm{Sat}(X_s(\Delta_{v_r}), \{\langle v_r, \Delta_{v_r}\rangle\} \cup g_{\langle 1,\varphi,u\rangle},$

$g_{\langle 2,\varphi,u\rangle}, F)\,[f \cup g_{\langle 1,\varphi,u\rangle},\ \mathrm{pr}_1\mathrm{As}(g) \cup g_{\langle 2,\varphi,u\rangle}]$

which proves the theorem for case (b).

(c) The proof of cases $\varphi = \neg\psi$ and $\varphi = \psi \vee \theta$ is easy.

(d) If $\varphi = \exists X_k\, \psi$ then

$$\langle u \cup C_{n_0}^{\langle u,h\rangle}, C_{n_0}^{\langle u,h\rangle}, h\rangle \models \varphi[f,g] \leftrightarrow f \in {}^{\mathrm{fv}_1\Delta_\varphi}u,\ g \in {}^{\mathrm{fv}_2\Delta_\varphi}C_{n_0}^{\langle u,h\rangle}$$

and there exists $b \in C_{n_0}^{\langle u,h\rangle}$ such that

$\langle u \cup C_{n_0}^{\langle u,h\rangle}, C_{n_0}^{\langle u,h\rangle}, h\rangle \models \psi[f, g \cup \{\langle b, \Delta_{x_k}\rangle\}]$;

this is:

$f \in {}^{fv_1 \Delta_\varphi}u$, $g \in {}^{fv_2 \Delta_\varphi} C_{n_0}^{\langle u,h\rangle}$ and there exists $k_b \in form_{n_0}(\mathcal{L}(F))$

and a finite sequence of parameters in u, $\vec{z}$ such that

(.1) $\langle u \cup C_{n_0}^{\langle u,h\rangle}, C_{n_0}^{\langle u,h\rangle}, h\rangle \models \psi[f, g \cup \{\langle b, \Delta_{x_k}\rangle\}]$ where

$b = \{x : x \in u \wedge St(u, k_b, \vec{z}, h)\}$.

By inductive hypothesis, (.1) is equivalent to:

$\langle u,h\rangle \models \psi^*_{n_0}[f \cup g_{\langle 1,\varphi,u\rangle}, pr_1 As(g) \cup \{\langle k_b, \Delta_{x_k}\rangle\} \cup g_{\langle 2,\varphi,u\rangle}]$;

but this is

$$\langle u,h\rangle \models \exists X\ for_{n_0}(\mathcal{L}(F))\ \exists f \exists g\ \psi^*(X)[f \cup g_{\langle 1,\varphi,u\rangle}, pr_1 As(g) \cup g_{\langle 2,\varphi,u\rangle}]$$

Theorem 2.2.4 is thus proved. ■

This theorem is still correct replacing $For_n(\mathcal{L}(F))$ by Fm and $C_n^{\langle u,h\rangle}$ by $C^{\langle u,h\rangle} = \{y : \exists m(m \in fm \wedge \exists u_1 \ldots u_q \in u \wedge$

$$y = \{x : x \in u \wedge St(u, m, x, u_1, \ldots, u_q, h\})\}.$$

The reduction theorem shows in a certain sense that M2 is in fact a second order set theory because quantification over classes can be reduced to quantification over formulas.

The applicability of the reduction theorem is ilustrated in the following theorems.

THEOREM SCHEMA 2.2.5. (M2C) *For each* n *let* $\varphi(v_{i_0}, X_{i_0}, \ldots, X_{i_k}) \in For_n(\mathcal{L}(F))$ *and let* $g \in {}^{\{\Delta_{x_{i_0}}, \ldots, \Delta_{x_{i_k}}\}} C_{n_0}^{\langle u,h\rangle}$ $(n_0 > n)$ *where* u *is a supertransitive set,* $\omega \in u$ *and* $Lim(\rho \cdot u)$. *If* $g \cdot \Delta_{x_{i_0}} = A_{i_0}, \ldots, g \cdot \Delta_{x_{i_k}} = A_{i_k}$ *then:*

$$\{x\colon \varphi(\begin{smallmatrix}F\\h\end{smallmatrix})^{\langle u,\, C^{\langle u,h\rangle}_{n_0}\rangle}(x, A_{i_0},\dots, A_{i_k})\} = \{x\colon \varphi(\begin{smallmatrix}F\\h\end{smallmatrix})^{*u}(x, y_{i_0},\dots, y_{i_k}, \vec{z})\}$$

where $y_{i_j} = \mathrm{pr}_1\mathrm{As}(g)`\Delta_{x_{i_j}}$ $(j = 0,\dots, k)$ *and* $\vec{z}$ *is a finite sequence of new parameters belonging to* u.

PROOF: By definition of the class abstractor $\{:\}$ we have

$$b \in \{x\colon \varphi(\begin{smallmatrix}F\\h\end{smallmatrix})^{\langle u, C^{\langle u,h\rangle}_{n_0}\rangle}(x, A_{i_0},\dots, A_{i_k}) \leftrightarrow \varphi(\begin{smallmatrix}F\\h\end{smallmatrix})^{\langle u, C^{\langle u,h\rangle}_{n_0}\rangle}(b, A_{i_0},\dots, A_{i_k})\}.$$

By Theorem 2.1.5 we have

$$\varphi(\begin{smallmatrix}F\\h\end{smallmatrix})^{\langle u, C^{\langle u,h\rangle}_{n_0}\rangle}(b, A_{i_0},\dots, A_{i_k}) \leftrightarrow \langle u,\, C^{\langle u,h\rangle}_{n_0}, \Delta_{\varphi(\begin{smallmatrix}F\\h\end{smallmatrix})}, f, g, h\rangle \in \mathrm{St}$$

where $f \in {}^{\{\Delta_{v_{i_0}}\}}u$ and $f`\Delta_{v_{i_0}} = b$.

Hence, by Reduction Theorem 2.2.4, we get

$$\mathrm{St}(u, C^{\langle u,h\rangle}_{n_0}, \Delta_{\varphi(\begin{smallmatrix}F\\h\end{smallmatrix})}, f, g, h) \leftrightarrow \mathrm{St}(u, \Delta_{\varphi(\begin{smallmatrix}F\\h\end{smallmatrix})^*}, f \cup g_{\langle 1,\varphi,u\rangle},$$
$$\mathrm{pr}_1\mathrm{As}(g) \cup g_{\langle 2,\varphi,u\rangle}).$$

Therefore using Theorem 2.1.5 we have

$$\varphi(\begin{smallmatrix}F\\h\end{smallmatrix})^{\langle u,\, C^{\langle u,h\rangle}_{n_0}\rangle}(b, A_{i_0},\dots, A_{i_k}) \leftrightarrow \varphi(\begin{smallmatrix}F\\h\end{smallmatrix})^{*u}(b, y_{i_0},\dots, y_{i_k}, \vec{z})$$

where $y_{i_j} = \mathrm{pr}_1\mathrm{As}(g)`\Delta_{x_{i_j}}$ $(j = 0,\dots,k)$ and $\vec{z}$ is a finite sequence of parameters in u. This completes the proof of the Theorem 2.2.5. ∎

An instance of the axiom of class specification for a given $\mathcal{L}(F)$-formula φ with at most n quantifiers becames formalized by the $\mathcal{L}(F)$-formula $CE_{(n,\varphi)}$ defined as follows:

DEFINITION SCHEMA 2.2.6. For every $\varphi \in \mathrm{For}_n(\mathcal{L}(F))$ if $\mathrm{Fv}\,\varphi = \{X_0,\dots, X_{k-1}\}$ let $CE_{(n,\varphi)} \in \mathrm{For}_{n+k+2}(\mathcal{L}(F))$ be the formula:

$$CE_{(n,\varphi)}: \forall X_0 \ldots \forall X_{k-1}\, \exists A\, \forall x (x \in A \leftrightarrow \varphi(x, X_0, \ldots, X_{k-1})).$$

The next theorem gives a simple sufficient condition for a structure being a model of $CE_{(n,\varphi)}$.

THEOREM SCHEMA 2.2.7. (M2C) *Let* $\langle u \cup v, v, h\rangle$ *be a structure such that* u *is a supertransitive set,* $\omega \in u$ *and* $\mathrm{Lim}(\rho \cdot u)$. *For each* n *and for every* $\varphi \in \mathrm{For}_n(\mathcal{L}(F))$ *if* $v = C_{n_0}^{\langle u, h\rangle}$ $(n_0 > \mathrm{Qn}(\varphi^*))$ *then*

$$\langle u \cup v, v, h\rangle \models CE_{(n,\varphi)}.$$

PROOF: We must show that $\{x: \varphi(^F_h)^{\langle u,\, C_{n_0}^{\langle u,h\rangle}\rangle}\} \in v$. By Theorem 2.2.5, $\{x: \varphi(^F_h)^{\langle u,\, C_{n_0}^{\langle u,h\rangle}\rangle}\} = \{x: \varphi(^F_h)^{*u}\}$. Because $\omega \in u$ and Definition 2.2.1 of the class $C_{n_0}^{\langle u,h\rangle}$ we have $\{x: \varphi(^F_h)^{*u}\} \in C_{n_0}^{\langle u,h\rangle} = v$, from which the theorem can be deduced. ■

The extension of the above theorem to a class of structures is a follows:

THEOREM SCHEMA 2.2.8. (M2C) *For each* n *let* $\mathcal{K}_n$ *be a class of structures of the form* $\langle u \cup v, v, h\rangle$ *such that* u *is a supertransitive set* $\omega \in u$ *and* $\mathrm{Lim}(\rho \cdot u)$ *and* $v = C_{n_0}^{\langle u,h\rangle}$ $(n_0 > \mathrm{Qn}(\varphi^*))$. *Then for every* $\varphi \in \mathrm{For}_n(\mathcal{L}(F))$, $\mathcal{K}_n \models CE_{(n,\varphi)}$.

Moreover, under these conditions, $\mathcal{K} \models$ M2 ~ { *Axiom schema of class specification and replacement axiom*}.

PROOF: The first part of the theorem follows from Theorem 2.2.7. As is proved in [15], $\mathrm{Stran}(u) \leftrightarrow u = R(\alpha)$. Then from various results from Montague-Vaught [12] the second part of the theorem can be obtained. ■

3. MODEL THEORETICAL RESULTS.

In the previous section we found sufficient conditions for a structure to be a model of a finite number of axioms of M2E without replacement. In order to have models for replacement we use Gaifman's Theorem (see [8], p. 259).

3.1 THE CONSTRUCTION OF THE MODELS.

We use German capital letters $\mathfrak{A}, \mathfrak{B}, \ldots$ to denote structures; we shall understand once for all, that all the structures are of the form $\langle u \cup v, v, h \rangle$ where h is a function.

DEFINITION 3.1.1. Let $\mathfrak{A} = \langle u \cup v, v, h \rangle$ be a structure. The set u is called the *support of* $\mathfrak{A}$ and is denoted by $Sup\ \mathfrak{A}$. v is called the *class structure of* $\mathfrak{A}$ and is denoted by $Cls\ \mathfrak{A}$. It is associated with $\mathfrak{A}$ an ordinal, called *rank of the support of* $\mathfrak{A}$, denoted $Rs\ \mathfrak{A}$, and which is, in fact, $\rho\text{`}Sup\ \mathfrak{A}$. If $v_0 \subseteq v$ then the structure of the form $\langle u \cup v_0, v_0, h \rangle$ is called the *restriction of the class structure of* $\mathfrak{A}$ *to* v_0 and is denoted by $\mathfrak{A} \upharpoonright v_0$. By the *universe* of a structure $\mathfrak{A}$, denoted $Un\ \mathfrak{A}$, it will be understood $u \cup v$. Function h is called the *choice set function of* $\mathfrak{A}$ and is denoted by $h^{\mathfrak{A}}$.

DEFINITION 3.1.2. Let $n \in N$.

(i) Let $\mathfrak{A}, \mathfrak{B}$ be structures with support $R(\alpha)$, $R(\beta)$ ($\alpha \leqslant \beta$) respectively. Then

(a) $\mathfrak{A} <_n \mathfrak{B}$ iff $\mathfrak{A} \upharpoonright R(\alpha) \models \varphi(\vec{y}) \leftrightarrow \mathfrak{B} \upharpoonright R(\beta) \models \varphi(\vec{y})$ for every $\varphi \in For_n(\mathcal{L}(F))$ and $\vec{y} = \langle y_0, \ldots, y_{n-1} \rangle$, $y_0 \in R(\alpha)$, $\ldots$, $y_{k-1} \in R(\alpha)$;

(b) $\mathfrak{A} \subset \mathfrak{B}$ iff $\alpha \leqslant \beta \wedge Cls\ \mathfrak{A} \subseteq Cls\ \mathfrak{B} \wedge h^{\mathfrak{A}} \subseteq h^{\mathfrak{B}}$.

(ii) Let $\mathcal{K}$ be a class of structures. Then:

(a) $\mathcal{K}$ is called *n-model complete* iff $\forall \mathfrak{A}, \mathfrak{B} \in \mathcal{K}\ (\mathfrak{A} \subset \mathfrak{B} \rightarrow \mathfrak{A} <_n \mathfrak{B})$

(b) $\mathcal{K}$ is called *closed* iff $\mathcal{K}$ is closed under unions of ascending sequences.

METALEMMA 3.1.3. *Let $\mathcal{K}$ be a class of structures such that:*

(*i*) *For evey α there exists $\beta > \alpha$ such that*
$i_{(\alpha,\beta)}$: *for every $\mathfrak{A} \in \mathcal{K}$, $Rs\,\mathfrak{A} = \alpha$ there exists $\mathfrak{B} \in \mathcal{K}$, $Rs\,\mathfrak{B} = \beta$ such that $\mathfrak{A} \subset \mathfrak{B}$;*

and

(*ii*) *$\mathcal{K}$ is closed.*

Then there exists an operation Ω defined over OR *such that for every α, $\Omega(\alpha)$ is a closed unbounded class of ordinals with the following properties:*

(*a*) *α is the least element of $\Omega(\alpha)$*

(*b*) $\beta, \gamma \in \Omega(\alpha) \wedge \beta \leqslant \gamma \rightarrow i_{(\beta,\gamma)}$.

PROOF: [8], pg. 259.

Let T_0 be the following class of structures:

$$T_0 = \{\mathfrak{A} : \exists\alpha(\alpha = \cup\alpha \wedge \alpha \neq 0 \wedge \alpha > \omega \wedge Sup\,\mathfrak{A} = Cls\,\mathfrak{A} = R(\alpha) \wedge \wedge E(h^{\mathfrak{A}})^{R(\alpha)})\}.$$

If a structure of T_0 is denoted by a given German letter, say $\mathfrak{A}$, we shall understand that the corresponding lower case Greek letter, α, denotes the rank of the support of $\mathfrak{A}$, i.e. $Sup\,\mathfrak{A} = R(\alpha)$. When $u = R(\alpha)$ we abbreviate $C_n^{\langle u,h\rangle}$ by $C_n^{\langle \alpha,h\rangle}$.

The following theorem is due to Gaifman ([8] pg. 259).

THEOREM 3.1.4. (Gaifman). *For each* $n \in N$ *there exists proper classes of* M2, $T_0,\ldots,T_n$ *such that*

(1) $T_0 \supset T_1 \supset \ldots T_n \supset T_{n+1} \ldots$

(2) *For evry α there exists $\beta > \alpha$ such that for every $\mathfrak{A} \in T_n$ there exists $\mathfrak{B} \in T_{n+1}$ and $\mathfrak{A} \subset \mathfrak{B}$;*

(3) T_n *is* n*-model complete*;

(4) T_n *is closed.*

If we call $M2E^{\sim}$ the theory obtained by the axioms of M2E without the axioms of replacement and class specification, then from various results in [12], $M2C \vdash (T_0 \models M2E^{\sim})$.

Theorems 2.2.8 and 3.1.4 lead to the formulation of the next theorem.

THEOREM SCHEMA 3.1.5. *For each* $n \in N$ *there exists proper classes of* M2, $\mathcal{K}_0, \ldots, \mathcal{K}_n, \ldots$ *such that:*

$$\mathcal{K}_n \models (M2E^{\sim} + CE_{(n,\varphi)}) \quad \textit{for each } \varphi \in For_n(\mathcal{L}(F)).$$

PROOF: Define the classes $\mathcal{K}_n$ by

$$\mathcal{K}_n = \{ (Rs\,\mathfrak{A} \cup C_{n_0}^{\langle \alpha, h^{\mathfrak{A}} \rangle}, h^{\mathfrak{A}}) : \mathfrak{A} \in T_n \}$$

Then by Theorems 2.2.8 and 3.1.4 the theorem can be derived. ■

In the following German letters that denote structures, will denote structures of some of the classes $\mathcal{K}_n$.

As it might be expected we have :

THEOREM SCHEMA 3.1.6. *If* $\mathfrak{A} \in \mathcal{K}_n$ *then* $\mathfrak{A}$ *is the least model with support* $R(\alpha)$ *that satisfies* $CE_{(n,\varphi)}$ *for every* $\varphi \in For_n(\mathcal{L}(F))$.

THEOREM SCHEMA 3.1.7. *Let* Σ *be a finite subset of the axioms of* M2E. *Then there exists* n_1, $n_1 \in N$, *such that for every* $n \geq n_1$

$$M2C \vdash (\mathcal{K}_n \models \Sigma).$$

PROOF: If Σ does not contain the replacement axiom nor instances of the axiom schema of class specification then, by 3.1.5, $M2C \vdash (\mathcal{K}_0 \models \Sigma)$.

Assume Σ contains a finite number of instances of the axiom schema of class specification

$$CE_{(n_0,\varphi_{i_0})}, \ldots, CE_{(n_k,\varphi_{i_k})}.$$

Then by Theorem 3.1.5 we have:

$$\mathcal{K}_{n_j} \models CE_{(n_j,\varphi_{i_j})} \qquad j = 0, \ldots, k.$$

Let $n_1 = \max\{n_j : j \in k+1\}$. Then for every $n \geqslant n_1$,

$$M2C \vdash (\mathcal{K}_n \models \Sigma).$$

If in addition Σ contains the replacement axiom, let n_1 be such that $M2C \vdash (\mathcal{K}_{n_1} \models \Sigma \sim \{\text{replacement axiom}\})$.

We must show that given a function g, then for all large enough n, we can prove in M2C that if $\mathfrak{A} \in \mathcal{K}$ and if

(.1) $\mathfrak{A} \models \forall v_1 \in x\, \exists v_2\, (g^* v_1 = v_2)$, where $x \in R(\alpha)$;

then there exists $y \in R(\alpha)$ such that

(.2) $\mathfrak{A} \models \forall v_1 \in x\, \exists v_2 \in y (g^* v_1 = v_2)$.

But by Definition 2.2.1 of $C_n^{\langle \alpha, h^{\mathfrak{A}} \rangle}$ and the construction of the classes $\mathcal{K}_{n_1}$ this means that g is a function defined by a formula

$\varphi \in For_n(\mathcal{L}(F))$ with parameters in $R(\alpha)$, i. e.

$v_2 = g^* v_1 \leftrightarrow \mathfrak{A} \models \varphi[v_1, v_2, a_0, \ldots, a_{j-1}]$ $v_1 \in R(\alpha)$, $a_i \in R(\alpha)$.

Then we must prove in M2C that if $\mathfrak{A} \in \mathcal{K}_n$ and if

(.3) $\mathfrak{A} \models \forall v_1 \in x\, \exists v_2\, \varphi[v_1, v_2, a_0, \ldots, a_{j-1}]$ where $x \in R(\alpha)$,

$a_i \in R(\alpha)$

then

(.4) $\mathfrak{A} \models \exists v_3\, \forall v_1 \in x\, \exists v_2 \in v_3\, \varphi[v_1, v_2, a_0, \ldots, a_{j-1}]$.

Let $k = Qn(\varphi)$ and $n \geqslant k + n_1 + 3$. If $\mathfrak{A} \in \mathcal{K}_n$ and if

$\mathfrak{A} \models \varphi[v_1, v_2, a_0, \ldots, a_{j-1}]$ $v_1 \in R(\alpha)$, $a_i \in R(\alpha)$ then using (1) and (2) of 3.1.4 we get $\mathfrak{B} \in \mathcal{K}_n$ such that $\mathfrak{A} \restriction R(\alpha) \subset \mathfrak{B} \restriction R(\beta) \wedge \alpha < \beta$.

Since T_n is n-model complete then

(.5) $\mathfrak{A} \restriction R(\alpha) <_n \mathfrak{B} \restriction R(\beta)$.

On the other hand because $R(\alpha) \in Sup\,\mathfrak{B}$ then

$$\mathfrak{B} \restriction R(\beta) \models \forall v_1 \in x\, \exists x_2 \in R(\alpha)\, \varphi[v_1, x_2, a_0, \ldots, a_{j-1}]$$

which implies

$$\mathfrak{B} \models \exists v_3 \forall v_1 \in x\ \exists v_2 \in v_3\ \varphi[v_1, v_2, a_0, \ldots, a_{j-1}]$$

and finally by (.5) we obtain (.4) and the theorem is proved. ∎

Using this theorem it is possible to show that M2 is not finitely axiomatizable (for another proof see [2]).

3.2 CONSERVATIVE RESULTS EXTENSION.

We give now a forcing-free proof for: M2E *is a conservative extension of* M2C *with respect to sets*. It was proved by Felgner [5], that NBGE is a conservative extension of NBGC with respect to sets using forcing. This same proof can be modified to yield our result (as pointed out in [6]).

As was mentioned earlier, Gaifman [8] gave a forcing-free proof of: ZFE *is a conservative extension of* ZFC. Our proof is based on this one. This proof give some additional information about models for finite sets of axioms.

DEFINITION 3.2.1.

(i) $M2E^* = \{\varphi : \varphi \in \mathrm{Pred} \wedge \varphi \in \mathrm{Fmw.o.} \wedge M2E \vdash \varphi\}$.

(ii) $M2C^* = \{\varphi : \varphi \in \mathrm{Fmw.o.} \wedge \varphi \in \mathrm{Pred} \wedge M2C \vdash \varphi\}$.

THEOREM 3.2.2. $M2E^* = M2C^*$.

PROOF: Clearly $M2C^* \subseteq M2E^*$. In order to prove $M2E^* \subseteq M2C^*$ let $\varphi \in$ Fmw.o.. We must show: if $M2E \vdash \varphi^V$ then $M2C \vdash \varphi^V$. By Reflection Theorem (see [13]) there is a closed unbounded class Γ of ordinals such that

(.1) $\varphi V \leftrightarrow R(\alpha) \models \varphi^V$ for $\alpha \in \Gamma$.

Let Σ be the finite set of axioms from which φ^V is derived. By 3.1.5 there is an n such that $\mathcal{K}_n \models \Sigma$. Using 3.1.3 and 3.1.4 we obtain a closed unbounded class of ordinals Ω, such that for each $\beta \in \Omega$ there is an $\mathfrak{A} \in \mathcal{K}$ such that $Sup\, \mathfrak{A} = R(\beta)$. Take $\alpha \in \Omega \cap \Gamma$ and $\mathfrak{A} \in \mathcal{K}_n$ with $Sup\, \mathfrak{A} = R(\alpha)$. Then $\mathfrak{A} \models \Sigma$. Thus, $\mathfrak{A} \restriction R(\alpha) \models \varphi^V$. By (.1) we obtain φ^V. ∎

It is proved in [2], that M2 with stronger versions of the axiom of choice (for instance WO_C) is a conservative extension of M2E with respect to sets. Hence, these theories are also conservative extensions of M2C with respect to sets.

REFERENCES.

[1] R. Chuaqui, *Forcing for the impredicative theory of classes*, Journal of Symbolic Logic, 37 (1972), 1-18.

[2] R. Chuaqui, *Internal and forcing models for the impredicative theory of classes*, to appear in Dissertationes Mathematicae.

[3] R. Chuaqui, *Bernays' class theory*, this volume.

[4] S. Feferman, *Arithmetization of metamethematics in a general setting*, Fundamenta Mathematica, 49 (1960), 35-92.

[5] U. Felgner, *Comparison of the axioms of local and universal choice* , Fundamenta Mathematica, 71 (1971), 43-62.

[6] U. Felgner, *Choice functions on sets and classes;* in Sets and Classes: on the work of Paul Bernays (Ed. G. H. Müller), North-Holland, Amsterdam, 1976.

[7] A. A. Fraenkel, *Über die Gleichheitsbeziehung in der Mengenlehre*, Journal für die reine und angewandte Mathematik, 157 (1927), 79-81.

[8] H. Gaifman, *Global and local choice functions*, Israel Journal of Mathematics, 22 (1975), 257-265.

[9] K. Gödel, *Über formal unentscheidbare Sätze der Principia Mathematica und verwandter Systeme, I*, Monatshefte für Mathematik und Physik, 38 (1931), 173-198.

[10] K. Gödel, The consistency of the axiom of choice and of the generalized continuum hypothesis with the axioms of set theory, seventh printing, Princeton University Press, Princeton, 1966.

[11] J. L. Kelley, General Topology, Van Nostrand, New York, 1955.

[12] R. Montague and R. L. Vaught, *Natural models of set theory*, Fundamenta Mathematica, 47 (1959), 219-242.

[13] A. Mostowski, Constructible sets with applications, North-Holland, Amsterdam, 1969.

[14] H. Rubin and J. E. Rubin, Equivalents of the axiom of choice, (2nd edition) North-Holland, Amsterdam, 1970.

[15] J. C. Shepherdson, *Inner models for set theory, II*, The Journal of Symbolic Logic, 17 (1952), 225-237.

Universidad de Chile
Facultad de Ciencias
Santiago, Chile

and

Universidad Católica de Chile
Instituto de Matemática
Santiago 8, Chile.

An Algebraic Study of a Propositional System of Nelson.

by *MANUEL M. FIDEL.*

ABSTRACT. In this paper we introduce a new method for a semantical study of Nelson's logic of constructible falsity (D. Nelson, *Constructible falsity*, The Journal of Symbolic Logic 14 (1949), 247-257) through algebraic models formed by couples (a_1, a_2) where a_1 and a_2 are elements of a Heyting algebra. The method introduced here shows the advantage of using models based on n-tuples of elements of other known algebras to the study of certain logical systems or other algebras. For instance, our method can be generalized to the study of Fitch's system (cf. F. B. Fitch, Symbolic Logic, Ronald Press, New York, 1952).

1. Introduction.

In this work we introduce a new way to study the semantics for Nelson's logic of constructible falsity [3], by algebraic methods.

From our point of view the most important property of that system is the non-validity of the rule

$$\frac{A \equiv B}{\neg A \equiv \neg B}$$

where A and B are formulas, $\equiv$ is the logical equivalence and $\neg$ is the negation. Thus, the equivalence relation needed for the algebraic study of that system, as defined on the basis of the logical equivalence $\equiv$, is not compatible with negation.

Such a problem has been solved by Rasiowa in [4] by defining the following relation between formulas of the system:

$$A \sim B \text{ iff } A \equiv B \text{ and } \neg A \equiv \neg B \text{ are theses,}$$

introducing thereby the concept of **N**-lattice (or quasi-pseudo-Boolean algebra, see [5] Chap. V, and [4]).

We shall see in this paper however that it is possible to make an albraic study of the system using the equivalence relation based on $\equiv$, which provides us moreover with many technical advantages.

As a preview of our algebraic models we observe that they are formed by couples (a_1, a_2) where a_1 and a_2 are elements of a Heyting algebra. (For the concept and properties of Heyting algebra or pseudo - Boolean algebra see, e. g., [6]).

As the algebraic counterpart of the concept of constructivity is that of a Heyting algebra we may interpret a_1 as the positive constructive value of a formula, and a_2 as the corresponding negative constructive value.

This corresponds to Nelson's idea of giving a constructive sense both to A and to $\neg A$. In Section 6 we shall define semantic models and certain valuations v^+ and v^- which are to verify conditions analogous to those for the concept of **P**-realizability and **N**-realizability to be found in Nelson [3].

From the algebraic viewpoint we have that the simplicity of the defition of our models as well as the simplicity of its properties are in great measure reducible to those of the Heyting algebras. The construction of the Lindenbaum canonical model and also the topological representation theorems turn out to be much simpler in our models than in **N**-lattices.

We also prove that there is a certain equivalence between our models and **N**-lattices.

We shall restrict ourselves here to the study of the propositional fragment of Nelson's system but we remark that it is possible to extend the treatment to predicate logic by standard methods.

This work is but a first example of the advantages gained by using models based on n-tuples of elements of known algebras to study certain logical systems or other algebras. For instance, our methods can be generalized to study Fitch's system [1] (see, e.g., [7]) with which we shall occupy ourselves on another occasion.

2. The logical system.

We repeat here the axiom-schemata and the rule for Nelson's system (see [3]):

i) All the axioms of the positive intuitionistic logic for the connectives $\supset$ (*implication*), $\vee$ (*disjunction*), $\wedge$ (*conjunction*). The rule is *modus ponens*.

ii) Additional axioms for $\neg$ (*negation*):

$$A \wedge \neg A \supset B,$$

$$\neg(A \wedge B) \equiv \neg A \vee \neg B,$$

$$\neg(A \vee B) \equiv \neg A \wedge \neg B,$$

$$\neg(A \supset B) \equiv A \wedge \neg B,$$

$$\neg\neg A \equiv A,$$

where A and B are formulas and $A \equiv B =_{df} (A \supset B) \wedge (B \supset A)$

3. The algebraic models.

Let H be a Heyting algebra with respect to the operations $\supset$ (*implication*), $\wedge$ (*infimum*), $\vee$ (*supremum*), and with first (0) and last element (1).

If $x \in H \times H$, let

$$x^+ = proj_1(x) \qquad \text{and} \qquad x^- = proj_2(x)$$

be the projections of x onto the first and second coordinates respectively. Thus,

$$x = (x^+, x^-).$$

A *Nelson algebraic model* is a non-void subset X of $H \times H$ such that $(0,1)$ belongs to X and:

i) For every x in X, $x^+ \wedge x^- = 0$;

ii) For every couple of elements x, y of X there are associated elements $x \supset y$, $x \wedge y$, $x \vee y$, $\neg x$ in X satisfying:

$$(x \supset y)^+ = x^+ \supset y^+, \qquad (x \supset y)^- = x^+ \wedge y^-,$$

$$(x \wedge y)^+ = x^+ \wedge y^+, \qquad (x \wedge y)^- = x^- \vee y^-,$$

$$(x \vee y)^+ = x^+ \vee y^+, \qquad (x \vee y)^- = x^- \wedge y^-,$$

$$(\neg x)^+ = x^-, \qquad (\neg x)^- = x^+.$$

Let us see some examples of such models.

i) Let $B = \{0,1\}$ be the Heyting algebra (actually a Boolean algebra) with two elements. The subset X of $B \times B$ formed by the couples

$$0 = (0,1), \qquad a = (0,0), \qquad 1 = (1,0)$$

is an algebraic Nelson model where

$\supset$	0	a	1
0	1	1	1
a	1	1	1
1	0	a	1

$\wedge$	0	a	1
0	0	0	0
a	0	a	a
1	0	a	1

$\vee$	0	a	1
0	0	a	1
a	a	a	1
1	1	1	1

	$\neg$
0	1
a	a
1	0

ii) The above example is a particular case of the following more general one:

Let H be an Heyting algebra and consider the following subset of H x H:

$$C = \{(h_1, h_2) \mid h_1 \wedge h_2 = 0\}.$$

It is easy to see that C is a Nelson model. We indicate as an example that given $x = (x_1, x_2)$, $y = (y_1, y_2)$ in C we have

$$\begin{aligned}(x_1 \supset y_1) \wedge (x_1 \wedge y_2) &= x_1 \wedge (x_1 \supset y_1) \wedge y_2 \\ &= x_1 \wedge y_1 \wedge y_2 \\ &= 0\end{aligned}$$

using properties of Heyting algebras. The remaining conditions offer no difficulties.

iii) We shall construct now the Lindenbaum model for Nelson's system.

Let A and B be formulas. It is well known that from the axioms of intuitionistic logic one can prove that

$$A \sim B \text{ iff } A \equiv B \text{ is a thesis}$$

is an equivalence relation compatible with the connectives $\supset$, $\wedge$ and $\vee$.

Let H_L be the set of equivalence classes $[A]$ where A is a formula.

It is well known that in H_L one can define the operations:

$$[A] \supset [B] = [A \supset B]$$
$$[A] \wedge [B] = [A \wedge B]$$
$$[A] \vee [B] = [A \vee B]$$

with respect to which H_L is a Heyting algebra with last element $1 = [T]$, where T is a thesis.

Moreover one sees that $[A] = 1$ iff A is a thesis.

From the axiom $A \supset \neg A \supset B$ it follows that $0 = [A \wedge \neg A]$ is the first element of H_L.

Let us denote by L the subset of $H_L \times H_L$ formed by all the ordered couples having the form: $([A], [\neg A])$ where A is a formula.

In order to see that L is a Nelson model we show as an example that given $a = ([A], [\neg A])$ and $b = ([B], [\neg B])$ in L, $a \supset b$ exists and is well-defined. This results from the equality

$$([A] \supset [B],\ [A] \wedge [\neg B]) = ([A \supset B],\ [\neg (A \supset B)]).$$

Analogously for the remaining conditions. For the element $(1,0)$ it suffices to consider the thesis

$$T \supset (\neg T \supset A),$$

where T is a thesis.

4. The valuations.

We proceed now to relate our models to the logic through adequate valuations.

Call an ordered couple (a^+, a^-) of functions from the set of propositional variables into an algebraic model X, an *assignment*. By induction this assignment can be extended in a unique fashion to a *valuation* consisting of an ordered pair (v^+, v^-) of functions from the set of formulas into the model X in the following manner:

i) $v^+(P) = a^+(P)$ $\qquad$ $v^-(P) = a^-(P)$

where P is a propositional variable;

ii) Let A and B be formulas. Suppose $a = (v^+(A), v^-(A))$ and $b = (v^+(B), v^-(B))$ are defined. Then set

$$\begin{aligned} \upsilon^+(A \supset B) &= (a \supset b)^+, & \upsilon^-(A \supset B) &= (a \supset b)^-, \\ \upsilon^+(A \wedge B) &= (a \wedge b)^+, & \upsilon^-(A \wedge B) &= (a \wedge b)^-, \\ \upsilon^+(A \vee B) &= (a \vee b)^+, & \upsilon^-(A \vee B) &= (a \vee b)^-, \\ \upsilon^+(\neg A) &= (\neg a)^+, & \upsilon^-(\neg A) &= (\neg a)^-. \end{aligned}$$

Some useful and easy to demonstrate properties of valuations are the following:

$$\upsilon^+(A \equiv B) = 1 \quad \text{iff} \quad \upsilon^+(A) = \upsilon^+(B) \quad ,$$

$$\upsilon^+(A) \wedge \upsilon^-(A) = 0,$$

$$\begin{aligned} \upsilon^+(A \supset B) &= \upsilon^+(A) \supset \upsilon^+(B), & \upsilon^-(A \supset B) &= \upsilon^+(A) \wedge \upsilon^-(B), \\ \upsilon^+(A \wedge B) &= \upsilon^+(A) \wedge \upsilon^+(B), & \upsilon^-(A \wedge B) &= \upsilon^-(A) \vee \upsilon^-(B), \\ \upsilon^+(A \vee B) &= \upsilon^+(A) \vee \upsilon^+(B), & \upsilon^-(A \vee B) &= \upsilon^-(A) \wedge \upsilon^-(B), \\ \upsilon^+(\neg A) &= \upsilon^-(A), & \upsilon^-(\neg A) &= \upsilon^+(A). \end{aligned}$$

We say that a formula A is valid in an algebraic model X if for every valuation (υ^+, υ^-) on X we have $\upsilon^+(A) = 1$. We say that A is valid if it is valid in every algebraic model.

As can be easily seen all theses are valid. Indeed, let X be an algebraic model constructed from a Heyting algebra **H**. The validity of the negationless axioms and of the modus ponens rule follows from the properties of υ^+ and from the fact that **H** is a Heyting algebra. As to the remainning axioms we give only the following example:

$$\begin{aligned} \upsilon^+(\neg(A \supset B)) &= \upsilon^-(A \supset B) \\ &= \upsilon^+(A) \wedge \upsilon^-(B) \\ &= \upsilon^+(A) \wedge \upsilon^+(\neg B) \\ &= \upsilon^+(A \wedge \neg B) \end{aligned}$$

which entails the validity of the axiom

$$\neg(A \supset B) \equiv A \wedge \neg B.$$

We conclude by induction on the length of demonstrations that all theses are valid.

In order to prove the completeness of the logic with respect to these models we must show that if A is valid then A is a thesis.

Suppose A is valid. Therefore it is valid in every algebraic model and particularly in the Lindenbaum model L. Let us define the following assignment on L:

$$(a^+(P), a^-(P)) = ([P], [\neg P])$$

for every propositional variable P.

We can prove by induction on the length of formulas that this assignment can be extended to a valuation satisfying

$$v^+(A) = [A] \quad \text{and} \quad v^-(A) = [\neg A]$$

for every formula A. As an example we offer the inductive step corresponding to implication. Suppose the property hold for A and B. Then

$$\begin{aligned} v^+(A \supset B) &= v^+(A) \supset v^+(B) \\ &= [A] \supset [B] \\ &= [A \supset B] \end{aligned}$$

and

$$\begin{aligned} v^-(A \supset B) &= v^+(A) \wedge v^-(B) \\ &= [A] \wedge [\neg B] \\ &= [\neg(A \supset B)] \end{aligned}$$

consequently, if A is valid

$$v^+(A) = [A] = 1$$

and A is a thesis.

5. ALGEBRAIC PROPERTIES.

We can outline now some algebraic properties of our Nelson models.

It is clear that if X is an algebraic Nelson model on H and if $x = (x_1, x_2)$ belongs to X then

$$\neg(x_1, x_2) = (x_2, x_1).$$

Let us prove that $proj_1(X) = proj_2(X)$.

In fact, $x_1 \in proj_1(X)$ means that there exists x in X such that $proj_1(x) = x_1$. Since $\neg x$ is in X and $x_1 = proj_2(\neg x)$ it follows that $x_1 \in proj_2(X)$. The other inclusion is analogous.

It can be seen that

$$H' = proj_1(X)$$

is a subalgebra of H. As an example let us verify that if $x_1, y_1 \in H'$ then there exist $x, y \in H$ such that $proj_1(x) = x_1$ and $proj_1(y) = y_1$. Since $x \supset y$ belongs to X, $x_1 \supset y_1 = proj_1(x \supset y)$ belongs to H'. Analogously for the remaining conditions.

It is a consequence of what was said above that in any Nelson model we can always suppose that

$$H = proj_1(X)$$

because if it were otherwise one could consider X as a subset of $H' \times H'$ which would entail $H' = proj_1(X)$.

Thus call H the *base algebra* of X.

We say that Y is a submodel of X if Y is a subset of X such that:

i) $(1,0) \in Y$;

ii) If x, y are in Y then also are $x \supset y$, $x \wedge y$, $x \vee y$, $\neg x$.

It is clear that a submodel of a Nelson model is itself a Nelson model.

Now, every Nelson model X with base algebra H is a submodel of the model Y given by

$$Y = \{(h_1, h_2) \in H \times H : h_1 \wedge h_2 = 0\}.$$

In fact, X is a subset of Y since if $x = (x_1, x_2)$ belongs to X then $x_1 \wedge x_2 = 0$ so that x is in Y. The remaining conditions are immediate.

We could refer ro model Y as a *complete* model.

Consider now two Nelson models X and X' with base algebras H and H' respectively. A homomorphism of X into X' is a function h from X into X' such that:

$$h(1,0) = (1,0),$$

$$h(x \supset y) = h(x) \supset h(y)$$

$$h(x \wedge y) = h(x) \wedge h(y)$$

$$h(x \vee y) = h(x) \vee h(y)$$

$$h(\neg x) = \neg h(x)$$

for every x, y in X.

Moreover, h is an isomorphism if h is a homomorphism and also a bijection.

We are going now to establish a bijection between the homomorphic images of a Nelson model X and the homomorphic images of its base Heyting algebra H. This will allow us to reduce the theory of homomorphisms for Nelson's models to the corresponding theory for Heyting algebras.

Let X and X' be two Nelson models, H and H′ their corresponding base algebra, and g a homomorphism of H into H′.

Define a function h from X into $H' \times H'$ by:

$$h(x_1,x_2) = (g(x_1),g(x_2)) \text{ for every } (x_1,x_2) \text{ in } X.$$

Let X' be the image of X by h. It is easily seen that h is a Nelson homomorphism of X into X' if we take into account the fact that g is a Heyting homomorphism.

Moreover

$$H' = proj_1(X')$$

obtains.

In fact, pick $x_1' \in H'$. Since g is onto there exists x_1 in H such that $g(x_1) = x_1'$. Since $H = proj_1(X)$ there exists x in X such that $proj_1(x) = x_1$. Also since $g(x_1) = proj_1(h(x))$ and $h(x)$ is in X' we conclude that x_1' is in $proj_1(X')$.

We have thus obtained a Nelson image of X.

Reciprocally let X' be a Nelson image of X with base algebra H′ and h the respective homomorphism.

We are going to see that H′ is a Heyting image of H by defining a function g from H into H′ as follows:

Pick $x_1 \in H$. There exists x in X such that $proj_1(x) = x_1$. We then set

$$g(x_1) = proj_1(h(x)).$$

Let us see that this definition is independent of x. For this purpose we use the abbreviation

$$x \equiv y \quad \text{for} \quad (x \supset y) \wedge (y \supset x).$$

Now take x, y in H such that $proj_1(x) = proj_1(y)$. Let $z = (x \equiv y)$. According to our assumption $z = (1,0)$. Because h is a homomorphism we have

$$h(z) = (h(x) \equiv h(y)) = (1,0),$$

so that

$$proj_1(h(x)) = proj_1(h(y)).$$

With arguments similar to those we have used one can prove that g is a homomorphism of H into H′ using the fact that h is a homomorphism.

Finally, using g to form the image X'' of X we see that X'' is isomorphic to X'. To establish this let us first prove that

$$h(x) = (g(x_1), g(x_2)) \text{ for every } x = (x_1, x_2) \text{ in } X.$$

We know that

$$g(x_1) = proj_1(h(x)).$$

We define now

$$g'(x_2) = proj_2(h(x)).$$

From $g'(x_2) = proj_1(\neg h(x)) = proj_1(h(\neg x))$ it follows that the definition of g' is, likewise that of g, independent of x. Moreover g' is a homomorphism of H into H′ and

$$h(x) = (g(x_1), g'(x_2)) \text{ for every } x = (x_1, x_2) \text{ in } X.$$

Let us see that $g = g'$.

Indeed, pick $x_1 \in H$. There exists x in X such that $proj_1(x) = x_1$. Because h is a homomorphism

$$h(\neg x) = \neg h(x)$$

and therefore

$$proj_2(h(\neg x)) = proj_2(\neg h(x)),$$

that is $g'(x_1) = g(x_1)$.

Then it is clear that the identity is the isomorphism between X'' and X'.

We now give a topological representation theorem for Nelson models which reduces to the representation theorem for Heyting algebras. We proceed to state it:

Every Nelson model is isomorphic to a submodel of a complete model having the Heyting algebra of all the open sets in a topological space T *as its base algebra.*

For the demonstration it suffices to use the well-known Representation Theorem for Heyting algebras and apply the construction of an homomorphic image of X, using the isomorphism between the bases.

To construct free models we need the concept of a *generated model*, particularly when the construction is effected upon a Heyting algebra.

We write

$$H = \overline{G}$$

if H is a Heyting algebra generated by the set G.

Consider now a set G of couples of elements of a Heyting algebra H such that

$$g_1 \wedge g_2 = 0 \quad \text{for every} \quad g = (g_1, g_2) \quad \text{in} \quad G.$$

The Nelson model generated by G, in the Heyting algebra H is the smallest Nelson model, formed by elements of $H \times H$, containing G. We denote it by $\overline{G}$.

Such a model always exists since the complete model on H is a model which contains G.

One easily sees that $\overline{G}$ is the set of couples of $H \times H$ which are obtained from the elements of G applying the operations $\supset$, $\wedge$, $\vee$, $\neg$ a finite number of times.

Let be

$$G_1 = proj_1(G) \quad \text{and} \quad G_2 = proj_2(G),$$

one easily verifies that:

If $H = \overline{G_1 \cup G_2}$ then $H = proj_1(\overline{G})$.

Now let us see how to construct free Nelson models based on Heyting algebras.

Let G be a set of couples such that:

i) $g_1 \wedge g_2 = 0$ for every (g_1, g_2) in G;

ii) If $G_1 = proj_1(G)$ and $G_2 = proj_2(G)$ then $G_1 \cap G_2 = \emptyset$;

iii) If $(g_1, g_2) \neq (g_1', g_2')$ then $g_1 \neq g_1'$ and $g_2 \neq g_2'$ for every (g_1, g_2) and (g_1', g_2') in G.

Consider a Heyting algebra H_L such that:

i) $H_L = \overline{G_1 \cup G_2}$

ii) Any function f from $G_1 \cup G_2$ into a Heyting algebra H' satisfying

$$f(g_1) \wedge f(g_2) = 0 \quad \text{for every} \quad g_1 \in G_1 \text{ and } g_2 \in G_2$$

can be extended to a homomorphism g of H_L into H'.

Under such conditions the Nelson model generated by G in H_L is the free model on the set G of generators.

It is clear that

$$L = \overline{G} \quad \text{and} \quad H_L = proj_1(L).$$

Let f be a function from G into a Nelson model X'. According to conditions ii) and iii) previouly given on G, f naturally induces a function, also denoted by f, from $G_1 \cup G_2$ into the base algebra H' of X'.

According to condition ii) of H_L there exists a homomorphism g of H_L into H' which extends f. From what was said about homomorphisms we conclude that there is a homomorphism h of L into X' which extends f since

$$\begin{aligned} h(g_1,g_2) &= (g(g_1),g(g_2)) \\ &= (f(g_1), f(g_2)) \\ &= f(g_1,g_2) \end{aligned}$$

for every (g_1,g_2) in G.

The existence of H_L results from considering the Lindenbaum algebra for the intuitionistic theory possessing two sets of mutually distinct propositional variables P_i and Q_i and having as axioms:

$$P_i \wedge Q_i \supset A \quad \text{for every } i.$$

One can also see that the Lindenbaum model is the free model on the following set of generators:

$$G = \{([P],[\neg P]) : P \text{ is a propositional variable}\}$$

We are going to prove now a kind of equivalence between Nelson models and **N**-lattices. This equivalence could also be heuristically explored to obtain properties of **N**-lattices.

Thus let X be a Nelson model. We define the following constant, operation and relations on X.

$$\begin{aligned} 1 &= (1,0), \\ x \supset y &= (x_1 \supset y_1, x_1 \wedge y_2), \\ x \vee y &= (x_1 \vee y_1, x_2 \wedge y_2), \end{aligned}$$

$$x \wedge y = (x_1 \wedge y_1, x_1 \wedge y_2),$$

$$\neg x = (x_2, x_1),$$

$$x \prec y \text{ iff } x_1 \leqslant y_1,$$

$$x \subset y \text{ iff } x_1 \leqslant y_1 \text{ and } y_2 \leqslant x_2,$$

for $x = (x_1, x_2)$, $y = (y_1, y_2)$ in X.

One easily verifies that

$$x \prec y \quad \text{iff} \quad x \supset y = 1.$$

It is easy to see that the relation $\prec$ is reflexive and transitive.

Furthermore, one can prove that X provided with $\wedge$ and $\wedge$ as operations is a distributive lattice with last element 1.

The relation $\subset$ satisfies

$$x \subset y \quad \text{iff} \quad x \prec y \quad \text{and} \quad y \prec x$$

and coincides with the order relation of the lattice X.

Moreover, with the operation $\neg$ aggregated to it X turns out to be a De Morgan algebra (or quasi-Boolean algebra, see [5]).

The laws:

$$\text{if } x \prec z \text{ and } y \prec z \text{ then } x \vee y \prec z,$$

$$\text{if } z \prec x \text{ and } z \prec y \text{ then } z \prec x \wedge y,$$

$$x \prec y \supset z \quad \text{iff} \quad x \wedge y \prec z,$$

are consequences of the fact that $proj_1(X)$ is a Heyting algebra.

The laws:

$$\neg(x \supset y) \prec x \wedge \neg y$$

$$x \wedge \neg y \prec \neg(x \supset y)$$

results from the equality of the corresponding first coordinates.

Finally,

$$x \wedge \neg x \prec y$$

is a consequence of $x_1 \wedge x_2 = 0$.

We have thus verified (except for not having considered the intuitionistic negation operator) the definition of **N**-lattice given by Rasiowa in

[4].

As to the reciprocal we are going to prove a Representation Theorem according to which each **N**-*lattice is represented as a subset of a two-fold product of Heyting algebras*.

For the demonstration we use the concept of a *first class prime filter* or *prime-sffk* (see [5]). A *prime-sffk* is a prime filter q of a **N**-lattice X such that:

If $x, y \in X$, $x \in q$ and $x \prec y$ then $y \in q$.

According to [5], pg. 93, a prime-sffk q also satisfies:

If x, $x \supset y$ are in q then $y \in q$.

We mention two important properties of prime-sffk's:

i) If $x \prec y$ does not hold then there exists a prime-sffk q such that $x \in q$ and $y \notin q$.

ii) If p is a prime-sffk and $x \supset y \notin p$ then there exists a prime-sffk p' such that $p \subseteq p'$, $x \in p'$ and $y \notin p'$.

As to property i) see [5] pg. 100. Property ii) is analogous to a corresponding property of Heyting algebra and for its demonstration it suffices to consider the sffk generated by $p \cup \{x\}$ and to note that y is not in this generated filter using the law

$$z \wedge x \prec y \text{ iff } z \prec x \supset y \quad \text{(see [5] pg. 94).}$$

Now let X be an **N**-lattice. We are going to consider for each x a function also denoted by x, from P, the set of all prime-sffk's of X, into the two-element Boolean algebra $\{0,1\}$ defined in the following manner:

$$x(p) = \begin{cases} 1 & \text{if } x \in p \\ 0 & \text{if } x \notin p \end{cases} \qquad \text{for each } p \text{ in } P.$$

Since $\{0,1\}$ is a Boolean alaebra we can define, on the set of all functions x, the operations $\wedge$, $\vee$, $-$ elementwise.

We also define:

$$(x \supset y)(p) = \bigwedge_{p \subseteq p'} (-x \vee y)(p')$$

which means that $x \supset y$ at p is the infimum of the values $-x(p') \vee y(p')$

for every $p \subseteq p'$. Such infimum always exists being either 0 or 1.

One can show that the set of functions x provided with the operations $x \supset y$, $x \wedge y$, $x \vee y$ and with the constant functions $1(p) = 1$ and $0(p) = 0$ constitutes a Heyting algebra which we denote by **H**.

For each x in X we consider a couple of functions (x^+, x^-) such that:

$$x^+(p) = x(p) \quad \text{and} \quad x^-(p) = \neg x(p) \quad \text{for every } p \text{ in } P.$$

Then one sees, using properties of prime-sffk's, that:

$$(x \supset y)^+ = x^+ \supset y^+, \qquad (x \supset y)^- = x^+ \wedge y^-,$$

$$(x \wedge y)^+ = x^+ \wedge y^+, \qquad (x \wedge y)^- = x^- \vee y^-,$$

$$(x \vee y)^+ = x^+ \vee y^+, \qquad (x \vee y)^- = x^- \wedge y^-,$$

$$(\neg x)^+ = x^-, \qquad (\neg x)^- = x^+,$$

for any functions x, y. As an example let us see the demonstrations of the first two equalities:

$$\begin{aligned}(x \supset y)^+(p) = 1 \ &\text{iff} \ x \supset y \in p \\ &\text{iff for every } p' \supseteq p,\ x \notin p' \text{ or } y \in p' \\ &\text{iff for every } p' \supset p,\ x^+(p') = 0 \text{ or } x^+(p') = 1 \\ &\text{iff } (x^+ \supset y^+)(p) = 1\end{aligned}$$

$$\begin{aligned}(x \supset y)^-(p) = 1 \ &\text{iff} \ \neg(x \supset y) \in p \\ &\text{iff } x \wedge \neg y \in p \\ &\text{iff } x \in p \text{ and } \neg y \in p \\ &\text{iff } x^+(p) = 1 \text{ and } y^-(p) = 1 \\ &\text{iff } (x^+ \wedge y^-)(p) = 1.\end{aligned}$$

Now let X' be the complete model having **H** as base algebra. We are going to see that X is isomorphic to a submodel of X'. For this purpose we define a function f such that:

$$f(x) = (x^+, x^-) \quad \text{for every } x \text{ in } X.$$

Using what was seen above it is easy to see that f is a homomorphism of **X** onto $f(X)$. For instance,

$$f(1) = (1, 0)$$

is a consequence of

$$1^{+}(p) = 1 \quad \text{iff} \quad 1 \in p$$

$$1^{-}(p) = 0 \quad \text{iff} \quad \neg 1 = 0 \notin p$$

for every p in P.

Finally let us prove that f is a bijection. Suppose x and y are in X and $x \neq y$. Then, either not $x \prec y$, or not $\neg y \prec \neg x$, or not $y \prec x$, or not $\neg x \prec \neg y$.

If not $x \prec y$ holds then there is a prime-sffk q such that $x \in q$ and $y \notin q$. Thus $x^{+}(q) = 1$ and $y^{+}(q) = 0$. Consequently $(x^{+}, x^{-}) \neq (y^{+}, y^{-})$ and $f(x) \neq f(y)$. Analogously for the remaining cases.

Furthermore it is clear that

$$H = proj_1(f(X)).$$

This completes the demonstration of the equivalence.

6. SEMANTICAL MODELS.

Now we introduce semantical models for Nelson's Logic analogous to those of Kripke [2] for intuitionistic logic. Such models differ from those given by Thomason [8] and Routley [7]. We use the algebraic results stated in Section 4 to prove the completeness relative to these models.

We proceed to define the concept of a Nelson *semantical model*. Consider a non-void set X ordered by an order relation R. An assigment is a couple (a^{+}, a^{-}) of functions from the set of propositional variables of Nelson's logic and from the set S into the set $C = \{(0,1), (0,0), (1,1)\}$ such that:

if $a^{+}_{s}(P) = 1$ and sRs' then $a^{+}_{s'}(P) = 1$,

if $a^{-}_{s}(P) = 1$ and sRs' then $a^{-}_{s'}(P) = 1$,

for any propositional variable P and s in S.

By induction this assignment can be extended in a unique manner to a valuation, that is, to a couple (v^{+}, v^{-}) of functions from the set of formulas and from the set S into C in the following way:

i) $v^{+}_{s}(P) = a^{+}_{s}(P)$ and $v^{-}_{s}(P) = a^{-}_{s}(P)$ for every propositional variable P and for every s in S.

ii) $\upsilon_s(A \supset B) = 1$ iff for every s' such that sRs', $\upsilon^+_{s'}(A) = 0$ or $\upsilon^+_{s'}(B) = 1$. Otherwise, $\upsilon^+_s(A \supset B) = 0$;

$\upsilon^-_s(A \supset B) = 1$ iff $\upsilon^+_s(A) = 1$ and $\upsilon^-_s(B) = 1$. Otherwise, $\upsilon^-_s(A \supset B) = 0$;

$\upsilon^+_s(A \wedge B) = 1$ iff $\upsilon^+_s(A) = 1$ and $\upsilon^+_s(B) = 1$. Otherwise, $\upsilon^+_s(A \wedge B) = 0$;

$\upsilon^-_s(A \wedge B) = 1$ iff $\upsilon^-_s(A) = 1$ or $\upsilon^-_s(B) = 1$. Otherwise, $\upsilon^-_s(A \wedge B) = 0$;

$\upsilon^+_s(A \vee B) = 1$ iff $\upsilon^+_s(A) = 1$ or $\upsilon^+_s(B) = 1$. Otherwise, $\upsilon^+_s(A \vee B) = 0$;

$\upsilon^-_s(A \vee B) = 1$ iff $\upsilon^-_s(A) = 1$ and $\upsilon^-_s(B) = 1$. Otherwise, $\upsilon^-_s(A \vee B) = 0$;

$\upsilon^+_s(\neg A) = 1$ iff $\upsilon^-_s(A) = 1$. Otherwise, $\upsilon^+_s(\neg A) = 0$;

$\upsilon^-_s(\neg A) = 1$ iff $\upsilon^+_s(A) = 1$. Otherwise, $\upsilon^-_s(\neg A) = 0$;

(for every formula A and B and for every s in S).

Intuitively the elements of S represent the stages of a construction by which one attempts to prove that A is true and at the same time that A is false. This is represented by the valuations υ^+ and υ^- whose values are 0 if one does not know whether A is true or false at this stage but falls under either one of the possibilities, and 1 if one has proved either that A is true or that A is false.

By induction on the length of formulas one can prove that:

if $\upsilon^+_s(A) = 1$ *and* sRs' *then* $\upsilon^+_{s'}(A) = 1$;

if $\upsilon^-_s(A) = 1$ *and* sRs' *then* $\upsilon^-_{s'}(A) = 1$,

where A *is a formula and* s *and* s' *are in* S.

We say that a formula A is valid in a semantical model S if $\upsilon^+_s(A) = 1$ for every s in S. We say that A is valid if it is valid in every semantical model.

One can show by induction on the length of demonstrations that *all theses of Nelson's logic are valid.*

As to the reciprocal, that is, to prove *completeness*, we consider the following canonical semantical model constructed from the algebraic Lindenbaum model L.

Let S be the set of all prime filters of the base algebra $\mathbf{H_L}$ of $\mathbf{L}$ and let R be the inclusion relation.

We define on S the following assignment:

$$a^+_s(P) = \begin{cases} 1 & \text{if } [P] \in s \\ 0 & \text{if } [P] \notin s \end{cases} \qquad a^-_s(P) = \begin{cases} 1 & \text{if } [\neg P] \in s \\ 0 & \text{if } [\neg P] \notin s \end{cases}$$

for every propositional variable P and for every s in S.

By induction on the length of formulas one can prove the possibility of extending this assignment to a valuation (v^+, v^-) satisfying:

$$v^+_s(A) = \begin{cases} 1 & \text{if } [A] \in s \\ 0 & \text{if } [A] \notin s \end{cases} \qquad v^-_s(A) = \begin{cases} 1 & \text{if } [\neg A] \in s \\ 0 & \text{if } [\neg A] \notin s \end{cases}$$

for every formula A and every s in S.

For the demonstration we use nothing more than the properties of prime filters of a Heyting algebra. Let us see as an example the inductive step corresponding to implication.

Suppose $[A \supset B] \in s$, and let us prove that for every s' such that sRs', if $v^+_{s'}(A) = 1$ then $v^+_{s'}(B) = 1$. By inductive hypothesis, if $v^+_{s'}(A) = 1$ then $[A] \in s'$. From $[A \supset B] = [A] \supset [B] \in s'$ and $[A] \in s'$ it follows that $[B] \in s'$, therefore $v^+_s(B) = 1$. According to the definition of valuation $v^+_s(A \supset B) = 1$.

Suppose now that $[A \supset B] \notin s$. Consequently there exists a prime filter s' containing s and such that $[A] \in s'$ and $[A \supset B] \notin s'$. Therefore $[B] \notin s'$.

By inductive hypothesis $v^+_{s'}(A) = 1$ and $v^+_{s'}(B) = 0$ follow from $[A] \in s'$ and $[B] \notin s'$. Using the definition of valuation, $v^+_{s'}(A \supset B) = 0$.

Suppose $[\neg(A \supset B)] \in s$. Since $[\neg(A \supset B)] = [A] \wedge [\neg B]$ and s is a filter. This is equivalent to $[A] \in s$ and $[\neg B] \in s$. By inductive hypothesis this is equivalent to $v^+_s(A) = 1$ and $v^-_s(B) = 1$. Again by the definition of valuation this amounts to $v^-_s(A \supset B) = 1$.

To prove the completeness of Nelson's logic relative to these semantical models suppose A is valid. In particular A is valid in the canonical model referred to above, that is, if s is a prime filter of $\mathbf{H_L}$,

$$[A] \in S$$

since $\{[T]\}$, where T is a thesis, is the intersection of all prime filter of H_L we have: $[A] = [T]$, and $A \equiv T$ is a thesis. Therefore, A is a thesis.

REFERENCES.

[1] F. B. Fithch, Symbolic Logic, Ronald Press, New York, 1952.

[2] S. A. Kripke, *Semantical analysis of intuitionistic logic, I,* Formal Systems and Recursive Fuctions (Ed. J. Crossley and M. Dummett) North-Holland Publs. Co., Amsterdam, 1965, 206-220.

[3] D. Nelson, *Constructible falsity,* The Journal of Symbolic Logic, 14 (1949), 16-26.

[4] H. Rasiowa, *N-lattices and constructive logic with strong negation,* Fundamenta Mathematica, 46 (1958), 61-80.

[5] H. Rasiowa, An algebraic Approach to Non-Classical Logics, North-Holland Publs. Co., Amsterdam, 1974.

[6] H. Rasiowa and R. Sikorski, The Mathematics of Metamathematics. PWM, Warszawa, 1963.

[7] R. Routley, *Semantical analysis of propositional systems of Fitch and Nelson,* Studia Logica, 33 (1974), 283-298.

[8] R. H. Thomason, *A semantical study of constructible falsity,* Zeitschr. für Math. Logik und Grundl. der Math., 15 (1969), 247-257.

Instituto de Matemática
Universidad Nacional del Sur
8.000 Bahía Blanca, Argentina.

ADDED IN PROOF: we have founded an algebraic construction for the free Nelson models using only the existence and properties of the free Heyting algebras.

Some Remarks in Set Theory.

by *MARCEL GUILLAUME*.

ABSTRACT. In this partially expository paper we present some remarks on the theory ZF' obtained by dropping the power set axiom from the theory ZF of Zermelo-Fraenkel.

We first show in ZF' that given a set a the class of subsets of x definable in a and with parameters in a is a set. Therefore we have in ZF' a notion of sets constructible from the set x.

Our main result is that adding to ZF' the sentence

$$(\mathrm{PL}) \quad (\forall x\, \exists y\, \forall z\, (\, z \subset x \wedge z \text{ constructible from } x \rightarrow z \in y))$$

as an axiom, we can prove in ZF' + (PL) that $L \models$ ZF.

As a corollary we obtain the equiconsistency of ZF and ZF' + (PL).

Work in set theory showed, since Cohen, that outside any countable model of ZF to which it belongs, an infinite set always admits parts, so that from an "absolute" point of view - if one there may be, but we are always tempted to do so- the situation may be viewd as though the parts of an infinite set form a *class*, not a set, hypothesis which amount to drop the Power Set Axiom.

I wish here to present some remarks on the theory ZF' so obtained from the theory ZF. This theory bears all the defects common to all systems in which the Peano arithmetic can be expressed; from this point of view, it is not worse than ZF. My idea is that, while the study of ZF cannot be substituted for that of ZF', it may yet jointly with that of the other axiomatizations, bring some light over foundational problems. Such is the case for theories close to ZF, whose interest semms sometimes much smaller, as Z, the initial theory of Zermelo, ZF deprived from the axiom of extensionality, or from the axiom of foundation, etc. ...

The principal problem of ZF' is that the reals form a class, and that one loses, in a word, the freedom afforded to develop mathematics by the hypothesis of a unique process of collectivization (it must be said, in passing, that this hypothesis seems no longer tenable). Then, what analysis, what mathematics can be done in ZF'? What sets of reals will be at our disposal? Eventually, would it not be reasonable to admit that, if *all* parts of an infinite set cannot constitute a set, the parts satisfying given conditions (which?) must do? (Therefore, to pass from ZF' to ZF' + an axiom of type

$$\mathrm{P'} \qquad \forall x[P(x) \to \exists y\, \forall z\, [\, z \subset x \;\wedge\; Q(z) \to z \in y]]$$

P and Q being one-free variable formulas to be determined; if for $P(x)$ and $Q(x)$ one takes $x = x$, one recovers the Power Set Axiom of ZF, and ZF (note that in ZF' the usual principle of separation still holds, so that in the formulation of P' the second $\to$ can alternatively be substituted by an $\leftrightarrow$).

I have more questions than answers. The situation is known for some people which worked on ZF' and related theories and have not written on the subject. My hope is to show that the study of such theories deserves interest.

Amongst these theories is that which is obtained from ZF' writing that all parts of an infinite set always form a proper class, as follows:

$$\mathrm{PC} \qquad \forall x[x \textit{ infinite} \to \forall y\, \exists z\, [\, z \subset x \;\wedge\; z \notin y]]$$

- an hypothesis on power set, still more radical than those of Takeuti [9]. The consistency of ZF entails that of ZF' + PC, as proved by a model of ZF' presented by Paul Cohen in his book [2] in order to prove the independence of the Power Set Axiom from the other axioms of ZF; this model consists of the hereditarily denumerable sets from any model of ZF; it shows that it is consitent with ZF + PC to admit the axiom of countability, postulating the countability of all sets. But that model is in a way disappointing; starting from that the parts of an infinite set must form a great class, we reach a realization consisting in supressing great sets (an analogous situation is pointed out by J. Friedmann [3] with respect to the general continuum hypothesis, hypothesis minorating the cardinals of power sets, but equivalent to a principle maximizing the cardinal numbers of infinite sets closed under pairing, union and replacement, assuming that all such sets contain as elements all their parts of cardinality less than that of the set itself).

"Little" as it seems, that model does *not* however totally forbid mathematicians's evolutions; it permits to talk - without talking of their collections - about real numbers, functions of real variables, and so on.

More generally, putting

$$\text{rank}(x) = \sup_{y \in x} \text{rank}(y),$$

for all sets x - this is permitted, since the ranks of the $y \in x$ form a set by the replacement schema - we can very well introduce the notions of sets of rank at most $\omega + 1$, $\omega + 2$, etc ..., notwithstanding that $R_{\omega+1}$, $R_{\omega+2}$ etc ... cannot be introduced as sets of ZF' (we are using freely the notations of the Morse theory of sets, taken as our metatheory, R_α denoting the class of sets of rank at most α). All that leads us to see that the theory of the von Neumann ordinals works in ZF', as in ZF in most respects: principle of transfinite induction, definitions by transfinite induction, normal functions ans their critical values, and so on ... up to the introduction of cardinals (the Cantor-Bernstein Theorem remains available, as seen in Quine [7]); it is of course *not* possible to establish the existence of ω^+ (the first cardinal after ω) since we know a transitive model whose ordinals form just ω^+ (in that model, the Hartog's function $\aleph(\alpha)$ *cannot* be defined, for $\alpha \geqslant \omega$).

Likewise ZF' remains *strongly semantically complete* in the sense of Montague [5]: the product of two sets is defined without recourse of the Power Set Axiom, as is, therefore, the set of all applications from a *finite* ordinal into a set, the set $\text{Seq}(u)$ of all *finite* sequences of elements of a set u, the transitive closure $\text{TC}(u)$ of the latter, the set of all the *finite* parts of a set (so we need no special axiom of type P' in order to introduce the power set of any finite set), and then R_0, R_1, R_2, ..., R_n, ..., and $R_\omega = \bigcup_{n \in \omega} R_n$. So we fall here upon another important feature of ZF': since we have available R_ω as *set*, and the definitions by induction, we have available at once, as in ZF, "Gödel's codes" for formulas, free variables, and so on ..., hence we can encode the notion of satisfaction (by a given sequence, in a given set, of a given code of formula); so for any given once for all such coding (cf. e. g. Mostowski

[6], or Jensen [4]), we can introduce in ZF', and for any set E:

the set Var of all (codes of) variables;

the set Fml of all (codes of) formulas;

the set $\mathrm{Vl}(\varphi)$ of the (codes of) variables occurring free in the (code of) formulas φ; the set $\mathrm{I}(E) = \bigsqcup_{s \in \mathrm{Seq}(\mathrm{Var})} {}^{s}E$ of all interpretations of finite sequences of variables in E; the set $\mathrm{Fml} \times \mathrm{Var} \times \mathrm{I}(E)$; its part $\mathrm{J}(E)$ of the triples $\langle \varphi, x, a \rangle$ such that there exists an $s \in \mathrm{Seq}(\mathrm{Var})$ such that $a \in {}^{s}E$ and $\mathrm{Vl}(\varphi) = s^\frown x$ (obtained by concatenation of x to s) and finally, the set $\mathrm{Def}(E)$ of all parts of E definable in E with parameters in E, which is the image of the set $\mathrm{J}(E)$ by the function which to the triple $\langle \varphi, x, a \rangle$, with $\mathrm{Vl}(\varphi) = \langle x_1, \ldots, x_k, x \rangle$ and $a = \langle a_1, \ldots, a_k \rangle$, associates the set $\{u \in E;\ E \models \varphi[a_1, \ldots, a_k, u]\}$. So we have:

THEOREM 1. $\mathrm{ZF'} \vdash \forall x \exists y \forall z$ ($z \subset x \wedge z$ *definable in* E *with parameters in* $E \leftrightarrow z \in y$).

One can therefore, in ZF', from any set x, define the hierarchy of sets L^x_α such that $L^x_0 = \mathrm{TC}(x)$, $L^x_{\alpha+1} = \mathrm{Def}(L^x_\alpha)$ for each ordinal α, $L^x_\lambda = \bigcup_{\alpha<\lambda} L^x_\alpha$ for each ordinal λ, and the notion of sets *constructible from the set* x. In the metatheory, we write $L^x = \bigcup_{\alpha \in \mathrm{On}} L^x_\alpha$ and $L = L^\phi$, as usual.

THEOREM 2. $\mathrm{ZF'} \models (L \models \mathrm{ZF'} + V = L$ (*here* $V = L$ *means* $\forall x \exists \alpha [\alpha$ *ordinal* $\wedge\ x \in L_\alpha]$ *and* $L \models \varphi$ *denotes the relativization of the formula* φ *to the formula* $\exists \alpha [\alpha$ *ordinal* $\wedge\ x \in L_\alpha]$).

PROOF: As in ZF, but the principle of reflection needs some adaptation. The point at which the proof of the ordinary principle for ZF breaks down is the following:

Given the formula $\exists x\, \psi(x, x_1, \ldots, x_k)$ and the sequence $\langle a_1, \ldots, a_k \rangle$ of members of R_α, we *cannot* use the fact that the class

$$\{u :\ \psi[u, a_1, \ldots, a_k] \wedge \mathrm{rank}(u) \text{ is minimal for that property}\}$$

is a *set*. That is no longer true for ZF'. But one has in ZF' the following *Principle of Reflection* for the hierarchy L^X:

For any formula φ and any ordinal α, there exists an ordinal $\beta>\alpha$ such that any interpretation of the sequence of the free variables of φ satisfies φ in L^X_β if and only if it satisfies the relativization of φ to the formula expressing the constructibility of x, because, if not empty, the class

$$\{u : u \in L^X \wedge L^X \models \varphi[u, a, \ldots, a_k] \wedge \text{the first } \gamma_0 \text{ of the ordinals } \gamma \text{ such that } u \in L^X_{\gamma_0} \text{ is minimal for that property}\},$$

is in $\mathrm{Def}(L^X_{\gamma_0})$ and hence is a set. ∎

REMARK. on a similar account, the definition of $\mathrm{OD}(x)$ – x is ordinal-definable as $\exists\alpha[\alpha$ ordinal and $\{x\} \in \mathrm{Def}(R_\alpha)]$, is no longer available in ZF'.

Therefore, even on our "little" model, we dispose already of certain sets of reals, of functions of reals variables, etc ..., covering all the same, a great part of the ones needed for the other sciences. One may well think that if the coding technique were available in his time, Emile Borel would have been very interested in the theory ZF'. Now, the means of expression included in any model of ZF' are in fact still richer.

THEOREM 3. *Let κ be the first admissible ordinal after ω; it belongs to any model of* ZF'.

PROOF: According to Barwise, Gandy, Moschovakis [1],

$$\kappa = \sup\{ O(\Gamma) : \Gamma \text{ first-order positive operator without parameters}\}.$$

The operator in question cerries subsets of $A = R_\omega$ into subsets of A; such a Γ is called *first-order positive* if there is a formula $\varphi(U, x, x_1, \ldots, x_k)$ with free variables amongst the $U, x, x_1, \ldots, x_k$ and only positive accurrences of U, and a sequence $\langle a_1, \ldots, a_k\rangle$ of members of A, such that $\forall x\,[x \in \Gamma(U) \leftrightarrow \varphi(U, x, a_1, \ldots, a_k)]$.

Taking $\varphi \in \mathrm{Fml}$, since, owing to the coding, a formula of ZF' can easy be written that the code φ satisfies if and only of this code has the property of positivity of the occurrences of U, we can reason as before over the *set* of all positive "definitions" of first-order positive operators, each such "definitions" being compounded from a suitable formula and a suitable sequence of members of A: Then the first-order positive operators (without parameters) form a set, and so does their *closure ordinals* from which κ is the supremum. So the sole point under discussion is if, given a first-order positive operator Γ, one can define its coosure ordinal $O(\Gamma)$ *within* ZF'. Now, first-order positive operators such as Γ are monotone, increasing:

$$U_1 \subset U_2 \rightarrow \Gamma(U_1) \subset \Gamma(U_2);$$

so, we can define the *increasing* hierarchies

$$\Gamma^{\alpha} = \Gamma(\bigcup_{\beta < \alpha} \Gamma^{\beta})$$

in ZF', and speak metatheoretically of

$$\Gamma^{\infty} = \bigcup_{\alpha \in \mathrm{On}} \Gamma^{\alpha}.$$

If any, the first ordinal α such that $\Gamma^{\alpha} = \Gamma^{\infty}$ (in which case the latter is obviously a set) is called the *closure ordinal* of Γ.

In fact, this ordinal always exists in ZF'; because, Γ^{∞} being a *part* of A, as it can easily be seen, is a set C (i. e., the set of the x's belonging to A such that there exists an ordinal α such that $x \in \Gamma^{\alpha}$); for each $x \in C$ one can define $\gamma(x)$ = *the first ordinal* α *such that* $x \in \Gamma^{\alpha}$; it is easily seen that $\delta = \sup(C)$ is the closure ordinal of Γ. So the proof is complete. ■

Can it be said, as a corollary of Theorem 1, that the *constructible parts* of an infinite set constitute in ZF' a set? Our previous model, if extracted from the minimal model of ZF, shows us the contrary (the set of constructible parts of ω cannot be countable in the constructibles). so we get an axiom of type P' which can be viewed as a very reasonable candidate for being adjoined to ZF':

PL $\forall x \forall y \forall z (z \subset x \wedge z$ constructible from $x \rightarrow z \in y)$.

This principle is in fact a powerfull tool in creating new ordinals, as shown by

THEOREM 4. ZP' + PL $\vdash$ (L $\models$ ZF).

PROOF: Let x be constructible. Then, as $\emptyset$ is included in $\mathrm{TC}(x)$ and is a member of it, all constructible sets are constructible from x. Conversely, as $\mathbf{TC}(x)$ is constructible, so included in some L_α and a member of it, all sets constructible from x are constructible. Then the (new) set $C = \{z : z \subset x \wedge z$ constructible from $x\}$ is also the set of all parts of x which are constructible *tout court*. So we get the relativization to the constructibles of the Power Set Axiom if we prove that C is constructible. Now, this is just the case, if for each $z \in C$, we denote by $\mathrm{od}(z)$ the first ordinal α such that $z \in L_\alpha$, these ordinals form the *set* $\mathrm{od}(C)$, and putting $\delta = \sup \mathrm{od}(C)$ we obtain $C \subset L_\delta$; and clearly $C \in \mathrm{def}(L_\delta) = L_{\delta+1}$. ∎

COROLLARY. *ZF and ZF' + PL are equiconsistent.*

So the theory **ZF' + PL** is in a sense as "reasonable" as **ZF**.

REMARK. The author wishes to thank M. Boffa, for pointing out another model of **ZF'**, which is even a model of ZF' + ($L \models$ ZF) + the Axiom of Countability: the one constructed by Dana Scott in [8].

It is not clear what is the exact status of the axiom of choice with respect to ZF'. The usual proof of equivalence between the ordinary (local) axiom, Zorn's Lemma, and the well-ordering principle, breaks down in ZF'.

References.

[1] K. J. Barwise, R. O. Gandy and Y. N. Moschovakis, *The next admissible set,* The Journal of Symbolic Logic, 36 (1971), 108-120.

[2] P. J. Cohen, Set Theory and the Continuun Hypothesis, Benjamin, New York, 1966.

[3] J. I. Friedmann, *The generalized continuun hypothesis is equivalent to the generalized maximization principle,* The Journal of Symbolic Logic 36 (1971), 39-54.

[4] R. B. Jensen, Modelle der Mengenlehre, Lecture Notes 37, Springer, Berlin, 1967.

[5] R. Montague, *Semantical closure and non-finite axiomatizability I*, Infinitistic Methods, Proc. of the Symposium on Foundations of Mathematics, Warsaw, 1959, 45-69.

[6] A. Mostowski, Modèles Transitifs de la Théorie des Ensembles de Zermelo-Fraenkel, Presses de L'Université de Montréal, 1967.

[7] W. O. Quine, Set Theory and its Logic, Harvard University Press, Cambridge, Mass., 1965.

[8] D. Scott, Lectures on Boolean Valued Models for Set Theory, Mimeographed Notes, 1967.

[9] G. Takeuti, *Hypothesis on power set*, Proceedings of AMS, part I, Symposium in Pure Mathematics held at the University of California, Los Angeles, 1967, 439-446.

Mathématiques Pures
Université de Clermont
63.170 Aubière, France.

ON THE PROBLEM OF JAŚKOWSKI AND THE LOGICS OF ŁUKASIEWICZ.

by *J. KOTAS and N. C. A. da COSTA.*

ABSTRACT. In this paper we extend the results of I. M. L. D'Ottaviano and N. C. A. da Costa, *Sur un problème de Jaśkowski*, C. R. Acad. Sc. Paris, 270 A (1970), 1349-1353; by means of the *generalized logics of Łukasiewicz*, new solutions of the so called Jaśkowski's problem are presented and discussed.

INTRODUCTION.

Let $\mathcal{L}$ denote a given language containing negation. Any set of formulas of $\mathcal{L}$ is called a propositional system (or simply, a system) of $\mathcal{L}$. The elements of a system S are called theses of S.

A system S is said to be inconsistent if it contains at least two theses, such that one is the negation of the other; in the opposite case, S is called consistent. S is said to be trivial (or overcomplete) if any formula of $\mathcal{L}$ is also a thesis of S; otherwise, S is nontrivial (or not overcomplete).

A propositional system which has an underlying logic, i. e., that is based on a logic, is called a deductive system. If in such system the rule: "From α and not-α, infer β" is permissible, then it is inconsistent if and only if it is trivial. This is precisely what happens with deductive systems based on classical logic and on several other categories of logics, as, for example, the intuitionistic.

Of course, trivial systems have no practical importance. But the situation is completely different in connection with consistent systems, which are of fundamental relevance from the theoretical as well as from the practical points of view. Nonetheless, there are relatively few systems of which we really know that they are consistent. We cannot give absolute proofs of consistence even for certain rather elementary mathematical systems. The situation is still worse when we consider systems based on results of experimental research. These facts and other stronger reasons motivated Jaśkowski to formulate in [8] the problem: To construct logics satisfying the following conditions: 1) when they are employed as underlying logics of inconsistent proposi ionaltsystems, inconsistency does not necessarily imply triviality; 2) they are rich enough to make possible most common inferences.

In [8] and [9], Jaśkowski presented a solution for his problem at the level of the propositional calculus: in effect, he introduced the so called *discussive propositional calculus* $\mathbf{D}_2$, defined by means of an appropriate interpretation in the modal system **S5** of Lewis. Therefore, Jaśkowski's solution depends on standard modal logic. Results by Furmanowski (see [7]), Perzanowski (see [12] and Błaszczuk and Dziobiak (cf. [1], [2] and [3]) showed that we obtain very interesting solutions to Jaśkowski's problem using other modal systems instead of **S5** . In [5], a completely distinct solution to the problem is described, where a hierarchy of logical systems not restricted to the propositional calculus is studied and developed up to the construction of inconsistent but apparently nontrivial set theories. D'Ottaviano and da Costa in [6] studied still another solution, the calculus $\mathbb{J}_3$, which is founded on Łukasiewicz three-valued logic.

In every-day life the process of assertion of sentences is very complex. We may suppose that we consider any sentence as true when our conviction about its truth is strong enough or, is other words, when a "logical value" sufficiently large corresponds to the sentence. Clearly, the assertion of a sentence as probable is made analogously. If we restrict ourselves to $\mathbf{D}_2$ or $\mathbb{J}_3$, then we have to consider as probable all sentences which have (in our conviction, of couse) a "logical value" greater than 0, and moreover we are constrained to assume that "to be probable" and "to be true" have practically the same meaning. This attitude, although convenient and full of interesting consequences, is only a crude approximation to the actual procedure. Also, we do not look at sentences which we believe to have sufficiently small "logical values" as probable. To introduce finer distinction than it is possible with the help of $\mathbf{D}_2$ and $\mathbb{J}_3$, we study in this paper some new logical calculi.

In fact, the aim of our work is to show that if we take as bases the finite or infinite logics of Łukasiewicz, then it is possible to define a class of logical calculi, some of which are solutions of Jaśkowski's problem. Therefore, this paper contains a generalization of D'Ottaviano – da Costa's results. It seems worthwhile to observe that the method of characterizing logical calculi employed here and applied to Lukasiewicz logics, can be extended to other logical calculi.

JAŚKOWSKI'S PROBLEM AND THE LOGICS OF ŁUKASIEWICZ.

Let M be a finite or infinite-valued matrix of Łukasiewicz, i. e., a matrix of the form:

$$\langle A, \{1\}, \Rightarrow, \neg \rangle,$$

where $A = \{0, 1/n, \ldots, n-1/n, 1\}$, $n \geqslant 1$, or A is the set of all real numbers x such that $0 \leqslant x \leqslant 1$, and the operations $\Rightarrow$ and $\neg$ are defined as usual. Let $M_{a,b}$, $0 < a \leqslant 1$, $0 < b \leqslant 1$, be the matrix obtained from M by the addition to its two operations, of the new operations $\cup$, $\cap$, and C_a defined as follows: for any $x,y \in A$, $x \cup y = \max(x,y)$, $x \cap y = \min(x,y)$, $C_a x = 0$ for $x < a$ and $C_a x = 1$ for $x \geqslant a$, and by replacing $\{1\}$ by the set of all $x \in A$ such that $x \geqslant b$. $M_{a,b}$ will be called a *generalized matrix of Łukasiewicz*, and may be a finite or an infinite-valued matrix.

We note that $\cup$ and $\cap$ are definable in terms of $\Rightarrow$ and $\neg$ in any matrix of Łukasiewicz, since for any $x,y \in A$, we have: $x \cup y = (x \Rightarrow y) \Rightarrow y$ and $x \cap y = \neg(\neg x \cup \neg y)$, but that C_a is definable only in finite-valued matrices.

Let p, q, r, ... be propositional variables and $\rightarrow$, $\vee$, $\wedge$, $\sim$, and $\diamond_a$, $0 < a \leqslant 1$, be the symbols of implication, disjunction, conjunction, negation and a-possibility, respectively. $\mathbb{F}$ will denote the set of formulas defined in the usual manner, and α, β, γ, ... will be variables whose values are formulas.

The set of all formulas valid in $M_{a,b}$, $0 < a \leqslant 1$, $0 < b \leqslant 1$, symbolized by $\mathfrak{L}_{a,b}$, will be called a *generalized logic of Łukasiewicz*. In the case that $M_{a,b}$ is a finite-valued matrix the notion just defined coincides with the concept of a generalized logic of Łukasiewicz, as introduced in [14] by Rosser and Turquete. If $M_{a,b}$ is a finite-valued matrix, then $\mathfrak{L}_{a,b}$ can also be defined as a *discussive* system, to wit, $\mathcal{L}_{a,b}$ can be interpreted as the set of all formulas α such that $\Diamond_b\alpha$ is valid in M.

It is easy to verify that in those systems in which all sentences having "logical values" at least equal to b, $0 < b \leqslant 1$, are considered as true, that is, as having distinguished values, the implication of Łukasiewicz ($\rightarrow$) does not have some of the fundamental properties commonly associated with the notion of implication; for instance, the rule of modus ponens is not valid. This rule can be applied in connection with $\rightarrow$, only if the sentences considered true have the value 1. Intuitively, this means that assertion has an absolute character: we accept as true only sentences about which we are absolutely confident that they are true. Obviously, such condition is almost never satisfied in empirical systems. In particular, concerning the logic $\mathcal{L}_{a,b}$, where $a < b$ and $M_{a,b}$ is finite-valued, it occurs that there are formulas α and β such that α and $\alpha\rightarrow\beta$ are valid, but β is not. To obtain the formulas α and β, we can make use of the criterion of the definability of functions in matrices of Łukasiewicz given by McNaughton in [11] or the criterion formulated by Prucnal in [13].

Since the rule of modus ponens plays an important role in deductive systems, the question arises, whether we can define in $\mathcal{L}_{a,b}$ a binary operation which could be accepted as an implication. We want, especially, that the rule of detachement (modus ponens), relative to such operation, when applied to formulas of $\mathcal{L}_{a,b}$, would always give formulas belonging to $\mathcal{L}_{a,b}$. For this purpose, we proceed precisely as Jaśkowski in [8], where he defines the *discussive implication*; we extend the language of $\mathcal{L}_{a,b}$ by the addition of the following operation, which it is natural to call a-discussive implication (or simply discussive implication):

DEFINITION 1. $\alpha \underset{a}{\rightarrow} \beta =_{\text{def}} \Diamond_a\alpha \rightarrow \beta$.

This definition is analogous to Jaśkowski's definition of discussive implication in $\mathbf{D}_2$, and has a similar meaning. If $\Diamond_a$ is interpreted as possibility, then $\underset{a}{\rightarrow}$ coincides with discussive implication.

The logic obtained from $\mathcal{L}_{a,b}$ by extending its language with the addition of $\underset{a}{\rightarrow}$, according the above definition, will still be denoted by the symbol $\mathcal{L}_{a,b}$.

THEOREM 1. *If $a \leqslant b$, then the following rules of inference are permissible in $\mathfrak{L}_{a,b}$:*

(1) *If* α *and* $\alpha \underset{a}{\rightarrow} \beta$, *then* β.

(2) *If* α, *then* $\beta \underset{a}{\rightarrow} \alpha$.

(3) *If* $\alpha \underset{a}{\rightarrow} \beta$ *and* $\beta \underset{a}{\rightarrow} \gamma$, *then* $\alpha \underset{a}{\rightarrow} \gamma$.

(4) *If* α *and* β, *then* $\alpha \wedge \beta$.

(5) *If* $\alpha \wedge \beta$, *then* α; *if* $\alpha \wedge \beta$, *then* β.

(6) *If* $\alpha \underset{a}{\rightarrow} \beta$ *and* $\alpha \underset{a}{\rightarrow} \gamma$, *then* $\alpha \underset{a}{\rightarrow} \beta \wedge \gamma$.

(7) *If* α, *then* $\alpha \vee \beta$; *if* β, *then* $\alpha \vee \beta$

(8) *If* $\alpha \underset{a}{\rightarrow} \gamma$ *and* $\beta \underset{a}{\rightarrow} \gamma$, *then* $\alpha \vee \beta \underset{a}{\rightarrow} \gamma$.

PROOF of (1): Let us suppose that $\alpha, \alpha \underset{a}{\rightarrow} \beta \in \mathfrak{L}_{a,b}$, i.e., for every valuation ϑ, we have $\vartheta(\alpha) \geqslant b$ and $\vartheta(\alpha \underset{a}{\rightarrow} \beta) \geqslant b$. Then, we have also $\vartheta(\beta) \geqslant b$, because $\vartheta(\alpha \underset{a}{\rightarrow} \beta) = \vartheta(\Diamond_a \alpha \rightarrow \beta) = \vartheta(\Diamond_a \alpha) \Rightarrow \vartheta(\beta) = C_a \vartheta(\alpha) \Rightarrow \vartheta(\beta) = 1 \Rightarrow \vartheta(\beta) = \vartheta(\beta)$.

The proofs of (2)-(8) are similar. ■

It is rather interesting that, if $a < b$, then the formulas corresponding to the rules of inference of Theorem 1 do not belong to $\mathfrak{L}_{a,b}$ (we assume, for example, that $(p \wedge (p \underset{a}{\rightarrow} q)) \underset{a}{\rightarrow} q$ corresponds to rule (1) and that $p \underset{a}{\rightarrow} (q \underset{a}{\rightarrow} p)$ corresponds to rule (2)). Usually those formulas are considered as characteristic of a good implication. Thus, the implication $\underset{a}{\rightarrow}$ apparently cannot find fundamental applications in deductive systems, because the set of formulas of $\mathfrak{L}_{a,b}$ in which $\underset{a}{\rightarrow}$ occurs is "poor". However, it is known that in the application of logical calculi to propositional systems, the rules of the calculi are more important than their theses. Then, Theorem 1 says that the implication $\underset{a}{\rightarrow}$ is actually not so weak, and that $\mathfrak{L}_{a,b}$ is apt to be used as underlying logic of deductive systems (in which the basic implication is $\underset{a}{\rightarrow}$ and not $\rightarrow$).

DEFINITION 2. *For any* $\alpha, \beta \in \mathbb{F}$,

$\alpha \approx \beta$ *if and only if* $\alpha \underset{a}{\rightarrow} \beta \in \mathfrak{L}_{a,b}$ *and* $\beta \underset{a}{\rightarrow} \alpha \in \mathfrak{L}_{a,b}$.

THEOREM 2. *$\approx$ is an equivalence relation if and only if $a = b$. Moreover, under this hypothesis, if $\alpha_1 \approx \beta_1$ and $\alpha_2 \approx \beta_2$, then $\alpha_1 \vee \alpha_2 \approx \beta_1 \vee \beta_2$, $\alpha_1 \wedge \alpha_2 \approx \beta_1 \wedge \beta_2$ and $\alpha_1 \underset{a}{\rightarrow} \alpha_2 \approx \beta_1 \underset{a}{\rightarrow} \beta_2$, but it is not true that $\alpha_1 \rightarrow \alpha_2 \approx \beta_1 \rightarrow \beta_2$ and $\sim\alpha_1 \approx \sim\alpha_2$.*

PROOF: If $a < b$, then $\approx$ is not reflexive, because $p \underset{a}{\rightarrow} p$ does not belong to $\mathfrak{L}_{a,b}$, and if $a > b$, then $\approx$ is not transitive, owing to the fact that it is not true that if $\alpha \underset{a}{\rightarrow} \beta \in \mathfrak{L}_{a,b}$ and $\beta \underset{a}{\rightarrow} \gamma \in \mathfrak{L}_{a,b}$, then $\alpha \underset{a}{\rightarrow} \gamma \in \mathfrak{L}_{a,b}$. (It follows from Prucnal's criterion that in the case of a finite-valued generalized Łukasiewicz logic we can find formulas α, β and γ such that $\alpha \approx \beta$ and $\beta \approx \gamma$, but not $\alpha \approx \gamma$.) Supposing $a = b$, $\alpha \approx \beta$ is true if and only if for any valuation ϑ one has $\vartheta(\alpha) < a$ and $\vartheta(\beta) < a$, or $\vartheta(\alpha) \geqslant a$ and $\vartheta(\beta) \geqslant a$. Therefore, it follows that in this case the relation $\approx$ has the properties required by the theorem. ■

We observe that $\alpha \underset{a}{\rightarrow} \beta \in \mathfrak{L}_{a,b}$, with $a \geqslant b$, if and only if, for any valuation ϑ, we have $\vartheta(\alpha) < a$ or $\vartheta(\beta) \geqslant b$. Consequently:

THEOREM 3. *If $a \geqslant b$, then the following formulas belong to $\mathfrak{L}_{a,b}$:*

(i) $p \underset{a}{\rightarrow} (q \underset{a}{\rightarrow} p)$,

(ii) $(p \underset{a}{\rightarrow} q) \underset{a}{\rightarrow} ((q \underset{a}{\rightarrow} r) \underset{a}{\rightarrow} (p \underset{a}{\rightarrow} r))$,

(iii) $((p \underset{a}{\rightarrow} q) \underset{a}{\rightarrow} p) \underset{a}{\rightarrow} p$,

(iv) $p \vee q \underset{a}{\rightarrow} ((p \underset{a}{\rightarrow} q) \underset{a}{\rightarrow} q)$,

(v) $((p \underset{a}{\rightarrow} q) \underset{a}{\rightarrow} q) \underset{a}{\rightarrow} p \vee q$,

(vi) $p \wedge q \underset{a}{\rightarrow} p$,

(vii) $p \wedge q \underset{a}{\rightarrow} q$,

(viii) $p \underset{a}{\rightarrow} (q \underset{a}{\rightarrow} p \wedge q)$.

$\mathfrak{L}_a$, $0 < a \leqslant 1$, will designate the set of all formulas of $\mathfrak{L}_{a,a}$ in which, besides propositional variables and parentheses, only the symbols $\neg$, $\vee$, $\wedge$, and $\underset{a}{\rightarrow}$ occur. Let $\mathfrak{L}_a^+$, $0 < a \leqslant 1$, be the subset of $\mathfrak{L}_a$ containing all formulas in which negation does not appear. $\mathfrak{L}_a^+$ will be called the positive part of $\mathfrak{L}_a$.

It is a consequence of Theorem 1 that the rule of detachement for $\underset{a}{\rightarrow}$ is permissible in $\mathfrak{L}_a^+$. Evidently, $\mathfrak{L}_a^+$ is a set of formulas closed under substitution (limited to the set of formulas of the language $\mathfrak{L}_a^+$). According to a result of Sobociński [16], the formulas listed in Theorem 3 together with substitution and detachement constitute an axiomatization for the classical positive propositional calculus, which, by Theorems 1 and 3, is contained in $\mathfrak{L}_a^+$. But if $\alpha \in \mathfrak{L}_a^+$, then α is a thesis of the classical propositional calculus and, eo ipso, of the classical positive propositional calculus, since for 0 and 1 the operations of $M_{a,b}$, corresponding to the logical connectives $\vee$, $\wedge$ and $\underset{a}{\rightarrow}$, have the same values as the appropriate operations of the two-valued classical matrix. Hence, we have proved the following:

THEOREM 4. $\mathfrak{L}_a^+$, $0 < a \leqslant 1$, *is the classical positive propositional calculus.*

The algebraic version of the classical positive propositional calculus is constituted by the notion of classical implicative lattice (as well as by the concept of a Boolean ring; see [4] and [15]). Denoting by t_a^+, $0 < a \leqslant 1$, the *algebra of formulas* $\mathfrak{L}_a^+$, we deduce from Theorems 2 and 4 that:

COROLLARY 1. *The quotient algebra* $t_a^+/\approx$, $0 < a \leqslant 1$, *is a classical implicative lattice.*

It is not difficult to prove the following proposition:

THEOREM 5. *Suppose that* $a < 1/2$; *then, the following formulas are not theses of* $\mathfrak{L}_a$ $(\alpha \underset{a}{\leftrightarrow} \beta =_{\text{def}} (\alpha \underset{a}{\rightarrow} \beta) \wedge (\beta \underset{a}{\rightarrow} \alpha))$:

$p \underset{a}{\rightarrow} (\sim p \underset{a}{\rightarrow} q)$, $\qquad (p \wedge \sim p) \underset{a}{\rightarrow} q$,

$(p \underset{a}{\rightarrow} q) \underset{a}{\rightarrow} ((p \underset{a}{\rightarrow} \sim q) \underset{a}{\rightarrow} \sim p)$, $\qquad (p \underset{a}{\rightarrow} q) \underset{a}{\rightarrow} (\sim q \underset{a}{\rightarrow} \sim p)$,

$(p \underset{a}{\leftrightarrow} q) \underset{a}{\rightarrow} (\sim p \underset{a}{\leftrightarrow} \sim q)$, $(p \vee (q \wedge \sim q)) \underset{a}{\leftrightarrow} p$,

$(p \underset{a}{\rightarrow} q) \underset{a}{\leftrightarrow} \sim p \vee q$, $(p \underset{a}{\rightarrow} q) \underset{a}{\leftrightarrow} \sim(p \wedge \sim q)$.

Theorem 4 shows that $\mathfrak{L}_a$, $0 \leqslant a \leqslant 1$, is a rather rich calculus, since its positive part coincides with the classical positive propositional calculus. However, $\mathfrak{L}_a$, $0 < a < 1$, is not rich enough so as to make it impossible to found on it inconsistent nontrivial systems; on the contrary, by Theorem 5, $\mathfrak{L}_a$, $0 < a < 1/2$, can be used as underlying logic for those systems. Hence, we have:

COROLLARY 2. *Every* $\mathfrak{L}_a$, $0 < a < 1/2$, *constitutes a solution to Jaśkowski's problem.*

We shall denote by $M_{a,b}^{(2k)}$ the $2k$-valued generalized matrices of Łukasiewicz; it is clear that 1/2 does not belong to the set of elements of $M_{a,b}^{(2k)}$. $\mathfrak{L}_{a,b}^{(2k)}$ symbolizes the set of valid formulas in $M_{a,b}^{(2k)}$.

LEMMA 1. (i) $p \underset{a}{\rightarrow} (\sim p \underset{a}{\rightarrow} q) \in \mathfrak{L}_{a,b}^{(2k)}$ *if and only if* $a \geqslant 1/2$;

(ii) $(\sim p \underset{a}{\rightarrow} p) \underset{a}{\rightarrow} p \in \mathfrak{L}_{a,b}^{(2k)}$ *if and only if* $b \leqslant a \leqslant 1/2$ *or* $a + b \leqslant 1$ *and* $a > 1/2$.

PROOF of (i): In this case $a \geqslant 1/2$. Hence, for any valuation ϑ, if $\vartheta(p) < a$, then $(p \underset{a}{\rightarrow} (\sim p \underset{a}{\rightarrow} q)) = 1$, and if $\vartheta(p) \geqslant a$, then also $\vartheta(p \underset{a}{\rightarrow} (\sim p \underset{a}{\rightarrow} q)) = 1$, because $\vartheta(\sim p) = 1 - \vartheta(p) < 1 - a \leqslant a$. Conversely, supposing that $a < 1/2$, it follows that for a valuation ϑ which satisfies the conditions $a \leqslant \vartheta(p) < 1/2$ and $\vartheta(q) = 0$, we have: $\vartheta(\sim p) = 1 - \vartheta(p) > 1/2 > a$ and $\vartheta(p \underset{a}{\rightarrow} (\sim p \underset{a}{\rightarrow} q)) = 0$; hence, $p \underset{a}{\rightarrow} (\sim p \underset{a}{\rightarrow} q)$ does not belong to $\mathfrak{L}_{a,b}^{(2k)}$.

PROOF of (ii): Admit that $b \leqslant a \leqslant 1/2$; then, we have for any valuation ϑ: $\vartheta((\sim p \underset{a}{\rightarrow} p) \underset{a}{\rightarrow} p) = 1$ for $\vartheta(p) < a$, and $\vartheta((\sim p \underset{a}{\rightarrow} p) \underset{a}{\rightarrow} p) \geqslant a \geqslant b$ for $\vartheta(p) \geqslant a$. If $a > 1/2$ and $a + b \leqslant 1$, then we have, for any valuation ϑ: $\vartheta((\sim p \underset{a}{\rightarrow} p) \underset{a}{\rightarrow} p) \geqslant b$ for $\vartheta(p) \leqslant 1 - a$, and $\vartheta((\sim p \underset{a}{\rightarrow} p) \underset{a}{\rightarrow} p) \geqslant 1 - a \geqslant a \geqslant b$ for $\vartheta(p) > 1 - a$. We can directly verify that $(\sim p \underset{a}{\rightarrow} p) \underset{a}{\rightarrow} p \notin \mathfrak{L}_{a,b}$ for values of a and b different from those especified in the lemma . ■

The proof of the following propositions offers no difficulty:

LEMMA 2. *The formulas*

(i) $p \vee q \underset{a}{\leftrightarrow} (\sim p \underset{a}{\rightarrow} q)$,

(ii) $p \wedge q \underset{a}{\leftrightarrow} \sim(p \underset{a}{\rightarrow} \sim q)$,

belong to $\mathfrak{L}^{(2k)}_{1/2,\,1/2}$.

It is known that $(p \underset{a}{\rightarrow} q) \underset{a}{\rightarrow} ((q \underset{a}{\rightarrow} r) \underset{a}{\rightarrow} (p \underset{a}{\rightarrow} r))$, $p \underset{a}{\rightarrow} (\sim p \underset{a}{\rightarrow} q)$ and $(\sim p \underset{a}{\rightarrow} p) \underset{a}{\rightarrow} p$, together with the rules of substitution and modus ponens, constitute an axiomatization for the classical propositional calculus with implication and negation as the sole primitive connectives. In order for the above formulas to belong to $\mathfrak{L}_{a,b}$ and for the rules of substitution and of modus ponens to be permissible in $\mathfrak{L}_{a,b}$, we must have: $a = b = 1/2$ and $M_{a,b}$ has $2k$ elements, as Theorems 1 and 3, and Lemma 1 show. Therefore, the classical propositional calculus with implication and negation as the sole primitive connectives is contained in $\mathfrak{L}^{(2k)}_{1/2,\,1/2}$. But from Lemma 2 we deduce that *all* classical propositional calculus is included in $\mathfrak{L}^{(2k)}_{1/2,\,1/2}$. Obviously, if $\alpha \in \mathfrak{L}^{(2k)}_{1/2,\,1/2}$, α is a thesis of the classical propositional calculus, since the operations of $M^{(2k)}_{1/2,\,1/2}$ which corresponds to $\vee, \wedge, \underset{a}{\rightarrow}$ and $\sim$ have the same values for 0 and 1 as the analogous operations of two-valued classical matrix. Consequently, $\mathfrak{L}^{(2k)}_{1/2,1/2}$ is contained in the classical propositional calculus. Thus, we proved the following:

THEOREM 6. $\mathfrak{L}^{(2k)}_{1/2,\,1/2}$ *is the classical propositional calculus.*

As we have already noted, the connective $\Diamond_a$, $0 < a \leqslant 1$, is definable in any finite-valued logic $\mathfrak{L}^{(n)}$ of Łukasiewicz. Then, if we have an axiomatization for $\mathfrak{L}^{(n)}$, we have also an axiomatization for this logic enriched by the definition of $\Diamond_a$: both are essentially the same. Rosser and

Turquette, in [14], were the first to give an axiomatization for the finite-valued generalized logics of Łukasiewicz. Now we present another, formally simpler, axiomatization of such logics.

A proof that the finite-valued logic of Łukasiewicz, $\mathfrak{L}$, which is the set of all formulas valid in the matrix $\langle [0,1], \{1\}, \Rightarrow, \neg \rangle$, was given by Wajsberg (see [10]).

The axioms of Wajsberg are:

A_1. $p \to (q \to p)$,

A_2. $(p \to q) \to ((q \to r) \to (p \to r))$,

A_3. $((p \to q) \to q) \to ((q \to p) \to p)$,

A_4. $(\sim p \to \sim q) \to (q \to p)$,

and the primitive rules are substitution and modus ponens.

Tokarz proved in [17] that for every natural number n, $n > 1$, there exists a formula α_n, which is called axiom of Tokarz, such that A_1-A_4, α_n and substitution and modus ponens form an axiomatics for $\mathfrak{L}^{(n)}$.

Let **A** be the following set of formulas:

(i) $\Diamond_1(p \to (q \to p))$,

(ii) $\Diamond_1((p \to q) \to ((q \to r) \to (p \to r))$,

(iii) $\Diamond_1(((p \to q) \to q) \to ((q \to p) \to p)))$,

(iv) $\Diamond_1((\sim p \to \sim q) \to (q \to p))$,

(v) $\Diamond_1 \alpha_n$, where α_n is the axiom of Tokarz,

and let **R** be the set of rules (1)-(4) bellow:

(1) Substitution,

(2) If $\Diamond_1 \alpha$ and $\Diamond_1 (\alpha \to \beta)$, then $\Diamond_1 \beta$,

(3) If $\Diamond_1 \alpha$, then α,

(4) If $\Diamond_a \alpha$, then α $(0 < a \leqslant 1)$.

THEOREM 7. *$\mathfrak{L}^{(n)}_{a,b}$ can be axiomatized by taking* **A** *as the set of axioms and* **R** *as the set of rules.*

PROOF: Obviously, the calculus based on the axioms of **A** and on the rules of **R** is contained in $\mathfrak{L}_{a,b}^{(n)}$. We have to prove that the converse inclusion holds. Assume that $\alpha \in \mathfrak{L}_{a,b}^{(n)}$; hence, there exists a finite sequence of formulas

$$\beta_1, \beta_2, \beta_3, \ldots, \beta_n,$$

which is a proof of $\Diamond_a \alpha$ in the axiomatics of Tokarz, referred to above. It is easy to see that the sequence

$$\Diamond_1\beta_1, \Diamond_1\beta_2, \ldots, \Diamond_1\beta_n, \beta_n, \alpha$$

is a proof of α in the calculus whose axioms belong to **A** and whose rules of inference are members of **R**. Therefore, the theorem is proved.

OPEN QUESTIONS.

Concluding our paper, we present some open problems:

PROBLEM 1. Is $\mathfrak{L}_{a,b}$, when it is based on the infinite-valued Łukasiewicz logic, axiomatizable?

PROBLEM 2. Are there axiomatizations of $\mathfrak{L}_{a,b}$ in which the sole primitive connectives are $\underset{a}{\rightarrow}$, $\rightarrow$, $\wedge$, $\vee$ and $\sim$? (In this problem, $\mathfrak{L}_{a,b}$ may be a finite or infinite-valued logic.)

PROBLEM 3. What results of this paper can be extended to generalized logics of Łukasiewicz with quantification ? (Evidently, *some* of our results can easily be adapted to the level of the predicate calculus.)

REFERENCES.

[1] J. Błaszczuk and W. Dziobiak, *Remarks on Perzanowski's modal system*, Bulletin of the Section of Logic, Polish Acad. of Sciences, 4 , nº 2 (1975), 57-64.

[2] J. Błaszczuk and W. Dziobiak, *Modal systems related to* $S4_n$ *of Sobociński*, Bulletin of the Section of Logic, Polish Acad. of Sci-

ences, 4, nº 5 (1975), 103-108.

[3] J. Błaszczuk and W. Dziobiak, *An axiomatization of M^n-counterparts for some modal calculi*, Reports of Math. Logic, 6, (1976), 3-6.

[4] H. B. Curry, Foundations of Mathematical Logic, McGraw-Hill, 1963.

[5] N. C. A. da Costa, *On the theory of inconsistent formal systems*, Notre Dame Journal of Formal Logic, 15, nº 4 (1974), 497-510.

[6] I. M. L. D'Ottaviano and N.C.A. da Costa, *Sur un problème de Jaśkowski*, C. R. Acad. Sc. Paris, 270 A (1970), 1349-1353.

[7] T. Furmanowski, *Remarks on discussive propositional calculus*, Studia Logica, 34 (1975), 39-43.

[8] S. Jaśkowski, *Rachunek zdań dla systemów dedukcyjnych sprzecznych*, Studia Societatis Scientiarun Torunensis, Sectio A, I, nº 5 (1948), 57-77.

[9] S. Jaśkowski, *O konjunkcji dyskusyjnej w rachunku zdań dla systemów dedukcyjnych sprzecznych*, Studia Societatis Scientiarum Torunensis, Sectio A, I, nº 8 (1949), 171-172.

[10] J. Łukasiewicz and A. Tarski, *Untersuchungen über den Aussagenkalkül*, Comptes Rendus des séances de la Societé des Sciences et des Lettres de Varsovie, 23 (1930), 30-50.

[11] R. McNaughton, *A theorem about infinite-valued sentential logic*, The Journal of Symbolic Logic, 16 (1951), 1-13.

[12] J. Perzanowski, *On M-fragmentas and L-fragments of normal modal propositional logics*, Reports on Math. Logic, 5 (1975), 63-72.

[13] T. Prucnal, *Kryterium definiowaluości funkcji w matrycach Łukasiewscza*, Studia Logica, 23 (1968), 71-77.

[14] J. B. Rosser and A. R. Turquette, Many-valued Logics, North-Holland, 1952.

[15] H. Rasiowa and R. Sikorski, The Mathematics of Metamathematics, PWN, 1968.

[16] B. Sobociński, *Aksjomatyzacja koniunkcyjno negacyjnej thoryi dedukcji*, Colletanea Logica, Warsaw, 1936.

[17] M. Tokarz, *A method of axiomatization of Łukasiewicz logics*, Studia Logica, 33 (1974), 333-338.

Uniwersytet Mikołaja Kopernika
Instytut Matematyki
87100 Toruń, Poland

and

Universidade de São Paulo
Instituto de Matemática
05.568 São Paulo, Brazil.

THE METHOD OF VALUATIONS IN MODAL LOGIC.

by *ANDRÉA LOPARIĆ.*

ABSTRACT. In this paper we present a decision method for Kripke's modal calculus **K**, obtained from a new 0-1 valuation semantics for this calculus which, by means of a weakening of the truth conditions concerning the modal operation of necessity, dispense with central notions of Kripkean semantics, such as the notion of an accessibility relation between worlds and of a plurality of worlds in the same model. The valuation semantics can be easily extended to many other calculi containing **K**, such as **T**, **S4**, etc. It can also be adapted to weaker modal calculi as well as to deontic and temporal logics. (More about this kind of semantics can be found, for example, in N. C. A. da Costa and E. H. Alves, *Une sémantique pour le calcul* C_1, C. R. Acad. Sc. Paris, 283 (1976), 729-731, and in A. Loparić, *Un étude sémantique de quelques calculs propositionnels*, C. R. Acad. Sc. Paris 284 (1977), 835-838.)

1. KRIPKE'S CALCULUS K.

In the following we use capital Roman letters as syntactical variables for formulas, and capital Greek letters for sets of formulas.

A basis for **K** is given by the six postulates:

P1. $A \supset (B \supset A)$.

P2. $(A \supset (B \supset C)) \supset ((A \supset B) \supset (A \supset C))$.

P3. $(\neg A \supset \neg B) \supset ((\neg A \supset B) \supset A)$.

P4. A , $A \supset B / B$. *(modus ponens)*

P5. $\Box(A \supset B) \supset (\Box A \supset \Box B)$.

P6. $A / \Box A$. *(Gödel's rule)*

The definitions of *theorem* and *proof* are usual; $\Gamma \vdash A$ (A is a consequence of the set Γ) if there is a sequence $D_1, \ldots, D_m$ (called a deduction of A from Γ) such that $D_m = A$ and for $1 \leqslant i \leqslant m$, a) $D_i \in \Gamma$; or b) D_i is an axiom; or c) there is a $j < i$ and a $k < i$ such that $D_k = D_j \supset D_i$; or d) there is a $j < i$ such that $D_i = \Box D_j$ and some subsequence of $D_1, \ldots, D_j$ is a proof of D_j. Remember that the following properties hold for **K**:

1.1. Deduction Theorem: $\Gamma \cup \{B\} \vdash A$ iff $\Gamma \vdash B \supset A$.

1.2. If $\Gamma \vdash A$, then $\Box\Gamma \vdash \Box A$ (where $\Box\Gamma = \{\Box B : B \in \Gamma\}$

2. SEMI-VALUATIONS AND VALUATIONS FOR K.

2.1. DEFINITION. B is a *subformula* of A if: a) $B = A$; or b) for some subformula C of A, i) $C = \neg B$; or ii) $C = \Box B$; or iii) for some D, $C = B \supset D$ or $C = D \supset B$.

2.2. DEFINITION. $A_1, \ldots, A_n$ is a Σ-*sequence* if:

1) For every B which is a subformula of A_i, $1 \leqslant i \leqslant n$, there is a $j \leqslant i$ such that $B = A_j$;

2) If $A_i = A_j$ then $i = j$, $1 \leqslant i \leqslant n$, $1 \leqslant j \leqslant n$.

2.3. DEFINITION. A *semi-valuation* s is a function from the set $\mathbb{F}$ of all formulas into $\{0, 1\}$ such that:

1) $s(\neg A) = 1$ iff $s(A) = 0$;

2) $s(A \supset B) = 1$ iff $s(A) = 0$ or $s(B) = 1$.

2.4. DEFINITION. Let Γ be a set of formulas and f, a function from a set Δ, containing Γ, into $\{0, 1\}$;

1) $\Gamma^{\Box} =_{def} \{F \in \Gamma : \text{for some } F', F = \Box F'\}$,

2) $\varepsilon(\Gamma^{\Box}) =_{def} \{F : \Box F \in \Gamma\}$,

3) $f \models \Gamma =_{def}$ for every $F \in \Gamma$, $f(F) = 1$,

4) $\Gamma_{f,u} =_{\text{def}} \{F \in \Gamma : f(F) = u\}$.

2.5. DEFINITION. ϑ is an $A_1, \ldots, A_n$-*valuation* if $A_1, \ldots, A_n$ is a Σ-sequence and:

1) $n = 1$ and ϑ is a semi-valuation;

2) $n > 1$, ϑ is an $A_1, \ldots, A_{n-1}$-valuation and if for some m we have $A_n = \Box A_m$, then:

   I) if $\vartheta(A_n) = 0$, then some $A_1, \ldots, A_{n-1}$-valuation ϑ' is such that $\vartheta'(A_m) = 0$ and $\vartheta' \models \varepsilon(\{A_1, \ldots, A_{n-1}\}^{\Box}_{\vartheta,1})$;

   II) if $\vartheta(A_n) = 1$ then, for every $p < n$ and every $q < p$ such that $A_p = \Box A_q$ and $\vartheta(A_p) = 0$, there is an $A_1, \ldots, A_{n-1}$-valuation ϑ_p such that $\vartheta_p(A_q) = 0$ and $\vartheta_p \models \varepsilon(\{A_1, \ldots, A_{n-1}\}^{\Box}_{\vartheta,1}) \cup \{A_m\}$.

2.6. DEFINITION. ϑ is a *valuation* if, for every Σ-sequence $A_1, \ldots, A_n$, ϑ is an $A_1, \ldots, A_n$-valuation.

2.7. DEFINITION. Let ϑ be an $A_1, \ldots, A_{n-1}$-valuation and $A_1, \ldots, A_n$ a Σ-sequence. Then, ϑ' is the canonical extension of ϑ to $A_1, \ldots, A_n$ if:

A) For every $m < n$, $A_n \neq \Box A_m$ and $\vartheta' = \vartheta$;

B) For some $m < n$, $A_n = \Box A_m$ and, for every F,

   1) If A_n is not a subformula of F, $\vartheta'(F) = \vartheta(F)$;

   2) If A_n is a subformula of F, we have:

      a) for $F = A_n$, $\vartheta'(F) = 0$ iff for some $A_1, \ldots, A_{n-1}$-valuation $\bar{\vartheta}$, $\bar{\vartheta}(A_m) = 0$ and $\bar{\vartheta} \models \varepsilon(\{A_1, \ldots, A_{n-1}\}^{\Box}_{\vartheta,1})$;

      b) for $F = \neg F'$, $\vartheta'(F) = 0$ iff $\vartheta'(F') = 1$;

      c) for $F = F' \supset F''$, $\vartheta'(F) = 0$ iff $\vartheta'(F') = 1$ and $\vartheta'(F'') = 0$;

      d) for $F = \Box F'$ and $F \neq A_n$, $\vartheta'(F) = \vartheta(F)$.

2.8. LEMMA. *Let* $A_1, \ldots, A_n$ *be a* Σ*-sequence,* ϑ *an* $A_1,\ldots,A_{n-1}$*-valuation,* ϑ' *the canonical extension of* ϑ *to* $A_1, \ldots, A_n$ *; assume further that for every* $p<n$, *every* $q<p$ *such that* $A_p = \Box A_q$ *and* $\vartheta(A_p) = 0$ *, there is some* $A_1,\ldots,A_{n-1}$*-valuation* ϑ_p *such that* $\vartheta_p(A_q) = 0$ *and* $\vartheta_p \models \varepsilon(\{A_1,\ldots,A_{n-1}\}^{\Box}_{\vartheta,1})$. *Then* ϑ' *is an* $A_1,\ldots,A_n$*-valuation.*

PROOF: If for every $m<n$, $A_n \neq \Box A_m$, then ϑ is an $A_1,\ldots,A_n$-valuation and since $\vartheta' = \vartheta$, ϑ' is an $A_1,\ldots,A_n$-valuation.
Suppose now that, for some $m<n$, $A_n = \Box A_m$. From conditions B1), B2b) and B2c) of 2.7 it follows that ϑ' is an $A_1,\ldots,A_{n-1}$-valuation. If $\vartheta'(A_n) = 0$ it follows from B2a) of 2.7, that there is an $A_1,\ldots,A_{n-1}$-valuation $\bar{\vartheta}$ such that $\bar{\vartheta}(A_n) = 0$ and $\bar{\vartheta} \models \varepsilon(\{A_1,\ldots,A_{n-1}\}^{\Box}_{\vartheta,1})$; by B1) of 2.7, for $1 \leqslant i \leqslant n-1$, $\vartheta'(A_i) = \vartheta(A_i)$; hence $\bar{\vartheta} \models \varepsilon(\{A_1,\ldots,A_{n-1}\}^{\Box}_{\vartheta',1})$, and by 2.5.2.I), ϑ' is an $A_1,\ldots,A_n$-valuation. Suppose now that $\vartheta'(A_n) = 1$; let $q<p<n$ be such that $A_p = \Box A_q$ and $\vartheta'(A_p) = 0$; by B1) of 2.7, $\vartheta(A_p) = 0$ and by hypothesis there is an $A_1,\ldots,A_{n-1}$-valuation ϑ_p such that $\vartheta_p(A_q) = 0$ and $\vartheta_p \models \varepsilon(\{A_1,\ldots,A_{n-1}\}^{\Box}_{\vartheta,1})$; now, $\vartheta(A_m) = 1$, otherwise, by B2a), $\vartheta'(A_n) = 0$; since, by B1), for $1 \leqslant i \leqslant n-1$, $\vartheta(A_i) = \vartheta'(A_i)$, $\vartheta_p \models \varepsilon(\{A_1,\ldots,A_{n-1}\}^{\Box}_{\vartheta',1}) \cup \{A_m\}$; hence, by 2.5.2.II, ϑ' is an $A_1,\ldots,A_n$-valuation. ∎

2.9. LEMMA. *Let* ϑ *be an* $A_1,\ldots,A_n$*-valuation; then for every* $p \leqslant n$, *every* $q<p$, *such that* $A_p = \Box A_q$ *and* $\vartheta(A_p) = 0$, *there is an* $A_1,\ldots,A_n$*-valuation* ϑ_p *such that* $\vartheta_p(A_q) = 0$ *and* $\vartheta_p \models \varepsilon(\{A_1,\ldots,A_n\}^{\Box}_{\vartheta,1})$.

PROOF: By induction on n. For $n=1$ the proposition is trivial. Let $n>1$ and assume that it holds for $n-1$. From the inductive hypothesis together with Lemma 2.8 we have:

a) for every $\bar{\vartheta}$, and $\bar{\bar{\vartheta}}$, if $\bar{\vartheta}$ is an $A_1,\ldots,A_{n-1}$-valuation and $\bar{\bar{\vartheta}}$ is the canonical extension of $\bar{\vartheta}$ to $A_1,\ldots,A_n$ then $\bar{\bar{\vartheta}}$ is an $A_1,\ldots,A_n$-valuation.

And from the fact that ϑ is an $A_1,\ldots,A_{n-1}$-valuation together with the inductive hypothesis it follows that:

b) for $p<n$, $q<p$, if $A_p = \Box A_q$ and $\vartheta(A_p) = 0$, there is an $A_1,\ldots,A_{n-1}$-valuation ϑ_p such that $\vartheta_p(A_q) = 0$ and $\vartheta_p \models \varepsilon(\{A_1,\ldots,A_{n-1}\}^{\Box}_{\vartheta,1})$.

For each p of b), let ϑ'_p be the canonical extension of ϑ_p to $A_1, \ldots, A_n$. Then since for $1 \leqslant i < n$, $\vartheta_p(A_i) = \vartheta'_p(A_i)$, we have:

c) for $p < n$, $q < p$, if $A_p = \Box A_q$ and $\vartheta(A_p) = 0$ there is an $A_1, \ldots, A_n$ - valuation ϑ'_p such that $\vartheta'_p(A_q) = 0$ and $\vartheta' \models \varepsilon(\{A_1, \ldots, A_{n-1}\}^{\Box}_{\vartheta,1})$.

Now consider the two cases:

I) $\vartheta(A_n) = 0$; then,

i) $\varepsilon(\{A_1, \ldots, A_n\}^{\Box}_{\vartheta,1}) = \varepsilon(\{A_1, \ldots, A_{n-1}\}^{\Box}_{\vartheta,1})$;

ii) by 2.I of definition 2.5, there is an $A_1, \ldots, A_{n-1}$ -valuation ϑ_n such that $\vartheta_n(A_m) = 0$ and $\vartheta_n \models \varepsilon(\{A \;, \ldots, A_{n-1}\}^{\Box}_{\vartheta,1})$

Let ϑ'_n be the canonical extension of ϑ_n; then, applying a) and B1 of 2.7 , from Ii) and Iii), it follows that:

1) for $p = n$, $(q = m)$ $A_p = \Box A_q$ and $\vartheta(A_p) = 0$, there is an $A_1, \ldots, A_n$-valuation ϑ'_p such that $\vartheta'(A_q) = 0$ and $\vartheta'_p \models \varepsilon(\{A_1, \ldots, A_n\}^{\Box}_{\vartheta,1})$;

And from c) and Ii) we get:

2) for $p < n$, $q < p$, if $A_p = \Box A_q$ and $\vartheta(A_p) = 0$, there is an $A_1, \ldots, A_n$-valuation ϑ'_p such that $\vartheta'_p(A_q) = 0$ and $\vartheta'_p \models \varepsilon(\{A_1, \ldots, A_n\}^{\Box}_{\vartheta,1})$.

Hence, the property holds, in case I, for $p \leqslant n$.

II) $\vartheta(A_n) = 1$; then,

i) $\varepsilon(\{A_1, \ldots, A_n\}^{\Box}_{\vartheta,1}) = \varepsilon(\{A_1, \ldots, A_{n-1}\}^{\Box}_{\vartheta,1}) \cup \{A_m\}$;

ii) by 2.II of definition 2.5 and from the fact that $\vartheta(A_n) = 1$, we have for each $p \leqslant n$, each $q < p$, if $A_p = \Box A_q$ and $\vartheta(A_p) = 0$, there is an $A_1, \ldots, A_{n-1}$- valuation ϑ_p such that $\vartheta_p(A_q) = 0$ and $\vartheta_p \models \varepsilon(\{A_1, \ldots, A_{n-1}\}^{\Box}_{\vartheta,1}) \cup \{A_m\}$;

For each p of IIii), let ϑ'_p be the canonical extension of ϑ_p. Then, by a) and II i) together with B1 of 2.7 we get from II ii):

- for each $p \leqslant n$, each $q < p$, if $A_p = \Box A_q$ and $\vartheta(A_p) = 0$, then there is an $A_1, \ldots, A_n$-valuation ϑ'_p such that $\vartheta'_p(A_q) = 0$ and $\vartheta'_p \models \varepsilon(\{A_1, \ldots, A_n\}^{\Box}_{\vartheta,1})$.

Hence, the property holds also in case II. ■

2.10. COROLLARY. *Let* $A_1, \ldots, A_n$ *be a* Σ*-sequence,* ϑ *an* $A_1, \ldots,$

A_{n-1}- valuation, and ϑ' the canonical extension of ϑ to $A_1, \ldots, A_n$. Then ϑ' is an $A_1,\ldots,A_n$- valuation and for $1 \leqslant i < n$, $\vartheta'(A_i) = \vartheta(A_i)$.

PROOF: Follows from 2.7, 2.8 and 2.9. ■

2.11. THEOREM. *ϑ is an $A_1,\ldots, A_n$- valuation iff $A_1,\ldots, A_n$ is a Σ-sequence, ϑ is a semi-valuation and for every $p \leqslant n$, every $q < p$ such that $A_p = \Box A_q$ and $\vartheta(A_p) = 0$, there is an $A_1,\ldots, A_n$- valuation ϑ_p such that $\vartheta_p(A_q) = 0$ and $\vartheta_p \models \varepsilon(\{A_1,\ldots, A_n\}^{\Box}_{\vartheta,1})$.*

PROOF: The implication from the left follows immediately from Definition 2.5 and Lemma 2.9. To prove the converse, assume the hypothesis and apply induction on n. From the fact that ϑ is a semi-valuation it follows that ϑ is an A_1-valuation. Suppose that every function that satisfies the hipothesis is an $A_1,\ldots, A_n$ -valuation - and in particular that ϑ is an A_n, $\ldots, A_{n-1}$- valuation. Clearly if, for every $m < n$, $A_n \neq \Box A_m$, ϑ is an $A_1,\ldots, A_n$ -valuation. Let then, for some m, $A_n = \Box A_m$. Remember that:

i) every $A_1,\ldots, A_n$-valuation is an $A_1,\ldots, A_{n-1}$ - valuation;

ii) for every function f, from $\mathbb{F}$ into $\{0,1\}$, if $f \models \varepsilon(\{A_1,\ldots,A_n\}^{\Box}_{\vartheta,1})$, then $f \models \varepsilon(\{A_1, \ldots, A_{n-1}\}^{\Box}_{\vartheta,1})$;

I) Let $\vartheta(A_n) = 0$ and take $p = n$, $q = m$. It follows from the hypothesis together with i) and ii) that there is an $A_1,\ldots, A_n$- valuation ϑ_n such that $\vartheta_n(A_m) = 0$ and $\vartheta \models \varepsilon(\{A \ ,\ldots, A_{n-1}\}^{\Box}_{\vartheta,1})$; hence, if $\vartheta(A_n) = 0$, ϑ is an $A_1,\ldots, A_n$ - valuation;

II) $\vartheta(A_n) = 1$; since $\varepsilon(\{A_1,\ldots, A_n\}^{\Box}_{\vartheta,1}) = \varepsilon(\{A_1,\ldots,A_{n-1}\}^{\Box}_{\vartheta,1}) \cup \{A_m\}$, applying i) we get from the hypothesis that for each $p \leqslant n$, each $q \leqslant n$ such that $A_p = \Box A_q$ and $\vartheta(A_p) = 0$, there is an $A_1,\ldots, A_{n-1}$- valuation ϑ_p such that $\vartheta_p(A_q) = 0$ and $\vartheta_p \models \varepsilon(\{A_1,\ldots, A_{n-1}\}^{\Box}_{\vartheta,1}) \cup \{A_m\}$. That is to say, if $\vartheta(A_n) = 1$ then ϑ is an $A_1,\ldots, A_n$-valuation. ■

3. CORRECTNESS.

3.1. LEMMA. *Let ϑ be an $A_1,\ldots, A_n$- valuation; then, for $1 \leqslant i \leqslant n$,* *if* A_i *is an axiom of* **K**, $\vartheta(A_i) = 1$.

PROOF: If A_i is an axiom of the classical propositional logic then $\vartheta(A_i) = 1$, for ϑ is a semi-valuation and a semivaluation respects the classical requirements for negation and implication (see Definition 2.3). Let A_i be of the form $\Box(A \supset B) \supset (\Box A \supset \Box B)$. Suppose $\vartheta(A_i) = 0$; then $\vartheta(\Box(A \supset B)) = (\Box A) = 1$ and $\vartheta(\Box B) = 0$; thus $A \supset B$, $A \in \varepsilon(\{A_1,\ldots, A_n\}^{\Box}_{\vartheta,1})$ and for some $p < n$, $\Box B = A_p$. But it is easy to see that no $A_1,\ldots,A_n$-valuation ϑ_p is such that $\vartheta_p(B) = 0$ and $\vartheta_p \models \varepsilon(\{A_1,\ldots,A_n\}^{\Box}_{\vartheta,1})$ - for if $\vartheta_p(B) = 0$, then $\vartheta_p(A) = 1$ iff $\vartheta_p(A \supset B) = 0$; hence, $\vartheta(A_i) = 1$. ■

3.2. THEOREM. *If* F *is an axiom of* **K** *and* ϑ *is a valuation, then* $\vartheta(F) = 1$.

PROOF: Arrange the subformulas of F in a sequence $A_1,\ldots, A_n$ which respects the length order: then $F = A_n$ and $A_1,\ldots, A_n$ is a Σ-sequence. Since every valuation is an $A_1,\ldots, A_n$-valuation, it follows from 3.1 that if F is an axiom and ϑ a valuation, $\vartheta(F) = 1$. ■

3.3. LEMMA. *For every* n, *if* ϑ *is an* $A_1,\ldots, A_n$ *- valuation and* $\vdash A_n$ *then* $\vartheta(A_n) = 1$.

PROOF: By induction on the number of rows of a proof for A_n. If $r = 1$ then A_n is an axiom and the property is already proved in 3.1. Let $r > 1$ and suppose that it holds for $r' < r$ (and every n'). Then, if A_n is not an axiom,

i) A_n was obtained by Modus Ponens from B and $B \supset A_n$, which precede A_n in the proof. Form $\alpha = \{C : C$ is a subformula of $B \supset A_n$ and $C \notin \{A_1,\ldots, A_n\}\}$. If $\alpha \neq \emptyset$, arrange the elements of α in some sequence $C_1,\ldots, C_n$ which respects the length order of the formulas. If $\alpha = \emptyset$, let $\sigma = A_1,\ldots, A_n$ and $\vartheta' = \vartheta$; otherwise, $\sigma = A_1,\ldots, A_n, C_1,\ldots, C_k$, form $\vartheta_0, \vartheta_1,\ldots, \vartheta_k$ where $\vartheta_0 = \vartheta$ and, for $1 \leqslant j \leqslant k$ let ϑ_j be the canonical extension of ϑ_{j-1} and take $\vartheta' = \vartheta_k$. Then σ is a Σ-sequence and ϑ' is a σ-valuation. Now, from the inductive hypothesis we have $\vartheta'(B \supset A_n) = \vartheta'(B) = 1$, hence $\vartheta'(A_n) = 1$, but since $\vartheta(A_n) = \vartheta'(A_n)$, $\vartheta(A_n) = 1$.

ii) for some $j < n$, $A_n = \Box A_j$ was obtained by Gödel's rule from A_j, which preceds A_n in the proof. Thus, by the inductive hypothesis , for every $A_1, \dots, A_n$-valuation ϑ', $\vartheta'(A_j) = 1$. Then taking $p = n$, $q = j$ and applying 2.9, we obtain that for every $A_1, \dots, A_n$-valuation $\vartheta, \vartheta(A_n) = 1$. ■

3.4. COROLLARY. *If* $\vdash F$ *then for every valuation* ϑ, $\vartheta(F) = 1$.

PROOF: Take a Σ-sequence $A_1, \dots, A_n$ with $A_n = F$. Since every valuation is an $A_1, \dots, A_n$-valuation, it follows from 3.3 that for every valuation ϑ, if $\vdash F$ then $\vartheta(F) = 1$. ■

3.5. THEOREM. *If* $\Gamma \vdash F$, *for every valuation* ϑ, *if* $\vartheta \models \Gamma$ *then* $\vartheta(F) = 1$.

PROOF: Suppose that $\Gamma \vdash F$ and let $D_1, \dots, D_m$ be a deduction of F from Γ. Then apply induction on m:

- if $m = 1$, $F \in \Gamma$ and there is nothing to prove, or F is an axiom and the property is already proved in 3.4;
- let $m > 1$; then, if $F \notin \Gamma$ and F is not an axiom,
  a) For some $j < m$, some $i < m$, $D_i = D_j \supset F$, that is to say, F was obtained by modus ponens from D_i and D_j. Hence, $\Gamma \vdash D_i$, $\Gamma \vdash D_j$ and by the inductive hypothesis, if $\vartheta \models \Gamma$ then $\vartheta(D_j) = \vartheta(D_i) = \vartheta(D_j \supset F) = 1$. Hence, $\vartheta(F) = 1$;
  b) For some $j < m$, $F = \Box D_j$ and was obtained by Gödel's rule from D_j; in this case, $\vdash D_j$ and $\vdash F$; hence, by 3.4, $\vartheta'(F) = 1$ for every ϑ'; a fortiori, if $\vartheta \models \Gamma$ then $\vartheta(F) = 1$. ■

4. F-SATURATED SETS; COMPLETENESS.

4.1. DEFINITION. Δ is an *F-saturated set* if $\Delta \nvdash F$ and for every $G \notin \Delta$, $\Delta \cup \{G\} \vdash F$.

4.2. LEMMA. *If* Δ *is an* F-*saturated set,*

a) $A \in \Delta$ *iff* $\Delta \vdash A$;

b) $\neg A \in \Delta$ *iff* $A \notin \Delta$;

c) $A \supset B \in \Delta$ *iff* $A \notin \Delta$ *or* $B \in \Delta$.

PROOF: a) if $A \in \Delta$, then $\Delta \vdash A$; for the converse, suppose that $\Delta \vdash A$; now, if $A \notin \Delta$ then $\Delta \cup \{A\} \vdash F$ and, by the Deduction Theorem $\Delta \vdash A \supset B$, but then $\Delta \vdash F$, by modus ponens; hence $A \in \Delta$.

b) $\Delta \vdash \neg A \supset (A \supset F)$; hence if both A, $\neg A \in \Delta$, $\Delta \vdash F$;on the other hand $\Delta \vdash (\neg A \supset F) \supset ((A \supset F) \supset F)$; but if both $\neg A, A \notin \Delta$, $\Delta \cup \{\neg A\} \vdash F$ and $\Delta \cup \{A\} \vdash F$; by the Deduction Theorem, $\Delta \vdash \neg A \supset F$, $\Delta \vdash A \supset F$, hence, $\Delta \vdash F$; so $\neg A \in \Delta$ iff $A \notin \Delta$.

c) If $A \supset B \in \Delta$ and $A \in \Delta$, then $\Delta \vdash A \supset B$, $\Delta \vdash A$ and, by Modus ponens, $\Delta \vdash B$; hence $B \in \Delta$. Suppose, now, that $A \notin \Delta$;then $\neg A \in \Delta$ and $\Delta \vdash \neg A$; since $\Delta \vdash \neg A \supset (A \supset B)$, $\Delta \vdash A \supset B$; hence $A \supset B \in \Delta$;finally let $B \in \Delta$; hence $\Delta \vdash B$ and, since $\Delta \vdash B \supset (A \supset B)$, $\Delta \vdash A \supset B$;therefore $A \supset B \in \Delta$. ∎

4.3. LEMMA. *If* $\Gamma \nvdash F$, *there is an* F-*saturated set* Δ *such that* $\Gamma \subset \Delta$.

PROOF: Take an enumeration $A_1, \ldots, A_i, \ldots$ of the formulas of **K**. Construct $\Delta = \cup \Gamma_i$, where $\Gamma_0 = \Gamma$ and, for $i > 0$, $\Gamma_i = \Gamma_{i-1}$ if $\Gamma_{i-1} \cup \{A_i\} \vdash F$, otherwise $\Gamma_i = \Gamma_{i-1} \cup \{A_i\}$. Then $\Gamma \subset \Delta$. It is easy to prove that if $\Gamma_{i-1} \nvdash F$ then $\Gamma_i \nvdash F$; since $\Gamma_0 \nvdash F$, by induction, for every $j \geqslant 0$, $\Gamma_j \nvdash F$. Now, if $\Delta \vdash F$, then for some finite $\alpha \subset \Delta$, $\alpha \vdash F$, and for some $i \geqslant 0$, $\alpha \subset \Gamma_i$; in this case $\Gamma_i \vdash F$; hence $\Delta \nvdash F$. Suppose that $G \notin \Delta$; since for some i, $G = A_i$, $\Gamma_{i-1} \cup \{G\} \vdash F$, hence $\Delta \cup \{G\} \vdash F$. Therefore Δ is an F-saturated set such that $\Gamma \subset \Delta$. ∎

4.4. COROLLARY. *If* $\Gamma \nvdash \Box F$, *there is an* F-*saturated set* Δ *such that* $\varepsilon(\Gamma^{\Box}) \subset \Delta$.

PROOF: If $\Gamma \nvdash \Box F$ then $\Gamma^{\Box} \nvdash \Box F$ and by 1.2., $\varepsilon(\Gamma^{\Box}) \nvdash F$; hence, by 4.3, there is an F-saturated set Δ such that $\varepsilon(\Gamma^{\Box}) \subset \Delta$. ∎

4.5. THEOREM. *If* Δ *is an* F-*saturated set and* $A_1, \ldots, A_n$ *is a* Σ-*sequence, the characteristic function* f, *of* Δ, *is an* $A_1, \ldots, A_n$-*valuation.*

PROOF: By induction on n. If $n=1$, it is easy to prove, using 4.2 b) and c), that f is a semi-valuation. Therefore f is an A_1- valuation. Suppose that for every Δ', every F' and every f', if Δ' is an F'-saturated set and

f' is the characteristic function of Δ', then f' is an $A_1,\ldots,A_{n-1}$-valuation. Then, in particular, f is an $A_1,\ldots,A_{n-1}$-valuation. Suppose that, for some $m < n$, $A_n = \Box A_m$; we have two cases:

I) $f(A_n) = 0$; then $A_n \notin \Delta$ and $\Delta \nvdash A_n$; by 4.4, there is an A_m-saturated Δ'' such that $\varepsilon(\Delta^{\Box}) \subset \Delta''$; but it follows from the inductive hypothesis that the characteristic function f'', of Δ'', is an $A_1,\ldots,A_{n-1}$-valuation; now $\{A_1,\ldots,A_{n-1}\}^{\Box}_{f,1} \subset \Delta^{\Box}$, therefore $\varepsilon(\{A_1,\ldots,A_{n-1}\}^{\Box}_{f,1}) \subset \varepsilon(\Delta^{\Box}) \subset \Delta''$; and since $A_m \notin \Delta''$, $f'' \models \varepsilon(\{A_1,\ldots,A_{n-1}\}^{\Box}_{f,1})$ and $f''(A_m) = 0$. Hence, if $f(A_n) = 0$, the characteristic function f, of Δ, is an $A_1,\ldots,A_n$-valuation.

II) $f(A_n) = 1$; then $A_m \in \varepsilon(\Delta^{\Box})$. Let $p < n$ and $q < p$ be such that $A_p = \Box A_q$ and $f(A_p) = 0$; then $A_p \notin \Delta$ and $\Delta \nvdash A_p$; from 4.4 it follows that there is an A_q-saturated Δ'' such that $\varepsilon(\Delta^{\Box}) \subset \Delta''$. Then $\varepsilon(\{A_1,\ldots,A_{n-1}\}^{\Box}_{f,1}) \cup \{A_m\} \subset \varepsilon(\Delta^{\Box}) \subset \Delta''$ and $A_q \notin \Delta''$; using the inductive hypothesis we obtain that f'', the characteristic function of Δ'', is an $A_1,\ldots,A_{n-1}$-valuation, $f'' \models \varepsilon(\{A_1,\ldots,A_{n-1}\}^{\Box}_{f,1}) \cup \{A_m\}$ and $f''(A_q) = 0$. That is to say, if $f(A_m) = 1$, f is an $A_1,\ldots,A_n$-valuation. ■

4.6. THEOREM. *ϑ is a valuation iff for some Δ and some F such that Δ is F-saturated, ϑ is the characteristic function of Δ.*

PROOF: Suppose that ϑ is a valuation; let $\Delta_1 = \{A: \vartheta(A)=1\}$ and $\Delta_0 = \{B: \vartheta(B)=0\}$; let $F \in \Delta_0$; then $F \notin \Delta_1$ and it is easy to prove that $\Delta_1 \nvdash F$; let $G \notin \Delta_1$; then $\vartheta(G) = 0$, $\vartheta(\neg G) = 1$ and $\neg G \in \Delta_1$; but, $\vartheta(\neg G \supset (G \supset F)) = 1$, hence $\neg G \supset (G \supset F) \in \Delta_1$; so $\Delta_1 \vdash G \supset F$, therefore $\Delta_1 \cup \{G\} \vdash F$. Hence, Δ_1 is an F-saturated set; and by construction of Δ_1, ϑ is the characteristic function of Δ_1. The converse is a direct corollary of 2.6 and 4.5. ■

4.7. DEFINITION. $\models F$ if for every valuation ϑ, $\vartheta(F) = 1$; $\Gamma \models F$ if for every valuation ϑ, if $\vartheta \models \Gamma$, then $\vartheta(F) = 1$.

4.8. THEOREM. *If $\Gamma \models F$ then $\Gamma \vdash F$.*

PROOF: Follows from 4.3, 4.5 and 4.7. ■

5. A SIMPLE EXTENSION: VALUATIONS FOR T.

The valuation method can be extended to most other modal calculi, as we will show elsewhere. To examplify, we indicate here how to adapt the results for the system **T**, of Feys-Von Wright. As it is well known, **T** can be obtained by adding to **K** the postulate: P7: $\Box A \supset A$. To obtain valuations for **T** it is sufficient to modify the semi-valuation notion by adding the condition: 3) if $s(A) = 0$ then $s(\Box A) = 0$. Adequacy theorems are obtained in the same way as in **K**, with obvious adaptations.

6. VALUATIONS AND KRIPKE MODELS.

Let M be a structure $\langle \omega, W, R \rangle$ where W is a non-empty set, ω is a member of W and R is a subset of $W \times W$; let $\mathbb{F}$ be the set of formulas of **K** and let I be a function from $\mathbb{F} \times W$ into $\{0,1\}$ such that, for A, $B \in \mathbb{F}$ and $a \in W$,

1) $I(\neg A, a) = 1$ iff $I(A, a) = 0$;

2) $I(A \supset B, a) = 1$ iff $I(A, a) = 0$ or $I(B, a) = 1$;

3) $I(\Box A, a) = 1$ iff for every $b \in W$, such that bRa, $I(A, b) = 1$.

Then the function I associated with the structure $\langle \omega, W, R \rangle$ is a Kripke model for **K**; and a formula A is true in this model iff $I(A, \omega) = 1$. (cf. [3].)

We say that two models $M = \langle I, \langle \omega, W, R \rangle \rangle$ and $M' = \langle I', \langle \omega', W', R' \rangle \rangle$ are equivalent $(M \sim M')$ iff, for every $A \in \mathbb{F}$, $I(A, \omega) = I'(A, \omega')$.

In this section we shall compare valuations and Kripke models showing a close connection between them.

Let $M_{\mathbf{k}}$ be the set of all Kripke models for **K**. For each $\langle I, \langle \omega, W, R \rangle \rangle \in M_{\mathbf{k}}$ and each $a \in W$, let Δ_a be the set $\{A \in \mathbb{F} : I(A, a) = 1\}$. Let $\mathcal{C}$ be the set $\{\Gamma \subset \mathbb{F}$: for some $\langle I, \langle \omega, W, R \rangle \rangle \in M_{\mathbf{k}}$ and some $a \in W$, $\Gamma = \Delta_a\}$. On the other hand, let $\mathcal{D}$ be the set $\{\Gamma$: for some $F \in \mathbb{F}$, Γ is an F-saturated set$\}$.

6.1. LEMMA. $\mathcal{C} = \mathcal{D}$.

PROOF: 1) Suppose that $\Gamma \in \mathcal{C}$; then for some $\langle I, \langle \omega\ W\ R \rangle \rangle \in M_{\mathbf{k}}$, and some $a \in W$, $\Gamma = \Delta_a$; it is easy to prove by induction on the length of a deduction, that $\Gamma \nvdash \neg(A \supset A)$, for every A; further, if $B \notin \Gamma$ then $\Gamma \cup \{B\} \vdash \neg(A \supset A)$, for every A, every B [for: a) $B \notin \Delta_a$, hence $I(B,a)$ and $I(\neg B,a)=1$; since $I(\neg B \supset (B \supset \neg(A \supset A)), a)=1$, $I(B \supset \neg(A \supset A)) = 1$, therefore $B \supset \neg(A \supset A) \in \Delta_a$, $B \supset \neg(A \supset A) \in \Gamma$ and $\Gamma \vdash B \supset \neg(A \supset A)$; so $\Gamma \cup \{B\} \vdash \neg(A \supset A)$]. Therefore for every A, Γ is a $\neg(A \supset A)$ - saturated set, that is to say, $\Gamma \in \mathcal{D}$.

2) To prove the converse, let $\Gamma \in \mathcal{D}$ and,

a) construct $\langle I, \langle \Gamma, \mathcal{D}, R\rangle\rangle$ such that: i) for every Γ', Γ'' in $\mathcal{D}$, $\Gamma'' R \Gamma'$ iff $\varepsilon(\Gamma'^{\Box}) \subset \Gamma''$; ii) $I(A,\Gamma')=1$ iff $A \in \Gamma'$, for every $A \in \mathbb{F}$ and every $\Gamma' \in \mathcal{D}$.

b) we prove now, for every $\Gamma' \in \mathcal{D}$ that $\varepsilon(\Gamma'^{\Box}) = \cap\{\bar{\Gamma} \in \mathcal{D} : \varepsilon(\Gamma'^{\Box}) \subset \bar{\Gamma}\}$. It is clear that $\varepsilon(\Gamma'^{\Box}) \subset \cap\{\bar{\Gamma} \in \mathcal{D} : \varepsilon(\Gamma'^{\Box}) \subset \bar{\Gamma}\}$. For the converse let $A \notin \varepsilon(\Gamma'^{\Box})$; then $\Box A \notin \Gamma'^{\Box}$, and $\Box A \notin \Gamma'$ and, since $\Gamma' \in \mathcal{D}$, $\Gamma' \nvdash \Box A$, hence $\Gamma'^{\Box} \nvdash \Box A$; by 1.2, $\varepsilon(\Gamma'^{\Box}) \nvdash A$ and, by 4.3 there is an A-saturated Γ^* such that $\varepsilon(\Gamma'^{\Box}) \subset \Gamma^*$; hence $A \notin \Gamma^*$ and $\Gamma^* \in \{\bar{\Gamma} \in \mathcal{D}: \varepsilon(\Gamma'^{\Box}) \subset \bar{\Gamma}\}$; therefore $A \notin \cap\{\bar{\Gamma} \in \mathcal{D}: \varepsilon(\Gamma'^{\Box}) \subset \bar{\Gamma}\}$.

c) From b) together with a.i), we have that: $A \in \varepsilon(\Gamma'^{\Box})$ iff $A \in \cap\{\bar{\Gamma} \in \mathcal{D}: \varepsilon(\Gamma'^{\Box}) \subset \bar{\Gamma}\}$, iff $A \in \cap\{\bar{\Gamma} \in \mathcal{D}: \bar{\Gamma} R \Gamma'\}$.

d) Now, we can prove that $\langle I, \langle \Gamma, \mathcal{D}, R\rangle\rangle$ is a Kripke model for **K**; in fact, $\mathcal{D}$ is a non-empty set, $\Gamma \in \mathcal{D}$, R is a subset of $\mathcal{D} \times \mathcal{D}$ and I is a function from $\mathbb{F} \times \mathcal{D}$ into $\{0,1\}$ such that, for $A, B \in \mathbb{F}$, and $\Gamma' \in \mathcal{D}$,

1) $I(\neg A\ \Gamma')=1$ iff $\neg A \in \Gamma'$ iff $A \notin \Gamma'$ (see 4.2.6) iff $I(A, \Gamma')=0$;

2) $I(A \supset B\ \Gamma') = 1$ iff $A \supset B \in \Gamma'$ iff $A \notin \Gamma'$ or $B \in \Gamma'$ (see 4.2.c) iff $I(A,\Gamma')=0$ or $I(B, \Gamma')=1$;

3) $I(\Box A, \Gamma\ ) = 1$ iff $\Box A \in \Gamma'$ iff $\Box A \in \Gamma'^{\Box}$ iff $A \in \varepsilon(\Gamma'^{\Box})$ iff $A \in \cap\{\bar{\Gamma} \in \mathcal{D}: \bar{\Gamma} R \Gamma'\}$ (see c), iff, for every $\bar{\Gamma} \in \mathcal{D}$, such that $\bar{\Gamma} R \Gamma'$, $A \in \bar{\Gamma}$ iff for every $\bar{\Gamma} \in \mathcal{D}$ such that $\bar{\Gamma} R \Gamma'$, $I(A, \bar{\Gamma})=1$.

Thus, $\langle I, \langle \Gamma, \mathcal{D}, R\rangle\rangle \in M_{\mathbf{k}}$.

e) Since $\Delta_{\Gamma} = \{A : I(A,\Gamma)=1\} = \Gamma$, $\Gamma \in \mathcal{C}$. ∎

6.2. THEOREM. *There is a bijective funtion h from the set V of all valuations into the set $M_{\mathbf{k}}/_{\sim}$ such that if $\vartheta \in V$, $h(\vartheta) = |M|$ and $M = \langle I, \langle \omega, W, R\rangle\rangle$ then for every $A \in \mathbb{F}$, $\vartheta(A) = I(A, \omega)$.*

PROOF: Consequence of 4.6 and 6.1. ∎

7. A DECISION METHOD FOR K, BASED ON VALUATIONS.

Let A be any formula; we see easily how to construct a Σ-sequence $A_1, \ldots, A_n$ such that $A_n = A$. Let $V(A_1,\ldots,A_n)$ be the class of all $A_1, \ldots, A_n$-valuations and, for $\vartheta, \vartheta' \in V(A_1,\ldots,A_n)$, define $\vartheta \sim \vartheta'$ iff, for $1 \leqslant i \leqslant n$; $\vartheta(A_i) = \vartheta'(A_i)$. Clearly $V(A_1,\ldots,A_n)/\sim$ is finite. Hence, if we exhibit an algorithm which permits us to produce, for every $|\vartheta| \in V(A_1,\ldots,A_n)/\sim$, the restriction $\bar{\vartheta}$ of some $\vartheta' \in |\vartheta|$ to the set $\{A_1,\ldots,A_n\}$, then we have obtained a decision method for K. We call such a construction the *tableau* for $A_1,\ldots,A_n$, in symbols, $T(A_1,\ldots,A_n)$.

Before giving the precise definition of $T(A_1,\ldots,A_n)$ let us illustrate the main ideas by an example. Let A be $\Box(p \supset \neg p) \supset \Box\neg p$ and $A_1,\ldots,A_6$ be the sequence $p, \neg p, p \supset \neg p, \Box\neg p, \Box(p \supset \neg p), \Box(p \supset \neg p) \supset \Box\neg p$. Then we begin constructing $T(A_1)$ and then extend it successively to $T(A_1, A_2), \ldots, T(A_1, A_6)$, in the following way:

	1	2	3	4	5	6
	p	$\neg p$	$p \supset \neg p$	$\Box\neg p$	$\Box(p \supset \neg p)$	$\Box(p \supset \neg p) \supset \Box\neg p$
1)	1	0	0	1	1	1
2)	0	1	1	1	1	1
3)	1	0	0	0	0	1
4)	0	1	1	0	0	1

As we see, $T(A_1)$, $T(A_1,A_2)$ and $T(A_1,A_2,A_3)$ are constructed as usual. But, in $T(A_1,\ldots,A_4)$ two new rows are constructed. The reason for this can be explained thus: $V(A_1,\ldots,A_3)/\sim$ has, as obviously, only two elements, say $|\vartheta_1|$ and $|\vartheta_2|$, and the members of $|\vartheta_1|$ and $|\vartheta_2|$, restricted to $A_1,\ldots,A_3$ are represented, respectively, by rows 1 and 2. Now, for $i \in \{1,2\}$, since $\{A_1,\ldots,A_3\}^{\Box} = \emptyset$, we have:

I) $\vartheta_1 \models \varepsilon(\{A_1,\ldots,A_n\}_{\vartheta_i,1})$ and $\vartheta_1(\neg p) = 0$; hence the necessary condition for the existence of a ϑ' and a ϑ'' in $V(A_1,\ldots,A_4)$ such that $\vartheta \in |\vartheta_1|$, $\vartheta' \in |\vartheta_2|$ and $\vartheta(A_4) = \vartheta'(A_4) = 0$, is satisfied; and, vacuously:

II) for each $p < 4$, each $q < p$ such that $A_p = \Box A_q$ and $\vartheta_i(A_p) = 0$ there is a $j \in \{1,2\}$ such that $\vartheta_j(A_q)$ and $\vartheta_j \models \varepsilon(\{A_1,\ldots,A_3\}_{\vartheta_i,1})$

$\cup \{A_2\}$. Hence, the necessary condition for the existence of a $\bar{\vartheta}$ and a $\bar{\vartheta}'$ in $V(A_1,\ldots,A_4)$ such that $\bar{\vartheta} \in |\vartheta_1|$, $\bar{\vartheta}' \in |\vartheta_2|$ and $\bar{\vartheta}(A_4) = \bar{\vartheta}'(A_4) = 1$, is also satisfied. Hence, it is plausible that $V(A_1,\ldots,A_4)/\sim$ has four members $|\vartheta_1|,\ldots,|\vartheta_4|$ and that, for $1 \leqslant i \leqslant 4$, the elements of $|\vartheta_i|$ restricted to $A_1,\ldots,A_4$, are represented by the lines 1-4. On the other hand, $V(A_1,\ldots,A_5)/\sim$ has, at most, $|\vartheta_1|,\ldots,|\vartheta_4|$ as its members, for:

I) If $i \in \{1,2\}$ and $\vartheta \in |\vartheta_i|$, $\varepsilon(\{A_1,\ldots,A_4\}_{\vartheta,1}) = \{\neg p\}$ and, for $1 \leqslant j \leqslant 4$, no $\vartheta' \in |\vartheta_j|$ is such that $\vartheta' \models \{\neg p\}$ and $\vartheta'(p \supset \neg p) = 0$; hence, for $i \in \{1,2\}$ and $\vartheta \in |\vartheta_i|$, $\vartheta(A_5) = 0$.

II) If $i \in \{3,4\}$ since $\Box\neg p = A_4 = \Box A_2$ and, for $\vartheta \in |\vartheta_i|, \vartheta(A_4) = 0$, $\vartheta(A_5) = 0$; for although $\varepsilon(\{A_1,\ldots,A_4\}_{\vartheta,1}) = \emptyset$, no $\vartheta' \in |\vartheta_j|$, for some j, $1 \leqslant j \leqslant 4$, is such that $\vartheta'(\neg p) = 0$ and $\vartheta' \models \emptyset \cup \{p \supset \neg p\}$; hence, for $i \in \{3,4\}$ and $\vartheta \in |\vartheta_i|$, $\vartheta(A_5) = 0$.

Let us now give a precise definition of a tableau for a Σ-sequence $A_1,\ldots,A_n$.

7.1. The tableau for a given Σ-sequence $A_1,\ldots,A_n$ is a function a_n: $I_n \times J(I_n) \longrightarrow \{0,1\}$, where $I_n = \{1,\ldots,n\}$ and:

1) for $n = 1$, $J(I_n) = \{1,2\}$, $a_1(1,1) = 1$ and $a_1(1,2) = 0$;

2) for $n > 1$, and $J(I_{n-1}) = \{1,\ldots,q\}$

a) If A_n is a propositional variable, then $J(I_n) = \{1,\ldots,2q\}$ and,

i) for $i < n$, $j \in J(I_{n-1})$, $a_n(i,j) = a_{n-1}(i,j)$;

ii) for $i < n$, $j' \in J(I_{n-1})$ and $j = q + j'$, $a_n(i,j) = a_{n-1}(i,j')$.

iii) for $i = n$, $j \in J(I_{n-1})$, $a_n(i,j)$;

iv) for $i = n$, $j' \in J(I_{n-1})$ and $j = q + j'$, $a_n(i,j) = 0$;

b) If $A_n = \neg A_k$, $J(I_n) = J(I_{n-1})$ and:

i) for $i < n$, $a_n(i,j) = a_{n-1}(i,j)$,

ii) for $i = n$, $a_n(i,j) \neq a_{n-1}(k,j)$;

c) If $A_n = A_k \supset A_e$, $J(I_n) = J(I_{n-1})$ and:

i) for $i < n$, $a_n(i,j) = a_{n-1}(i,j)$;

ii) for $i = n$, $a_n(i,j) = 1$ iff $a_{n-1}(k,j) = 0$ or $a_{n-1}(e,j) = 1$;

d) If $A_n = \Box A_k$, for each $j \in J(I_{n-1})$,

I) let $\alpha(j,n-1) = \{j' \in J(I_{n-1})$: $a_{n-1}(k,j') = 0$ and, for every r, $1 \leq r < n$, if $A_r = \Box A_s$ and $a_{n-1}(r,j) = 1$, then $a_{n-1}(s,j') = 1\}$;

II) for every $p < n$, every $q < p$ such that $A_p = \Box A_q$ and $a_{n-1}(p,j) = 0$, let $\beta(p,j,n-1) = \{j' \in J(I_{n-1})$: $a_{n-1}(q,j') = 0$, $a_{n-1}(k,j') = 1$ and for every r, $1 \leq r < n$, if $A_r = \Box A_s$ and $a_{n-1}(r,j) = 1$, then $a_{n-1}(s,j') = 1\}$.

III) Now let $\{j_1, \ldots, j_m\} \subseteq J(I_{n-1})$ be such that:

1) $j_{m'} < j_{m''}$ if $m' < m''$; 2) $j_{m'} \in \{j_1, \ldots, j_m\}$ if $\alpha(j_{m'}, n-1) \neq \emptyset$, and for each $p < n$, $\beta(p, j_{m'}, n-1) \neq \emptyset$.

Then, $J(I_n) = \{1, \ldots, q, \ldots, q+m\}$ and:

i) for $i < n$, $j \leq q$, $a_n(i,j) = a_{n-1}(i,j)$;

ii) for $i < n$, $j = q + m'$, $a_n(i,j) = a_{n-1}(i, j_{m'})$;

iii) for $i = n$, j such that $\alpha(j,n-1) = \emptyset$, $a_{n-1}(i,j) = 1$;

iv) for $i = n$, j such that $\alpha(j,n-1) \neq \emptyset$ and $\beta(p,j,n-1) = \emptyset$, for some $p < n$, $a_n(i,j) = 0$;

v) for $i = m$, j such that $\alpha(j\ n-1) \neq \emptyset$ and, for each $p < n$, $\beta(p,j,n-1) \neq \emptyset$, in which case, for some $m' \in \{1, \ldots, m\}$, $j = j_{m'}$ or $j = q + m'$,

1) if $j = j_{m'}$, then $a_n(1,j) = 1$;

2) if $j = q + j_{m'}$, then $a_n(i,j) = 0$.

Let a_n: $I_n \times J(I_n) \rightarrow \{0,1\}$ be the tableau of the Σ-sequence $A_1, \ldots, A_n$.

7.2. LEMMA. *For each* $j \in J(I_n)$ *there is a valuation* ϑ *such that*

$\vartheta(A_i) = a_n(i,j)$, *for* $1 \leqslant i \leqslant r$.

PROOF: Let $\{A_1,\ldots, A_n\}_{j,u} = \{A_i \in \{A_1,\ldots, A_n\}: a_n(i,j) = u\}$. By induction on n, we prove that if $A_i \in \{A_1,\ldots, A_n\}_{j,0}$, then $\{A_1, \ldots, A_n\}_{j,1} \nvdash A_i$. Then the proposition follows from 4.8. ■

7.3. LEMMA. *For every valuation* ϑ *there is a* $j \in J(I_n)$ *such that, for* $1 \leqslant i \leqslant n$, $\vartheta(A_i) = a_n(i,j)$.

PROOF: By induction on n. ■

7.4. THEOREM. $\vdash A_i$, *iff for every* $j \in J(I_n)$, $a_n(i,j) = 0$.

Thus, **K** is decidable by the tableau of valuations.

8. Final Remarks.

The resulsts obtained in Sections 6 and 7 are easily adapted to **T**. Valuations for **S4**, **S5** as well as for many other calculi, having or not Kripkean semantics, can be obtained by convenient modifications on the procedures here employed. They contribute also to the better understanding of Kripke models: as we have seen for the system **K**, the plurality of worlds in a given model and the relation between worlds can be considered as technical devices to simplify the truth conditions governing modal formulas.

REFERENCES.

[1] N. C. A. da Costa and E. H. Alves, *Une sémantique pour le calcul* C_1, C. R. Acad. Sc. Paris, 283 A (1976), 729-731.

[2] A. Loparić, *Un étude sémantique de quelques calculs propositionnels*, C. R. Acad. Sc. Paris, 284 A (1977), 835-838.

[3] S. A. Kripke, *Semantical analysis of modal logic, I, normal propositional calculus*, ZML, vol. 9 (1963), 67-96.

Centro de Lógica, Epistemologia e
História da Ciência
Universidade Estadual de Campinas
13.100 Campinas, SP., Brazil.

Discussive Versions of the Modal Calculi T, B, S4 and S5.

by *L. H. LOPES DOS SANTOS*.

ABSTRACT. In this paper we show that **T**, **B**, **S4** and **S5** can be specified by axiomatics having discussive conjunction as the sole non-extensional primitive notions, and having as postulates the discussive counterparts of some of the basic laws of classical conjunction in the classical propositional calculus. This will make clear not only that discussive conjunction in **T**, **B**, **S4** and **S5** shares several properties with classical conjunction, but also that a meager set of such properties is sufficient to characterize completely these logics. It seems natural to call a modal function elementary with respect to a certain modal logic when it is possible to characterize completely this logic by means of a simple axiomatics having that modal function as the sole non-extensional primitive notion. We present axiomatics which assure the elementariness of the modal function which defines discussive conjunction with respect to **T**, **B**, **S4** and **S5**.

1. Introduction.

Atenpting to find a logical basis for inconsistent but non-trivial theories, S. Jaškowski defined in [3] the discussive logic $\mathbf{D}_2$ as the set of all formulas p such that Mp is a thesis of the modal calculus **S5**, where M is the usual possibility operator. Following Jaškowski's suggestion, N. C. A. da Costa and L. Dubikajtis defined in [1] the *discussive implication* of a formula q by a formula p as the material implication of q by Mp, and the *discussive conjunction* of p and q as the classical conjunction of Mp and q. Taking these discussive operators as non-extensional primitive notions, they axiomatized $\mathbf{D}_2$ using a set of postulates which makes clear that discussive implication and discussive conjunction play in $\mathbf{D}_2$

the roles played by classical implication and conjunction in the positive propositional calculus, and some of the roles played by these classical operators in the classical propositional calculus.

What happens with the discussive operators in the modal logics **T**, **B** (the Brouwerian system), **S4** and **S5** themselves? It may be easily shown that several properties, which one would reasonably expect an implication operations to have, fail to be true of discussive implication in **T**, **B**, **S4** and **S5**; in particular, even the discussive counterparts of the basic laws usually employed in the specification of the intuitionist implicative calculus do not hold in these logics. In this paper, we show that discussive conjunction is not in the same situation.

More precisely, what will be shown is that **T**, **B**, **S4** and **S5** can be specified by axiomatics having discussive conjunction as the sole non-extensional primitive notion, and having as postulates the discussive counterparts of some of the laws of classical conjunction in the classical propositional calculus. This will make clear not only that discussive conjunction in **T**, **B**, **S4** and **S5** shares several properties with classical conjunction, but also that a meager set of such properties is sufficient to characterize completely these logics. It seems natural to call a modal function elementary with respect to a certain modal logic when it is possible to characterize completely this logic by means of a simple axiomatics having that modal function as the sole non-extensional primitive notion. The axiomatics presented below assure the elementariness of the modal function which defines discussive conjunction with respect to **T**, **B**, **S4** and **S5**.

2. THE AXIOMATICS **T**, **B**, **S4** AND **S5**.

In their regular version, the logics **T**, **B**, **S4** and **S5** will be formulated here in a language whose formulas are recursively built from propositional letters, the operators $\neg$ (*classical negation*), $\rightarrow$ (*material implication*) and M (*possibility*), and parentheses. We shall characterize them by means of axiomatics which will be also called **T**, **B**, **S4** and **S5** (concerning the ability of the axiomatics to characterize the logics, see for example [2], pp. 125, 126, 129). Throughout this paper, small latin letters will be employed as syntactical variables for formulas, the expression

"$\vdash_{\mathbf{A}} p$" will mean that p is a thesis of the axiomatics **A**, and the expression "$q_1, \ldots, q_n \vdash_{\mathbf{A}} p$" will mean that p is derivable from $q_1, \ldots, q_n$ in in the axiomatics **A**. We shall often omit the most external parentheses of a formula, as well as those unambiguously restorable according to the rule of association to the right.

Axiomatics **T** comprises the following definitions and postulates (axiom schemes and rules):

D1) $(p \vee q) =_{df} (\neg p \rightarrow q)$;

D2) $(p \equiv q) =_{df} \neg(\neg(p \rightarrow q) \vee \neg(q \rightarrow p))$;

D3) $\mathrm{L}p =_{df} \neg \mathrm{M} \neg p$;

D4) $(p \,\&\, q) =_{df} \neg(\neg \mathrm{M}p \vee \neg q)$;

P1) $p, q \vdash_{\mathbf{T}} r$, *if* r *is a tautological consequence of* p *and* q;

P2) $p \rightarrow \mathrm{M}p$;

P3) $\mathrm{M}(p \vee q) \equiv (\mathrm{M}p \vee \mathrm{M}q)$;

P4) $p \vdash_{\mathbf{T}} \mathrm{L}p$;

P5) $p \equiv q \vdash_{\mathbf{T}} \mathrm{M}p \equiv \mathrm{M}q$.

Now, let us consider the postulates:

P6) $\mathrm{ML}p \rightarrow p$;

P7) $\mathrm{MM}p \rightarrow \mathrm{M}p$.

Addind P6 to **T**, we obtain **B**; adding P7 to **T**, we obtain **S4**; adding both P6 and P7 to **T**, we obtain **S5**.

3. Axiomatics Td, Bd, S4d and S5d.

In their discussive version, the logics **T**, **B**, **S4** and **S5** will be formulated in a language whose formulas are recursively built from *propositional letters*, the operators $\neg$ (*classical negation*), $\rightarrow$ (*material implication*) and & (*discussive conjunction*), and parentheses.

We shall now introduce the axiomatics **Td**, **Bd**, **S4d** and **S5d**, which are to be thought as formulated in such a language. Axiomatics **Td** embraces the following definitions and postulates:

Dd1) $(p \vee q) =_{df} (\neg p \to q)$;

Dd2) $(p \equiv q) =_{df} \neg(\neg(p \to q) \vee \neg(q \to p))$;

Dd3) $Mp =_{df} (p \,\&\, (p \vee \neg p))$;

Dd4) $Lp =_{df} \neg M \neg p$;

Dd5) $Kp =_{df} (p \vee \neg p)$;

Pd1) $p, q \vdash_{\mathbf{Td}} r$, *if r is a tautological consequence of p and q*;

Pd2) $p \to q \to (p \,\&\, q)$;

Pd3) $(p \,\&\, q) \to q$;

Pd4) $((p \vee q) \,\&\, r) \to ((p \,\&\, r) \vee (q \,\&\, r))$;

Pd5) $r \to (p \,\&\, q) \to (p \,\&\, r)$;

Pd6) $\neg p \vdash_{\mathbf{Td}} \neg(p \,\&\, q)$;

Pd7) $p \to q \vdash_{\mathbf{Td}} (p \,\&\, r) \to (q \,\&\, r)$.

We get **Bd** from **Td** adding

Pd8) $\neg(p \,\&\, q) \vdash_{\mathbf{Bd}} \neg(q \,\&\, p)$;

we get **S4d** from **Td** adding

Pd9) $((p \,\&\, q) \,\&\, r) \to (p \,\&\, (q \,\&\, r))$;

axiomatics **S5** comprises Dd1-Dd5 and Pd1-Pd9.

4. Equivalences.

Any formula of the regular modal language of Section 2 is obviously translatable in the discussive language of Section 3 by means of Dd1-Dd5; similarly, any formula of the discussive language is translatable in the regular modal language by means of D1-D4. Now, let tr(p) denote the translation of the discussive formula p in the regular modal language, and let trd(q) denote the translation of the regular modal formula q in the discussive language. In order to show that **Td, Bd**, **Sd4** and **Sd5** are adequate axiomatics respectively for the logics **T**, **B**, **S4** and

S5, it is enough to show that they are equivalent respectively to the axiomatics **T**, **B**, **S4** and **S5**. In order to assure the equivalences, it is enough to show that: (i) p is a thesis of **Td** (**Bd**, **S4d**, **S5d**) if and only if tr(p) is a thesis of **T** (**B**, **S4**, **S5**); (ii) q is a thesis of **T** (**B**, **S4**, **S5**) if and only if trd(q) is a thesis of **Td** (**Bd**, **S4d**, **S5d**).

THEOREM 1. $\vdash_{\mathbf{Td}} p$, *if p is a tautology.*

PROOF: By Pd1 and any two axioms.

THEOREM 2. $\vdash_{\mathbf{Td}} p \rightarrow Mp$.

PROOF: (i) $\vdash_{\mathbf{Td}} p \rightarrow Kp \rightarrow (p \,\&\, Kp)$, by Pd2;

(ii) $\vdash_{\mathbf{Td}} p \rightarrow (p \,\&\, Kp)$, by (i) and Pd1;

(iii) $\vdash_{\mathbf{Td}} p \rightarrow Mp$, by (ii), Dd3 and Dd5.

THEOREM 3. $p \rightarrow q \vdash_{\mathbf{Td}} Mp \rightarrow Mq$.

PROOF: (i) $p \rightarrow q$;

(ii) $(p \,\&\, Kp) \rightarrow (q \,\&\, Kp)$, by (i) and Pd7;

(iii) $Kq \rightarrow (q \,\&\, Kp) \rightarrow (q \,\&\, Kq)$, by Pd5;

(iv) $(p \,\&\, Kp) \rightarrow (q \,\&\, Kq)$, by (ii), (iii) and Pd1;

(v) $Mp \rightarrow Mq$, by (iv), Dd3 and Dd5.

THEOREM 4. $p \equiv q \vdash_{\mathbf{Td}} Mp \equiv Mq$.

PROOF: By Pd1 and Theorem 3.

THEOREM 5. $\vdash_{\mathbf{Td}} M(p \vee q) \rightarrow (Mp \vee Mq)$.

PROOF: (i) $\vdash_{\mathbf{Td}} ((p \vee q) \,\&\, Kp) \rightarrow ((p \,\&\, Kp) \vee (q \,\&\, Kp))$, by Pd4;

(ii) $\vdash_{\mathbf{Td}} Kp \rightarrow ((p \vee q) \,\&\, K(p \vee q)) \rightarrow ((p \vee q) \,\&\, Kp)$, by Pd5;

(iii) $\vdash_{\mathbf{Td}} Kq \rightarrow (q \,\&\, Kp) \rightarrow (q \,\&\, Kq)$, by Pd5;

(iv) $\vdash_{Td} ((p \vee q) \;\&\; K(p \vee q)) \rightarrow ((p \;\&\; Kp) \vee (q \;\&\; Kq))$, by (i), (ii), (iii) and Pd1.

(v) $\vdash_{Td} M(p \vee q) \rightarrow (Mp \vee Mq)$, by (iv), Dd3 and Dd5.

THEOREM 6. $\vdash_{Td} (Mp \vee Mq) \rightarrow M(p \vee q)$.

PROOF: (i) $\vdash_{Td} p \rightarrow (p \vee q)$, by Theorem 1;

(ii) $\vdash_{Td} q \rightarrow (p \vee q)$, by Theorem 1;

(iii) $\vdash_{Td} Mp \rightarrow M(p \vee q)$, by (i) and Theorem 3;

(iv) $\vdash_{Td} Mq \rightarrow M(p \vee q)$, by (ii) and Theorem 3;

(v) $\vdash_{Td} (Mp \vee Mq) \rightarrow M(p \vee q)$, by (iii), (iv) and Pd1.

THEOREM 7. $\vdash_{Td} M(p \vee q) \equiv (Mp \vee Mq)$.

PROOF: By Pd1 and Theorems 5-6.

THEOREM 8. $p \vdash_{Td} Lp$.

PROOF: (i) p ;

(ii) $\neg\neg p$, by (i) and Pd1;

(iii) $\neg(\neg p \;\&\; K\neg p)$, by (ii) and Pd6;

(iv) Lp, by (iii), Dd3, Dd4 and Dd5.

THEOREM 9. $\vdash_{Td} (p \;\&\; q) \equiv \neg(\neg Mp \vee \neg q)$.

PROOF: (i) $\vdash_{Td} (p \;\&\; q) \rightarrow q$, by Pd3;

(ii) $\vdash_{Td} Kp \rightarrow (p \;\&\; q) \rightarrow (p \;\&\; Kp)$, by Pd5;

(iii) $\vdash_{Td} (p \;\&\; q) \rightarrow \neg(\neg(p \;\&\; Kp) \vee \neg q)$, by (i), (ii) and Pd1;

(iv) $\vdash_{Td} q \rightarrow (p \;\&\; Kp) \rightarrow (p \;\&\; q)$, by Pd5;

(v) $\vdash_{Td} \neg(\neg(p \;\&\; Kp) \vee \neg q) \rightarrow (p \;\&\; q)$, by (iv) and Pd1;

(vi) $\vdash_{Td} (p \;\&\; q) \equiv \neg(\neg(p \;\&\; Kp) \vee \neg q)$, by (iii), (v) and Pd1;

(vii) $\vdash_{Td} (p \;\&\; q) \equiv \neg(\neg Mp \vee \neg q)$, by (vi), Dd3 and Dd5.

THEOREM 10. $\vdash_{\mathbf{Bd}} \mathrm{ML}p \to p$.

PROOF: (i) $\vdash_{\mathbf{Bd}} \neg\mathrm{M}\neg p \vee \neg\neg\mathrm{M}\neg p$, by Theorem 1;

(ii) $\vdash_{\mathbf{Bd}} \neg(\neg p \;\&\; \neg\mathrm{M}\neg p)$, by (i), Theorem 9 and Pd1;

(iii) $\vdash_{\mathbf{Bd}} \neg(\neg\mathrm{M}\neg p \;\&\; \neg p)$, by (ii) and Pd8;

(iv) $\vdash_{\mathbf{Bd}} \neg\mathrm{M}\neg\mathrm{M}\neg p \vee \neg\neg p$, by (iii), Theorem 9 and Pd1;

(v) $\vdash_{\mathbf{Bd}} \mathrm{ML}p \to p$, by (iv), Pd1 and Dd4.

THEOREM 11. $\vdash_{\mathbf{S4d}} \mathrm{MM}p \to \mathrm{M}p$.

PROOF: (i) $\vdash_{\mathbf{S4d}} ((p \;\&\; \mathrm{K}p) \;\&\; \mathrm{K}(p \;\&\; \mathrm{K}p)) \to (p \;\&\; (\mathrm{K}p \;\&\; \mathrm{K}(p \;\&\; \mathrm{K}p)))$, by Pd9;

(ii) $\vdash_{\mathbf{S4d}} \mathrm{K}p \to (p \;\&\; (\mathrm{K}p \;\&\; \mathrm{K}(p \;\&\; \mathrm{K}p))) \to (p \;\&\; \mathrm{K}p)$, by Pd5;

(iii) $\vdash_{\mathbf{S4d}} ((p \;\&\; \mathrm{K}p) \;\&\; \mathrm{K}(p \;\&\; \mathrm{K}p)) \to (p \;\&\; \mathrm{K}p)$, by (i), (ii) and Pd1;

(iv) $\vdash_{\mathbf{S4d}} \mathrm{MM}p \to \mathrm{M}p$, by (iii), Dd3 and Dd5.

THEOREM 12. *Understanding & as a defined symbol of* **T, B, S4** *and* **S5** *postulates* Pd1-Pd7 *are valid in these axiomatics;* Pd8 *is valid in* **B** *and* **S5***;* Pd9 *is valid in* **S4** *and* **S5**.

PROOF: By means of the well-known semantical tests.

THEOREM 13. $\mathrm{M}p \equiv (p \;\&\; (p \vee \neg p))$ *is a valid scheme of* **T, B, S4** *and* **S5**.

PROOF: By means of the well-known semantical tests.

THEOREM 14. *If* p *is a thesis of* **T** (**B**, **S4**, **S5**), *then* trd(p) *is a thesis of* **Td** (**Bd**, **S4d**, **S5d**).

PROOF: By induction, with the help of Pd1 and Theorems 2, 4, 7, 8. 10 and 11.

THEOREM 15. *If* p *is a thesis of* **Td** (**Bd**, **S4d**, **S5d**), *then* tr(p) *is a thesis of* **T** (**B**, **S4**, **S5**).

PROOF: By induction, with the help of P1 and Theorem 12.

THEOREM 16. *$p \equiv$ tr(trd(p)) is a valid scheme of* **T, B, S4** *and* **S5**.

PROOF: By induction on the length of p, with the help of P1 and Theorem 13.

THEOREM 17. *$p \equiv$ trd(tr(p)) is a valid schema of* **Td, Bd, S4d** *and* **S5d**.

PROOF: By induction on the length of p, with the help of Pd1 and Theorem 9.

THEOREM 18. *If trd(p) is a thesis of* **Td** (**Bd, S4d, S5d**), *then p is a thesis of* **T** (**B, S4, S5**).

PROOF: If trd(p) is a thesis of **Td** (**Bd, S4d, S5d**), then tr(trd(p)) is a thesis of **T** (**B, S4, S5**), by Theorem 15; hence p is a thesis of **T** (**B, S4, S5**), by P1 and Theorem 16.

THEOREM 19. *If tr(p) is a thesis of* **T** (**B, S4, S5**), *then p is a thesis of* **Td** (**Bd, S4d, S5d**).

PROOF: Similar to the proof of Theorem 18, with the help of Pd1 and Theorems 14 and 17.

THEOREM 20. *Axiomatics* **T, B, S4** *and* **S5** *are equivalent respectively to* **Td, Bd, S4d** *and* **S5d**.

PROOF: By Theorems 14, 15, 18 and 19.

REFERENCES.

[1] N. C. A da Costa and L. Dubikajtis, *On Jaśkowski discussive logic*, Non-Classical Logics, Model Theory and Computability, Arruda, da Costa and Chuaqui (Eds.), North-Holland, Amsterdam, 1977, 37-56.

[2] G. E. Hughes ans M. J. Cresswell, An Introduction to Modal logic, Methuen, London, 1968.

[3] S. Jaśkowski, *Rachunek zdań dla systemów dedukcyjnych sprzecznych*, Studia Societatis Scientiarum Torunensis, A, I, 5 (1948), 55-77. (An English tranlation of this paper appeared in Studia Logica, XXIV (1969), 143-157.)

Centro de Lógica, Epistemologia e História da Ciência
Universidade Estadual de Campinas
13.100 Campinas, S.P., Brazil.

SUPERPOSITION OF STATES IN QUANTUM LOGIC FROM A SET THEORETICAL POINT OF VIEW.

by *CARLOS A. LUNGARZO.*

ABSTRACT. The aim of this paper is to study some connections between quantum logic and model theory of set theory, in the research line opened by Benioff and others. Our purpose is to analyze some characteristic operations in quantum logic from the point of view of their "absolutness" with respect to transitive models for ZFC theory (Zermelo-Freankel + axiom of choice). We begin by recalling some results by Benioff, especially those proving the existence of an isometric monomorphism between a set which is an algebra of bounded operators in a Hilbert space *inside* a ZFC model, and such an algebra *outside* the model. We prove that there is an injection between pure states inside and outside a model, when the algebra is that of finite dimensional or sodimensional operators. By resorting to Gleason's Theorem, we "translate" superposition inside the model to superposition outside the model, and prove that this weak version of the superposition principle is preserved.

1. MOTIVATION.

In [1] and [2] Benioff has undertaken the job of developing the basic tools for a comprehensive treatment of the mathematics of quantum mechanic, within the framework of axiomatic set theory (more specifically, his background is the transitive standard models of Zermelo-Fraenkel set theory plus axiom of choice). The basic idea of this and other scientists who are trying a set-axiomatic approach to "usual" mathematics is that sometimes the naive set theory in which standard analysis is constructed does not reflect the enormous complexity of the mathematical apparatus of physics. On one hand, axiomatic set theory restricts the "too large" Cantor universe, avoiding some (possibly) unnecessary generality. On the other hand, intuitive set theory, because of its own haziness and lack of precision, does not allow for some subtle problems that arise as soon as one deals with a specific axiomatized theory.

If several universes for the mathematics of quantum mechanics are to be studied, we believe that some basic notions, such as superposition of states, should be examined from this axiomatic set-theoretocal point of view. A still "raw" attempt in this direction is made in this paper. In Section 2, we give the basic definitions, just to fix notation and terminology. In particular, we explain in what sense the expressiom "quantum logic" is to be understood.

In Section 3, we formulate some general facts about the usual models for quantum logics, and state two main results by Benioff. Finally, we prove the existence of a map from pure states inside a transitive model into pure states outside the model.

2. General background.

We adopt here the definition of *quantum logic* as it appears for instance in Gudder [6] and [7] or Pulmanovā [9]; other definitions (for instance, Deliyannis [3]) also could work.

$(\mathbf{L}, \mathbf{M})$ is a *quantum logic* iff:

(i) $\mathbf{L}$ is a partially ordered set (poset).

(ii) There are a first (zero) element and a last (unit) element in $\mathbf{L}$, namely $\tilde{0}$ and $\tilde{1}$.

(iii) There is an orthocomplementation on $\mathbf{L}$:

(iii,1) $(a')' = a;$

(iii,2) $a \leqslant b$ implies $b' \leqslant a'$;

(iii,3) for all a, there exists the supremum of (a,a') denoted by $a \vee a'$) and it equals $\tilde{1}$.

(iv) If (a_n) is a sequence of disjoint elements, then there exists the supremum of (a_n) denoted by $\bigvee a_n$.

(v) If $a < b$ then $a \leftrightarrow b$ [we call this property "ortho (or weak) modularity". "$\leftrightarrow$" means compatibility]. (see [9].)

(The subtle differences among definitions in [6], [7] and [9] are of no relevance for our purposes).

The set **L** is called the set of *propositions* of the quantum logic under consideration.

Moreover, **M** is the set of *states*, which means:

If $m \in \mathbf{M}$, then:

(vi) $m : \mathbf{L} \to [0,1]$.

(vii) $m(\tilde{1}) = 1$.

(viii) m is σ-additive.

If $m \in \mathbf{M}$, it is called a *mixture* iff it is a strictly convex combinations of two others. $\boldsymbol{P}$ is the set of *pures states*, i. e., of those states which are not mixtures.

Define for every a in **L** and every m in $\boldsymbol{P}$, $\boldsymbol{P}(a) = \{m \in \boldsymbol{P} : m(a) = 1\}$ and $\mathbf{L}(m) = \{a \in \mathbf{L} : m(a) = 1\}$.

We finally state two last axioms:

(ix) $\boldsymbol{P}(a) \subset \boldsymbol{P}(b)$ entails $a \leqslant b$.

(x) $\mathbf{L}(m) \subset \mathbf{L}(n)$ entails $a = b$.

We speak of an "*additive*" *quantum logic* when (iv) is replaced by (iv'): if $(a_1, \ldots, a_n)$ is a (finite) sequence of disjoint propositions, then $a_1 \vee \ldots \vee a_n$ happens to exist, and (viii) is replaced by (viii'): m is additive.

We speak of a *logic* just when (i)-(viii) are fulfilled.

We recall the definition of *superposition* of states:

Let p, q, r be three states of **M**. We say that r is a *superposition* of p and q iff: for every a in $\mathbf{L}$, $p(a) = 0$ and $q(a) = 0$ imply $r(a) = 0$.

3. MODELS FOR QUANTUM LOGIC.

One of the best known examples of quantum logic is that provided by the complex Hilbert space formalism of conventional quantum mechanics. Henceforth, we always use the term "Hilbert space" as meaning "separable Hilbert space over the field of complex numbers", even though we believe analogous results could be obtained for quaternionic Hilbert spaces.

Let $(\mathbf{H}, +, \cdot, h)$ be a Hilbert space over C, with $+$ and $\cdot$ vector space operations and h the Hermitian form that defines the inner product. Let $\mathrm{PR}(\mathrm{H})$ be the set of all projection operators on $\mathbf{H}$. It is well known that $\mathbf{PR}(\mathrm{H})$ is a quantum logic, where $\leq$ is interpreted as set-theoretical inclusion and a' as $\mathrm{I} - a$, where I is the identity operator. See [8], p. 443

We aim to study the properties of superposition when they are looked "inside" and "outside" a certain kind of models. Following Benioff [1] we here restrict ourselves to transitive standard models for ZFC theory (Zermelo-Fraenkel + Axiom of Choice).

By $\mathcal{E}$ we denote always a model of this class, and by $\mathbf{E}$ the domain of $\mathcal{E}$. We state now the main results by Benioff:

PROPOSITION 1. *If* $\mathbf{H}_\mathrm{E}$ *is a Hilbert space inside* $\mathcal{E}$ *then there is a Hilbert space* $\mathbf{H}$, *outside* $\mathcal{E}$, *and an isometric monomorphism* $\varphi: \mathbf{H}_\mathrm{E} \to \mathbf{H}$.

We just recall the construction of $\mathbf{H}$. If H_E and $\mathbb{C}_\mathrm{E}$ are the sets of Cauchy sequences in H_E and $\mathbb{C}_\mathrm{E}$ respectively (where $\mathbb{C}$ is the set of complex numbers inside $\mathcal{E}$), then $\mathbf{H}$ is the quotient set $\mathrm{H}_\mathrm{E}/\sim$ where $\sim$ is the following equivalence relation:

$$(x_n) \sim (y_n) \quad \text{iff} \quad ((x - y)_n) \text{ converges to } 0.$$

(Notice that $\mathbf{H}_\mathrm{E}$ is a $\mathbb{C}_\mathrm{E}$-prehilbert space.)

$\varphi(x)$ is $[(x, x, \ldots x, \ldots)]$ where the square brackets denote "equivalent class" and $(x, x, \ldots x, \ldots)$ is the sequence identically equal to x.

$\mathbf{H}$ is very much like the classical completion of a metric space.

PROPOSITION 2. *Assume* $\mathbb{B}(\mathbf{H}_\mathrm{E})$ *is the set of linear bounded operators*

H_E, *inside* $\mathcal{E}$. *Then there is a Hilbert space* H, *outside* $\mathcal{E}$, *and an isometric monomorphism* V, *such that:*

$$V: \mathbb{B}(H_E) \to \mathbb{B}(H)$$

($\mathbb{B}(H)$ *is the set of bounded linear operators on* H).

PROPOSITION 3. *If* B *is a basis for* H_E *inside* $\mathcal{E}$, *then*

$$B^{\sim} = \{[(x, x, \ldots, x, \ldots)]: x \in B\}$$

is a basis for H.

PROPOSITION 4. *If* T *is a unitary (projection, density) operator on* H_E, *inside* $\mathcal{E}$, *then* $V(T)$ *is a unitary (projection, density) operator on* H *outside.*

PROPOSITION 5. *Let* a *and* b *be bounded linear operators on* H_E *such that* a *is a density operator and* b *is a projection operator,* H_E *is the trace, and* TR_E *is the trace in* $\mathcal{E}$. *Then:*

$$TR(V(a).V(b)) = TR_E(a.b).$$

Now we are not able to prove that superposition is an absolute property (i. e., to be preserved when looking from inside or from outside the model) when we consider states in their full generality; nevertheless, we can restrict ourselves to some "weak" structure on which the desired properties are preserved. We begin with the definition of some special class of projection operators.

DEFINITION 1. *We call a projection operator* a *on a Hilbert space a cof-operator iff the closed subspace associated to* a *has finite dimension or codimension.*

PROPOSITION 6. *The set* A *of all cof-operators on a Hilbert space, is an additive logic.*

PROOF: Clearly A is partially ordered by set theoretical inclusion. Moreover, if a subspace has finite dimension (codimension) then its orthocomplement has finite codimension (dimension).

If a and b belong to A, then their associated subspaces are (1) both of finite dimension, or (2) both of finite codimension, or (3) one of finite dimension and the other of finite codimension. In either case their closed linear hull belong to A.

Finally, if $a \leqslant b$, we have that $a \circ b = a$ (where "o" is the composition of operators) and this entails $a \circ b = b \circ a$; therefore $a \leftrightarrow b$. ∎

PROPOSITION 7. *The notations are the same as in Proposition 2. Let* $\mathbf{L}$ *be the additive logic of the cof-operator of* $\mathbb{B}(H_E)$ *inside* $\mathcal{E}$, *and* $\mathbf{L}^\sim$ *the additive logic of cof-operators of* $\mathbb{B}(H)$. *Then* $V(\mathbf{L}) = \mathbf{L}^\sim$.

PROOF: Let $\tilde{a}$ be an element in $V(\mathbf{L})$ and $a = V^{-1}(\tilde{a})$. Consider now the subspace S_a of H_E associated to the operator a. We chose a basis $\mathbf{A}$ for S_a *in* $\mathcal{E}$, and extend A to a basis $\mathbb{B}$ of H_E *in* $\mathcal{E}$. By the hypothesis, the set A (or, respectively $\mathbb{B} - A$) is finite.

By Proposition 3, $\mathbb{B}^\sim$ is a basis for H.

Define $A^\sim = \{[(x, x, \ldots, x, \ldots)] : x \in A\}$. Then $A^\sim$ (or, respectively, $B^\sim - A^\sim$) is finite. By the absolutness of "finite", $A^\sim$ is a basis for $S_a^\sim$, the subspace associated with $V(a)$. Then $V(a) \in \mathbf{L}^\sim$.

Conversely, assume $\tilde{a}$ in $\mathbf{L}^\sim$ and $S_a^\sim$ is the subspace associated with $\tilde{a}$. Let $A^\sim$ be a basis for $S_a^\sim$ and $B^\sim$ the extension of $A^\sim$ to a basis for H. By the hypothesis $A^\sim$ (or respectively $B^\sim - A^\sim$) is finite.

Define $A = \{x \in H_E : [(x, x, \ldots, x, \ldots)] \in A^\sim\}$, and

$$B = \{x \in H_E : [(x, x, \ldots, x, \ldots)] \in B^\sim\}.$$

Then A (or respectively $B - A$) is finite and spans the subspace S_a associated with $a = V^{-1}(\tilde{a})$. Therefore $V^{-1}(\tilde{a}) \in \mathbf{L}$. ∎

Note that his proposition fails to be true for more general classes of operators.

PROPOSITION 8. *Assume* $\mathbf{L}$ *and* $\mathbf{L}^\sim$ *are additive logics of cof-operators in* $\mathbb{B}(H_E)$ *and* $\mathbb{B}(H)$ *respectively, and* $\mathbf{P}$ *and* $\mathbf{P}^\sim$ *the sets of pure additive states on* $\mathbf{L}$ *and* $\mathbf{L}^\sim$. *Then, there is an injection* $\theta : \mathbf{P} \to \mathbf{P}^\sim$ *such that, for every* $\boldsymbol{a}$ *in* $\mathbf{L}$

$$p(a) = \theta(p)(V(a)).$$

PROOF: (1) Define $\theta(p) = p \circ V^{-1}$.

As $L^{\sim} = V(L)$ by Proposition &, and V being a monomorphism, $p \circ V^{-1}$ is well defined on $L^{\sim}$. Thus, for any $\tilde{a}$ in $L^{\sim}$ there is an a in L, such that $\tilde{a} = V(a)$. Therefore,

$$\theta(p)(\tilde{a}) = p \circ V^{-1}(V(a)) = p(a).$$

Moreover, $p \circ V^{-1} = q \circ V^{-1}$ entails $p = q$.

(2) $\theta(p)$ is a pure additive state on $L^{\sim}$.

As we always assume that $\mathbf{H}_{\mathbf{E}}$ is separable in &, we can apply Gleason's Theorem [5] inside the model. Them there is a projection operation having one-dimensional rande d, such that, *inside* & :

$$p(a) = \mathbf{TR}_{\mathbf{E}}(d \cdot a)$$

where $\mathbf{TR}_{\mathbf{E}}$ is a trace *inside* the model.

Given the assumptions on L and P, we know that the trace actually "collapses" for almost every term. Then we have just a *finite* sum eigenvalues. By absolutness of " = " and the finite sum, we conclude:

$$p(a) = \mathbf{TR}_{\mathbf{E}}(d \cdot a) \quad \textit{outside } \&. \quad (+)$$

By Proposition 5,

$$\mathbf{TR}_{\mathbf{E}}(d \cdot a) = \mathrm{TR}(V(d) \cdot V(a)). \quad (++)$$

From (+) and (++) we get, outside the model

$$p(a) = \theta(p)(V(a)) = \mathrm{TR}(V(d) \cdot V(a)).$$

But, d was a projection operator with one-dimensional range on $\mathbf{H}_{\mathbf{E}}$ inside the model, and a was a finite-cofinite projection operator on $\mathbf{H}_{\mathbf{E}}$ inside the model (we do not require that they still be outside the model). Then, by applying Proposition 3 (actually, d is a special case of the density operator), $V(d)$ and $V(a)$ have still the same properties on $\mathbf{H}$. Then we define $\tilde{a} = V(a)$ and $\tilde{d} = V(d)$, and $\theta(p)(\tilde{a}) = \mathrm{TR}(\tilde{d} \cdot \tilde{a})$.

By Gleason's Theorem, we can conclude that $\theta(p)$ is a pure state on $L^{\sim}$, since it is realized by a projection operator with one dimensional range. ■

PROPOSITION 9. *Let* L *and* $L^{\sim}$ *be like before. If* r *is a superposition of* p *and* q, *on* L (p, q *and* r *being pure states of* $\mathbf{P}$) *then* $\theta(r)$ *is a superposition of* $\theta(p)$ *and* $\theta(q)$ *on* $\mathbf{L}^{\sim}$.

PROOF: Assume $\theta(p)(V(a)) = 0$ and $\theta(q)(V(a)) = 0$. Then, by Proposition 8, $p(a) = 0$ and $q(a) = 0$. Therefore, by hypothesis, $r(a) = 0$, which means that $\theta(r)(V(a)) = 0$. Thus, $\theta(r)$ is a superposition of $\theta(p)$ and $\theta(q)$. ■

4. THE SUPERPOSITION OF STATES.

If the logics at issue are atomic orthocomplemented lattices, it is quite easy to state a "natural" principle of superposition of states (see, for example, Jauch [4], p. 106). In fact, the following assertion expresses accurately such a principle:

(S_1) *If* a_1 *and* a_2 *are two distinct atomic propositions, then there is a new atomic proposition* a_3 (*i.e.*, $a_3 \neq a_2$ *and* $a_3 \neq a_1$) *such that:*

$$a_1 \vee a_2 = a_1 \vee a_3 = a_2 \vee a_3.$$

In our more general case, S_1 does not necessarely make sense. We can however, use for instance Pulmannovā's definition:

(S_2) *For* p *and* q *two distinct pure states, there is a new pure state* r (*i. e.*, $r \neq p$ *and* $r \neq q$), *such that* r *is a superposition of* q *and* q .

It is easily shown that, for a logic which is an atomic orthocomplemented lattice, (S_1) entails (S_2). For (S_1) implies that the given quantum logic is irreducible, and this, with the usual axioms of "classical" quantum logic (i. e., Jauch-Piron's axioms, for instance), implies that (S_2) holds.

We turn now to the notations os Section 3.

PROPOSITION 10. *If* (S_2) *holds in* $\mathbf{L}$, *then it holds in* $\mathbf{L}^{\sim}$.

PROOF: By Proposition 9, if r is a superposition of p and q, $\theta(r)$ is a superposition of $\theta(p)$ and $\theta(q)$. By Proposition 8, θ is injective. ■

REFERENCES.

[1] P. Benioff, *Models of Zermelo-Fraenkel set theory as carries for the mathematics of physics, I*, J. Math. Physics, 17 (1976), 619-628.

[2] P. Benioff, *Models of Zermelo-Fraenkel set theory as carries for the mathematics of physics, II*, J. Math. Physics, 17 (1976), 629-640.

[3] P. Deliyannis, *Superposition of states and the structure of quantum logics*, J. Math. Physics, 17 (1976), 248-254.

[4] J. M. Jauch, Foundations of Quantum Mechanics, Addison-Wesley Readings, Mass., 1968.

[5] A. M. Gleason, *Measures on a closed subspace of a Hilbert space*, J. of Rat. Mech. and Analysis, 6 (1967), 885-912.

[6] S. P. Gudder, *A superposition principle in physics*, J. Math. Physics, 11 (1970), 1037-1040.

[7] S. P. Gudder, *Uniqueness ans existence properties of bounded observables*, Pacific Journal of Mathematics, 19 (1966), 81-93.

[8] C. Piron, *Axiomatique quantique*, Helv. Phys. Acta, 37 (1964), 439-468.

[9] S. Pulmannová, *A superposition principle in quantum logics*, Commun. Math. Physics, 49 (1976), 47-51.

Centro de Lógica, Epistemologia
e História da Ciência
Universidade Estadual de Campinas
13.100 Campinas, SP., Brazil.

I am indebted to Professors Linda Wessels, Rolando Chuaqui and Charles Pinter for helpful comments and criticisms.

CONSTRUCTIBILITY IN THE IMPREDICATIVE THEORY OF CLASSES.

by *M. V. MARSHALL* and *R. CHUAQUI*.

ABSTRACT. In *Internal and forcing models for the impredicative theory of classes* (to appear in Dissertationes Mathematicae) one of the authors defined the collection L of constructible classes for the impredicative theory of classes (which we call MT) and sketched a proof that L satisfies MT and that strong versions of the axiom of choice are valid in L. We shall study the model L further and prove that the generalized continuum hypothesis is valid in L, that all sets are constructed from ordinals, and that L is the minimal model which contains all ordinals.

1. INTRODUCTION.

In [1], one of the authors defined the collection L of constructible classes for the impredicative theory of classes (which we call MT) and sketched a proof that L satisfies MT and that strong versions of the axiom of choice are valid in L. We shall study the model L further and prove that the generalized continuum hypothesis is valid in L, that all sets are constructed from ordinals, and that L is the minimal model which contains all ordinals.

These facts are well-known for L, the class of constructible sets, which is a model of Zermelo-Fraenkel. The original proof for most of these facts appear in [4]. We shall base our proof, however, in the proof that appears in [3], which is due in essence to Lévy [5].

We shall use the same axioms for MT as those in [1] (see also [2], Section 1). Our notation, unless explicitly noted will also be the same as in

[1] (or [2]).

The proper structures for MT are collections of classes represented by a class $\mathfrak{A} = [A, [F(x): x \in A]]$ (a superclass) where A is the class of indices and F an operation (recall that $[F(x): x \in A] = \cup\{F(x) \times \{x\}: x \in A\}$, Definition 2.5 of [1]). y is an "element" of $\mathfrak{A}$, in symbols $\mathfrak{A}(y)$, if $\exists x(x \in A \wedge y = F(x))$. In [1], Section IV, satisfaction is defined for these type of structures. Section 2 of this paper contains a generalization of the theorems of Löwenheim-Skolem and Mostowski's Contraction Lemma. The proofs follow [6]. These two theorems are used later.

Section 3 introduces Δ_1^T-formulas for a theory T with classes. A slightly simplified definition of L is given and it is proved for this definition that it is Σ_1^{MTL} ($MTL = MT + V \simeq L$, i. e. MT with the axiom of constructibility). This proof is given with all its details. From it, it is easy to obtain the absoluteness of L for all models of MT.

These results are used in Section 4, to obtain the main results mentioned previously.

2. STRUCTURES FOR MT.

We shall use the notation of [1], in particular, DG domain of G, DG^{-1}: range of G; $G`a$: value of G at a; ${}^A B$ the class of function f with $Df = A$ and $Df^{-1} \subseteq B$; G^*A: the image of A under G; $\bar{\bar{A}}$: the cardinal of A; TC(X), the transitive closure of X.

The proper strucutures for MT are of the form $[A, [F(x): x \in A]]$, where $[F(x): x \in A] = \cup\{F(x) \times \{x\}: x \in A\}$ and $[A, B]$ is the pair of classes A,B. In [1], satisfaction is defined for these structures. We shall sketch the definitions leading to satisfaction. For further explanations see [1].

Let $\mathfrak{A} = [A, F(x): x \in A]]$. Then:

1. $C_{\mathfrak{A}} = \cup\{{}^c A . c \subseteq \omega \wedge c \text{ finite}\}$.

2. G is an assigment for the formula φ in $\mathfrak{A}$ iff $G \subseteq V \times V$, DG is a finite subset of ω and $(\forall i \in Fv\,\varphi)(\exists x \in A)\ (G^*\{i\} = F(x))$. ($Fv\,\varphi$ is the set of free variables of φ).

3. $H_{\mathfrak{A}}$ is the operation defined for each $x \in C_{\mathfrak{A}}$ by:

$$H_{\mathfrak{A}}(x) = [F(x \cdot i) : i \in x]$$

(i. e. $H_{\mathfrak{A}}(x)$ is the assignment coded by x).

4. For each formula φ the class of assignments for φ in $\mathfrak{A}$, $K_{\mathfrak{A}}(\ulcorner\varphi\urcorner)$, is defined by:

(i) $\ulcorner\varphi\urcorner = \ulcorner v_i \in v_j \urcorner \to K_{\mathfrak{A}}(\ulcorner\varphi\urcorner) = \{g : g \in {}^{\{i,j\}}A \wedge$
$H_{\mathfrak{A}}(g)(i) \in H_{\mathfrak{A}}(g)(j)\}$,

(ii) $\ulcorner\varphi\urcorner = \ulcorner v_i = v_j\urcorner \to H_{\mathfrak{A}}(\ulcorner\varphi\urcorner) = \{g : g \in {}^{\{i,j\}}A \wedge$
$H_{\mathfrak{A}}(g)(i) = H_{\mathfrak{A}}(g)(j)\}$,

(iii) $\ulcorner\varphi\urcorner = \ulcorner\neg\psi\urcorner \to K_{\mathfrak{A}}(\ulcorner\varphi\urcorner) = {}^{Fv\varphi}A \sim K_{\mathfrak{A}}(\ulcorner\psi\urcorner)$,

(iv) $\ulcorner\varphi\urcorner = \ulcorner\psi_1 \vee \psi_2\urcorner \to K_{\mathfrak{A}}(\ulcorner\varphi\urcorner) = \{g : g \in {}^{Fv\varphi}A \wedge$
$(g \upharpoonright Fv\psi_1 \in K_{\mathfrak{A}}(\ulcorner\psi_1\urcorner) \wedge g \upharpoonright Fv\psi_2 \in K_{\mathfrak{A}}(\ulcorner\psi_2\urcorner))\}$,

(v) $\ulcorner\varphi\urcorner = \ulcorner\exists v_i \psi\urcorner \wedge i \notin Fv\psi \to K_{\mathfrak{A}}(\ulcorner\psi\urcorner) = K_{\mathfrak{A}}(\ulcorner\varphi\urcorner)$,

(vi) $\ulcorner\varphi\urcorner = \ulcorner\exists v_i \psi\urcorner \wedge i \in Fv\psi \to K_{\mathfrak{A}}(\ulcorner\psi\urcorner) =$
$\{g : g \in {}^{Fv\varphi}A \wedge (\exists x \in A)(g \cup \{\langle x, i\rangle\} \in K_{\mathfrak{A}}(\ulcorner\psi\urcorner))\}$.

5. If G is an assignment for φ in $\mathfrak{A}$, we define:

$\mathfrak{A} \models \varphi[G]$ iff $(\exists g \in K_{\mathfrak{A}}(\ulcorner\varphi\urcorner))(\forall i \in Fv\varphi)(G^*\{i\} = H_{\mathfrak{A}}(g)(i))$.

It is easy to see that,

$\mathfrak{A} \models \varphi[H_{\mathfrak{A}}(x)]$ iff $x \in K_{\mathfrak{A}}(\ulcorner\varphi\urcorner)$.

Similarly, we have,

$\mathfrak{A} \models \varphi$ iff for every assignment G for φ in $\mathfrak{A}$, $\mathfrak{A} \models \varphi[G]$.

6. We define the cardinality of a structure $\mathfrak{A} = [A, [F(x) : x \in A]]$ as follows:

Let $R = \{\langle x, y\rangle : x, y \in A, F(x) = F(y)\}$, then the cardinality of $\mathfrak{A}$, is

$$\overline{\overline{\mathfrak{A}}} = \overline{\overline{A/R}} .$$

It is clear that $\overline{\overline{\mathfrak{A}}} \leqslant \overline{\overline{A}}$.

7. Let $\mathfrak{A} = [A, [F(x) : x \in A]]$ and $\mathfrak{B} = [B, [G(x) : x \in B]]$.

We say that $\mathfrak{A}$ is a substructure of $\mathfrak{B}$, in symbols $\mathfrak{A} \subseteq \mathfrak{B}$ iff $\forall X(\mathfrak{A}(X) \to \mathfrak{B}(X))$. $\mathfrak{A} \simeq \mathfrak{B}$ iff $\mathfrak{A} \subseteq \mathfrak{B}$ and $\mathfrak{B} \subseteq \mathfrak{A}$.

$\mathfrak{A}$ is an elementary submodel of $\mathfrak{B}$, in symbols $\mathfrak{A} \prec \mathfrak{B}$ iff for every formula φ and every assignment G for φ in $\mathfrak{A}$ we have

$$\mathfrak{A} \models \varphi[G] \quad \text{iff} \quad \mathfrak{B} \models \varphi[G].$$

H is an isomorphism of $\mathfrak{A}$ onto $\mathfrak{B}$, in symbols $\mathfrak{A} \cong_H \mathfrak{B}$ iff H is a one-to-one operation of $\mathfrak{A}$ onto $\mathfrak{B}$ such that,

$$\forall X \forall Y(\mathfrak{A}(X) \wedge \mathfrak{A}(Y) \to (X \in Y \leftrightarrow H(X) \in H(Y))).$$

8. A structure $\mathfrak{A}$ is transitive iff $\forall X(\mathfrak{A}(X) \to \forall y(y \in X \to \mathfrak{A}(y)))$.

We state now our version of the Löwenheim-Skolem Theorem. In the proof we use the Global Axiom of Choice (Axiom E of [4]).

THEOREM 2.1. (MTE) *Let* $\mathfrak{A} = [a, [F(x) : x \in a]]$ *with* $\overline{\overline{\mathfrak{A}}} \leqslant \kappa$ *and* κ *a cardinal,* $\kappa \geqslant \omega$. *Let* $\mathfrak{B} = [B, [G(x) : x \in B]]$ *and* $\mathfrak{A} \subseteq \mathfrak{B}$. *Then there is a structure* $\mathfrak{C} = [c, [H(x) : x \in c]]$ *such that* $\mathfrak{A} \subseteq \mathfrak{C}$, $\overline{\overline{c}} \leqslant \kappa$ *and* $\mathfrak{C} \prec \mathfrak{B}$.

PROOF: Let $a' \subseteq B$ be such that

$$\mathfrak{A}' = [a', [G(x) : x \in a']] \simeq \mathfrak{A},$$

(i. e. $\forall X(\mathfrak{A}'(X) \leftrightarrow \mathfrak{A}(X))$) and $\overline{\overline{a'}} \leqslant \overline{\overline{\mathfrak{A}}}$.

Let $u \in a'$. Define for each $n \in \omega$

(i) If $\varphi_n = \exists v_j \psi$ and $x \in C_{\mathfrak{B}}$ with $Dx = \mathrm{Fv}\,\varphi_n$,

let $g_n \cdot x$ = the first $z \in B$ such that $\mathfrak{B} \models \psi[H_{\mathfrak{B}}(x \cup \{\langle z, j\rangle\})]$, if there is one such ;

= u otherwise.

If $\varphi_n \neq \exists v_j \psi$, then $g_n \cdot x = u$ for every $x \in C_{\mathfrak{B}}$ with $Dx = \mathrm{Fv}\,\varphi_n$.

(ii) Let $a_0 = a' \subseteq B$

$$a_{n+1} = a_n \cup \{ g_m \cdot x : x \in C_{\mathcal{L}} \wedge Dx = \mathrm{Fv}\varphi_m \wedge Dx^{-1} \subseteq a_n \wedge m \in \omega \}.$$

Let $c = \cup \{ a_n : n \in \omega \}$.

It is easy to see using a similar proof as in [6], that

$$\mathcal{C} = [\, c, [\, G(x) : x \in c \,]\,]$$

satisfies the conclusion of the theorem. ■

THEOREM 2.2. (Mostowski's Contraction Lemma)
Let $\mathfrak{A} = [\, A, [\, F(x) : x \in A \,]\,]$ *be a structure which satisfies extensionality. Then there is a transitive structure* $\mathcal{B} = [\, A, [\, G(x) : x \in A \,]\,]$ *and an isomorphism* Π *which satisfies*:

(i) $\mathfrak{A} \cong_{\Pi} \mathcal{B}$,

(ii) $\mathfrak{A}(X) \wedge \forall y (y \in \mathrm{TC}(X) \rightarrow \mathfrak{A}(y)) \rightarrow \Pi(X) = X$.

PROOF: We define the operation Π by recursion on $\in$:

$$\Pi(X) = \{ \Pi(F(y)) : F(y) \in X \}.$$

The operation G is defined by:

$$G(x) = \Pi(F(x)) \quad \text{for each } x \in A.$$

Then Π and $\mathcal{B} = [\, A, [\, G(x) : x \in A \,]\,]$ satisfies the conclusion of the theorem. The proof is similar to that appearing in [6].

3. Constructible classes.

We define, as in [1], simultaneously by recursion on well-orderings an operation TW' that defines well-order types and a structure L_S for each well-ordering S. Our definition is a simplified version of 11.2 of [1]. The numbers in parentheses refer to the numbers of the corresponding definition, in [1].

(i) R_x is the initial segment of R determined by x.

(ii) (7.8) Let C be a collection of well orderings such that

$$C(R) \wedge C(S) \wedge R \cong S \rightarrow R = S.$$

Define

$$\inf \mathbf{C} = \text{the least } R \text{ such that } C(R).$$

(iii) (Simplified version of 6.5.)

Let $[F(x): x \in A]$ be a superclass of well orderings such that

$$\forall x \, \forall y \, (x, y \in A \wedge F(x) \cong F(y) \rightarrow F(x) = F(y)).$$

For $x, y \in A$, let xEy iff $F(x) = F(y)$.

Let $B = \{ds(E^*\{x\}): x \in A\}$ where $ds\,X$ is the set of elements of X of minimal rank. Let $G(x) = F(y)$ for $y \in x$ and $x \in B$. Let xRz iff $G(x) \leqslant G(z)$ for $x, z \in B$. Denote by

$$S = \sum_R [G(x): x \in B]$$

the ordered sum according to R of the $G(x)$'s. Define

$$\sup[F(x): x \in A] = \inf\{T: (T = S \vee \exists u(u \in DS \wedge T = S_u)) \wedge \forall x(x \in A \rightarrow F(x) \leqslant T)\}$$

2.(11.1) Let $\mathfrak{A} = [A, [F(x): x \in A]]$ be a structure. Then

(a) $\overline{C}_{\mathfrak{A}} = [{}^{Fv(\exists v_0 \varphi_m)}A: m \in \omega]$,

(b) $D_{\mathfrak{A}} = [\{y: \mathfrak{A}(y) \wedge \mathfrak{A} \models \varphi_m[y, H_{\mathfrak{A}}(g)]\}: \langle g, m \rangle \in \overline{C}_{\mathfrak{A}}]$,

(c) $D(\mathfrak{A}) = [\overline{C}_{\mathfrak{A}}, [D_{\mathfrak{A}}(y): y \in \overline{C}_{\mathfrak{A}}]]$.

2.(11.2) We define, for each well-ordering R, a well-ordering $TW'(R)$, a class B_R, an operation G_R, and a structure $\mathcal{L}_R = [B_R, [G_R(x): x \in B_R]]$.

(i) If $R \in V$, then $TW'(R)$ is the ordering of the ordinal corresponding to R.

(ii) If $R \notin V$, $R \cong S + 1$ for some S, and $TW'(S) \cong S$, then $TW'(R) = TW'(S) + 1$.

(iii) $B_0 = \{0\}$, $G_0(0) = 0$.

(iv) If $R \cong S + 1$ for some S, then $B_R = \overline{C}_{\mathcal{L}_S}$, $G_R(y) = D_{\mathcal{L}_S}(y)$ for $y \in \overline{C}_{\mathcal{L}_S}$.

(v) If $R \not\cong S + 1$ for every well-ordering S, then $B_R = [B_{R_x} : x \in DR]$

and $G_R(\langle y,x\rangle) = G_{R_x}(y)$ for $\langle y,x\rangle \in B_R$.

(vi) Let $R \notin V$, $R \not\cong S+1$ for every well-ordering S; suppose that TW' is defined for all $S<R$ and that $TW'(S) \cong S$ for every $S<R$; suppose futhermore that L_S is defined for every $S \leqslant R$.

1st STAGE. Define

1) $A(T,m,g) = \{y: L_R \models \varphi_m[y, H_{L_T}(g))]\}$, for $T<R$, $g \in {}^{Fv(\exists v_0 \varphi_m)}B_T$.

2) $\mu(x) = \inf\{T: T = TW'(T) \wedge L_T(x)\}$, for each x such that $L_R(x)$.

3) $B(T,m,g) = \sup[\mu(y): y \in A(T,m,g)]$.

4) $C(T) = \sup[B(T,m,g): m \in \omega \wedge g \in {}^{Fv(\exists v_0 \varphi_m)}B_T]$.

5) If there is a T such that $C(T) \cong R$, define $TW'(R) = C(T_0)$ where $T_0 = \inf\{T: T = TW'(T) \wedge C(T) \cong R\}$; otherwise pass to the second stage.

2 nd STAGE. Define

1) $J(T,m,g) = \inf\{T': T' = TW'(T') \wedge T' \geqslant T \wedge T' < R \wedge (\exists X(L_R(X) \wedge L_R \models \varphi_m[X, H_{L_T}(g)] \rightarrow \exists X(L_{T'}(X) \wedge L_R \models \varphi_m[X, H_{L_T}(g)]))\}$, for $T<R$, and $g \in {}^{Fv(\exists v_0 \varphi_m)}B_T$

2) $L(T) = \sup[J(T,m,g): m \in \omega \wedge g \in {}^{Fv(\exists v_0 \varphi_m)}B_T]$ for $T<R$ and $L(T) = T$ for $T \cong R$.

3) $M(T,0) = T$

$M(T,m+1) = L(M(T,m))$, for $T \leqslant R$ and $m \leqslant \omega$.

4) $N(T) = \sup[M(T,m): m \in \omega]$.

We have that, $N(T) \leqslant R$ for every $T \leqslant R$.

It can be proved that $L_{N(T)} \prec L_R$ for $T \leqslant R$.

If there is a $T < R$ such that $\mathrm{N}(T) \cong R$, then $\mathrm{TW}'(R) = \mathrm{N}(T_0)$ where $T_0 = \inf\{T\colon T = \mathrm{TW}'(T) \wedge \mathrm{N}(T) \cong R\}$. Otherwise, $\mathrm{TW}'(R) = 0$

(vi) If for some $S < R$ we have $\mathrm{TW}'(S) \not\cong S$, then $\mathrm{TW}'(R) = 0$, $\mathrm{B}_R = 0 = \mathrm{G}_R$.

(viii) $\mathcal{T}(R)$ iff $R = \mathrm{TW}'(R) \wedge (\forall S \leqslant R)(\mathrm{IN}\restriction \mathrm{OR} \leqslant S \rightarrow L_S \not\models \mathrm{MT})$, $L = \tilde{\bigcup}\{L_R\colon \mathcal{T}(R)\}$.

Let V be the collection of all classes. Then, the axiom of constructibility is expressed by

$$V \simeq L$$

or

$$\forall X \exists R(\mathcal{T}(R) \wedge L_R(X)).$$

The simplification obtained in (iv) with respect to 11.2 of [1], does not modify the proof of 11.3 (i) of [1], namely MT^L. Thus, L is a model of MT.

In order to prove 11.3 (ii), 11.4 and 11.5 of [1], i. e. that L satisfies the axiom of constructibility and strong versions of the axiom of choice, we have to prove that all types are in L, that is $\mathcal{T} \subseteq L$. In what follows, we shall prove that all definitions given above are $\Delta_1^{\mathrm{MT}L}$. By a similar, but simpler argument using Section 9 of [9], it is possible to prove that they are absolute with respect to models of MT. Thus, it is possible to prove by induction on well-orderings that $\mathcal{T} \subseteq L$. The rest of the proofs of 11.3 (ii), 11.4 and 11.5, then goes through.[1]

We now begin to study complexity of formulas, with respect to theories with classes. This study is based on that which appears in [3].

DEFINITION 3.1.

(i) A formula φ containing only primitive symbols is *bounded* iff all its quantifiers appear in the forms $\forall X(X \in Y \rightarrow \psi)$ or $\exists X(X \in Y \wedge \psi)$. As is usual, we shall abbreviate these formulas by $(\forall X \in Y)\psi$ and $(\exists X \in Y)\psi$, respectively.

(ii) Let T be a theory; we say that a formula φ is Δ_0^{T} iff $\mathrm{T} \vdash \varphi \leftrightarrow \psi$, where ψ is a bounded formula.

(iii) φ is Σ_1^T iff $T \vdash \varphi \leftrightarrow \exists X \psi$ where ψ is Δ_0^T ; φ is Π_1^T iff $T \vdash \varphi \leftrightarrow \forall X \psi$ where ψ is Δ_0^T; φ is Δ_1^T iff φ is Σ_1^T and Π_1^T.

(iv) A term τ is Δ_0^T, Σ_1^T, Π_1^T, or Δ_1^T iff the formula $\tau = X$ is of the corresponding kind.

(v) A unary operation F is Δ_0^T, Σ_1^T, Π_1^T, or Δ_1^T iff the formula $F(X) = Y$ is such.

The rest of this section deals with the proof that $L_R(X)$ and $TW'(R) = S$ are $\Delta_1^{MT + V \simeq L}$.

We shall show next some basic properties of Δ_0, Δ_1, Σ_1, and Π_1 formulas.

THEOREM 3.2.

(i) *φ is Π_1^T iff $\neg\varphi$ is Σ_1^T.*

(ii) *If φ and ψ are Σ_1^T (or Π_1^T), then $\varphi \vee \psi$ and $\varphi \wedge \psi$ are Σ_1^T (or Π_1^T).*

(iii) *If φ and ψ are Δ_0^T (or Δ_1^T), then $\neg\varphi$, $\varphi \vee \psi$, $\varphi \wedge \psi$, $\varphi \rightarrow \psi$, and $\varphi \leftrightarrow \psi$ are Δ_0^T (or Δ_1^T).*

THEOREM 3.3. *The following formulas are Δ_0^{MT}:*

(1) $X = 0 \leftrightarrow (\forall z \in X)(z \neq z)$.

(2) $X \subseteq Y \leftrightarrow (\forall z \in X)(z \in Y)$.

(3) $X = \{Y\} \leftrightarrow \forall z \in X\ (z = Y) \wedge Y \in X$.

(4) $X = \{Y, Z\} \leftrightarrow Y \in X \wedge Z \in X \wedge \forall u \in X\ (u = Y \vee u = Z)$.

(5) $X = Y \cup Z \leftrightarrow \forall u \in X (u \in Y \vee u \in Z) \wedge \forall u \in Y (u \in X) \wedge \forall u \in Z (u \in X)$.

(6) $X = Y \cap Z \leftrightarrow \forall u \in X (u \in Y \wedge u \in Z) \wedge \forall u \in Y (u \in Z \rightarrow u \in X)$.

(7) $X = \cup Y \leftrightarrow \forall z \in X\ \exists u \in Y (z \in u) \wedge \forall z \in Y\ \forall u \in z (u \in X)$

(8) $X = Y \sim Z \leftrightarrow \forall u \in X(u \in Y \wedge u \notin Z) \wedge \forall u \in Y(u \notin Z \to u \in X)$.

(9) $X = \{Y,0\} \leftrightarrow \exists u \in X(u = 0 \wedge X = \{y,u\})$.

(10) $X = 1 \leftrightarrow \exists u \in X(u = 0) \wedge \forall u \in X(u = 0)$.

(11) $x = \{y,1\} \leftrightarrow \exists u \in x(u = 1 \wedge x = \{y,u\})$.

(12) $Y = \langle x,0 \rangle \leftrightarrow \exists u \in Y\, \exists v \in Y(u = \{x\} \wedge v = \{x,0\} \wedge Y = \{u,v\})$.

(13) $Y = \langle x,1 \rangle \leftrightarrow \exists u \in Y\, \exists v \in Y(u = \{x\} \wedge v = \{x,1\} \wedge Y = \{u,v\})$.

(14) $\langle x,0 \rangle \in Y \leftrightarrow \exists u \in Y(u = \langle x,0 \rangle)$.

(15) $\langle x,1 \rangle \in Y \leftrightarrow \exists u \in Y(u = \langle x,1 \rangle)$

(16) $x = \langle y,z \rangle \leftrightarrow \exists u \in x\, \exists v \in x(x = \{u,v\} \wedge u \in \{y\} \wedge v = \{y,z\})$.

(17) $\Pi_1(x) = y \leftrightarrow \exists u \in x\, \exists z \in u(x = \langle y,z \rangle) \vee [\neg \exists u \in x\, \exists z \in u\, \exists v \in x$
$\exists w \in v(x = \langle w,z \rangle) \wedge y = 0]$.

(18) $\Pi_2(x) = y \leftrightarrow$ analogous for the second term of an ordered pair.

(19) $X = Y \times Z \leftrightarrow \forall u \in X\, \exists v \in Y\, \exists w \in Z(u = \langle v,w \rangle) \wedge$
$\forall v \in Y\, \forall w \in Z\, \exists u \in X(u = \langle v,w \rangle)$.

(20) $D(X) = Y \leftrightarrow \forall u \in Y\, \exists v \in X(u = \Pi_2(v)) \wedge \forall v \in X\, \exists u \in Y(u = \Pi_2(v))$.

(21) $D(X^{-1}) = Y$ analogous with Π_1 substituting Π_2.

(22) $\mathrm{REL}(X) \leftrightarrow (\forall a \in X)(\exists u \in a)(\exists v \in a)(\exists u' \in u)(\exists v' \in v)$
$(a = \langle u',v' \rangle)$.

(23) $\mathrm{FUN}(X) \leftrightarrow \mathrm{REL}(X) \wedge (\forall a \in X)(\forall b \in X)(\forall u \in a)(\forall v \in a)(\forall y \in u)$
$(\forall z \in v)(\forall u' \in b)(\forall v' \in b)(y' \in u')(z' \in v')$
$((a = \langle y,z \rangle \wedge b = \langle y',z' \rangle \wedge z = z') \to y = y')$.

(24) $1\mathrm{FUN}(X) \leftrightarrow \mathrm{FUN}(X) \wedge (\forall a \in X)(\forall u \in a)(\forall v \in a)(\forall y \in u)(\forall z \in v)$
$(\forall b \in X)(\forall u' \in b)(\forall v' \in b)(\forall y' \in u)(\forall z' \in v')$
$((a = \langle y,z \rangle \wedge b = \langle y',z' \rangle \wedge z = z') \to y = y')$.

(25) $X \cdot y = z \leftrightarrow \mathrm{FUN}(X) \wedge \exists a \in X(a = \langle z,y \rangle)$.

(26) $\mathrm{TRANS}(X) \leftrightarrow \forall y \in X(y \subseteq X)$.

(27) $\mathrm{OR}(x) \leftrightarrow \mathrm{TRANS}(x) \wedge \forall y \in x(\mathrm{TRANS}(y))$.

(28) $S(x) = y \leftrightarrow \forall z \in x (z \in y) \wedge \forall z \in y (z \in x \vee z = x)$.

(29) $\omega = x \leftrightarrow OR(x) \wedge x \neq 0 \wedge \neg \exists z \in x (OR(z) \wedge x = S(z)) \wedge \forall z \in x \exists y \in z (S(y) = z)$.

(30) $X = G \upharpoonright y \leftrightarrow X \subseteq G \wedge D(X) = y$.

(31) $X = R^* Y \leftrightarrow (\forall u \in X)(\exists v \in Y)(\langle u, v \rangle \in R)$.

THEOREM 3.4.

(i) *If* φ *is* Σ_1^{MT}, *then* $\exists X \varphi$ *is* Σ_1^{MT}.

(ii) *If* φ *is* Π_1^{MT}, *then* $\forall X \varphi$ *is* Π_1^{MT}.

PROOF:

(i) If φ is Σ_1^{MT}, then $MT \vdash \exists X \varphi \leftrightarrow \exists X \exists Y \psi(X, Y)$ where ψ is Δ_0^{MT}. Therefore, $MT \vdash \exists X \varphi \leftrightarrow \exists Z \psi(Z^*\{0\}, Z^*\{1\})$. Thus we have to show that $\psi(Z^*\{0\}, Z^*\{1\})$ is Δ_0^{MT}. In order to do this, we prove that the following formulas are Δ_0^{MT}.

1. $(\exists x \in Z^*\{0\})\theta \leftrightarrow (\exists u \in Z)(\exists x \in u)(\langle x, 0 \rangle \in Z \wedge \theta)$.

2. $(\forall x \in Z^*\{0\})\theta \leftrightarrow \neg(\exists x \in Z^*\{0\})\neg\theta$.
   similarly for $(\exists x \in Z^*\{1\})\theta$ and $(\forall x \in Z^*\{1\})\theta$.

3. $x \in Z^*\{0\} \leftrightarrow \langle x, 0 \rangle \in Z$.

4. $X = Z^*\{0\} \leftrightarrow (\forall u \in X)(u \in Z^*\{0\}) \wedge (\forall u \in Z^*\{0\})(u \in X)$.

5. $Z^*\{0\} \in X \leftrightarrow (\exists u \in X)(u = Z^*\{0\})$.
   similarly with $Z^*\{1\}$ instead of $Z^*\{0\}$.

6. $Z^*\{0\} = Z^*\{1\} \leftrightarrow (\forall x \in Z^*\{1\})(x \in Z^*\{0\})$
   $(\forall x \in Z^*\{0\})(x \in Z^*\{1\})$.

7. $Z^*\{0\} \in Z^*\{1\} \leftrightarrow (\exists x \in Z^*\{1\})(x = Z^*\{0\})$.
   similarly for $Z^*\{1\} \in Z^*\{0\}$.

(ii) If φ is Π_1^{MT}, then $\neg\varphi$ is Σ_1^{MT}. Hence, $\exists X \neg\varphi$ is Σ_1^{MT}, and, thus $\neg \exists X \neg \varphi$ is Π_1^{MT}. ■

THEOREM 3.5.

(i) *If* φ *is* Σ_1^{MTL}, *then* $(\exists x \in Y)\varphi$ *and* $(\forall x \in Y)\varphi$ *are* Σ_1^{MTL}.

(ii) *If* φ *is* Π_1^{MTL}, *then* $(\exists x \in Y)\varphi$ *and* $(\forall x \in Y)\varphi$ *are* Π_1^{MTL}.

(iii) *If* φ *is* Δ_1^{MTL}, *then* $(\exists x \in Y)\varphi$ *and* $(\forall x \in Y)\varphi$ *are* Δ_1^{MTL}.

PROOF:
(i) If φ is Σ_1^{MTL}, then by 3.4, $(\exists x \in Y)\varphi$ is also Σ_1^{MTL}. If φ is Σ_1^{MTL}, then

$$MTL \vdash (\forall x \in Y)\varphi \leftrightarrow (\forall x \in Y)\exists Z \psi \text{ where } \psi \text{ is } \Delta_0^{MTL}.$$

By 11.4 of [1], in MTL it is possible to define an operation Q such that, $\forall X \exists T (\mathcal{T}(T) \wedge Q(T) = X)$.

For $x \in y$ define an operation F by:

$$F(x) = Q(T_0) \text{ where } T_0 \text{ is the first type } T \text{ such that } Q(T) \text{ satisfies } \psi \text{, if there is one such;}$$

$$F(x) = 0 \text{, otherwise.}$$

Let $Z = [F(x): x \in Y]$. Then in MTL:

$$(\forall x \in Y)\exists Z \psi \leftrightarrow \exists Z(\forall x \in Y)\psi(x, Z^*\{x\}).$$

As in 3.4, we can show that $\psi(x, Z^*\{x\})$ is Δ_0^{MT}.

(ii) is obtained from (i), and (iii) from (i) and (ii). ■

In this proof we used only MT + WO_Q, which is a weaker theory than MTL (WO_Q is a strong version of the axiom of choice well ordering all classes). All the rest of this section could be done in MT + WO_Q instead of MTL.

THEOREM 3.6.

(i) *If* τ *is* Δ_1^{MTL} *and* $\varphi(X)$ *is* Δ_1^{MTL}, *then* $\varphi(\tau)$ *is* Δ_1^{MTL}. *In par-*

ticular, $(\exists x \in \tau)\varphi$ *and* $(\forall x \in \tau)\varphi$ *are* Δ_1^{MTL}; *and if* $\sigma(X)$ *is a* Δ_1^{MTL} *term, then* $\sigma(\tau)$ *is* Δ_1^{MTL}.

(ii) *If* τ, σ, *and* φ *are* Δ_1^{MTL}, *then* $\cup\{\tau : x \in \sigma \wedge \varphi\}$ *and* $\{\tau : x \in \sigma \wedge \varphi\}$ *are* Δ_1^{MTL}.

(iii) *If* τ *and* σ *are* Δ_1^{MTL}, *then* $[\tau, \sigma]$ *is* Δ_1^{MTL}.

(iv) *If* τ *and* σ *are* Δ_1^{MTL}, *then* $[\tau : x \in \sigma]$ *is* Δ_1^{MTL}.

PROOF:

(i) $$\varphi(\tau) \leftrightarrow \exists X(X = \tau \wedge \varphi(X))$$
$$\leftrightarrow \forall X(X = \tau \rightarrow \varphi(X)).$$

(ii) $$X = \cup\{\tau : x \in \sigma \wedge \varphi\} \leftrightarrow (\forall y \in X)(\exists x \in \sigma)(y \in \tau \wedge \varphi) \wedge (\forall x \in \sigma)(\forall y \in \tau)(\varphi \rightarrow y \in X)$$
$$\{\tau : x \in \sigma \wedge \varphi\} = \cup\{\{\tau\} : x \in \sigma \wedge \varphi\}.$$

(iii) $$[\tau, \sigma] = \tau \times \{0\} \cup \sigma \times \{1\}.$$

(iv) $$[\tau : x \in \sigma] = \cup\{\tau \times \{x\} : x \in \sigma\}.$$ ∎

THEOREM 3.7.

(i) *If* H *is a* Δ_1^{MTL} *operation and* F *is defined by recursion as*

$$F(R) = H([F(R_x) : x \in DR])$$

for R *well-orderings, then* **F** *is* Δ_1^{MTL}.

(ii) *If* H *is a function,* $H \cdot x = y$ *is* Δ_1^{MTL}, *and* $F \cdot x = H \cdot F^{\star} x$ *for every* x, *then* $F \cdot x = y$ *is* Δ_1^{MTL}.

PROOF: The proof is by induction using 3.6 (iv).

THEOREM 3.8. *If* $\tau(A) = \cup\{{}^{c}A : c \subseteq \omega \wedge c$ *finite*$\}$, *then* τ *is* Δ_1^{MTL}.

PROOF: Define by recursion for each $n \in \omega$,

$$\sigma(A, 0) = \emptyset$$

$$\sigma(A, n+1) = \sigma(A,n) \cup \{g \cup \{\langle c, i\rangle\} : g \in \sigma(A,n) \wedge c \in A \wedge i \in \omega\}.$$

Then, by 3.6 and 3.7, σ is Δ_1^{MTL}. But,

$$\sigma(A) = \bigcup\{\sigma(A,n) : n \in \omega\}.$$

Hence, by 3.6, σ is Δ_1^{MTL}.

THEOREM 3.9. *Let* A *and* F *be* Δ_1^{MTL} *operations and* $\mathfrak{A}(R) = [A(R), [F_R(x) : x \in A(R)]]$. *Then* $\mathfrak{A}(R)$, $C_{\mathfrak{A}(R)}$, $H_{\mathfrak{A}(R)}$ *and "G is an assignment for the formula* φ *in* $\mathfrak{A}(R)$", *are* Δ_1^{MTL}.

PROOF:

(i) $\mathfrak{A}(R)$ is Δ_1^{MTL} by 3.6. $C_{\mathfrak{A}(R)}$ is Δ_1^{MTL} by 3.8.

(ii) $G = H_{\mathfrak{A}(R)}(x) \leftrightarrow x \in C_{\mathfrak{A}(R)} \wedge G = [F_R(x`i) : i \in Dx]$.

(iii) G is an assignment for φ in $\mathfrak{A}(R) \leftrightarrow \mathrm{REL}(G) \wedge (\exists c \in \omega) DG \subseteq c \wedge (\exists x \in C_{\mathfrak{A}(R)} (\forall i \in \mathrm{Fv}\varphi)(G^*\{i\} = H_{\mathfrak{A}(R)}(x`i))$. ■

THEOREM 3.10. *Let* A *and* F *be* Δ_1^{MTL} *operations and* $\mathfrak{A}(R) = [A(R), [F_R(x) : x \in A(R)]]$. *Then* $x \in K_{\mathfrak{A}(R)}(\ulcorner\varphi\urcorner)$ *is* Δ_1^{MTL} *for each formula* φ.

PROOF: The proof is by induction on formulas.

a) If φ is $v_i \in v_j$, then

$$g \in K_{\mathfrak{A}(R)}(\ulcorner\varphi\urcorner) \leftrightarrow \mathrm{FUN}(g) \wedge Dg = \{i,j\} \wedge Dg^{-1} \subseteq A(R) \wedge H_{\mathfrak{A}(R)}(g)(i) \in H_{\mathfrak{A}(R)}(g)(j).$$

b) If φ is $v_i = v_j$, then

$$g \in K_{\mathfrak{A}(R)}(\ulcorner\varphi\urcorner) \leftrightarrow \mathrm{FUN}(g) \wedge Dg = \{i,j\} \wedge Dg^{-1} \subseteq A(R) \wedge H_{\mathfrak{A}(R)}(g)(i) = H_{\mathfrak{A}}(g)(j).$$

c) If φ is $\neg\psi$, then

$$g \in K_{\mathfrak{A}(R)}(\ulcorner\varphi\urcorner) \leftrightarrow \mathrm{FUN}(g) \wedge Dg = \mathrm{Fv}\varphi \wedge Dg^{-1} \subseteq A(R) \wedge$$

$$g \notin K_{\mathfrak{A}(R)}(\ulcorner \psi \urcorner).$$

d) If φ is $\psi_1 \vee \psi_2$, then

$$g \in K_{\mathfrak{A}(R)}(\ulcorner \varphi \urcorner) \leftrightarrow \text{FUN}(g) \wedge Dg = \text{Fv}\,\varphi \wedge Dg^{-1} \subseteq A(R) \wedge g \restriction \text{Fv}\psi_1 \in K_{\mathfrak{A}(R)}(\ulcorner \psi_1 \urcorner) \vee g \restriction \text{Fv}\psi_2 \in K_{\mathfrak{A}(R)}(\ulcorner \psi_2 \urcorner).$$

e) If φ is $\exists v_i \psi$, and $i \notin \text{Fv}\psi$, then

$$g \in K_{\mathfrak{A}(R)}(\ulcorner \varphi \urcorner) \leftrightarrow g \in K_{\mathfrak{A}(R)}(\ulcorner \psi \urcorner).$$

f) If φ is $\exists v_i \psi$, and $i \in \text{Fv}\,\psi$, then

$$g \in K_{\mathfrak{A}(R)}(\ulcorner \varphi \urcorner) \leftrightarrow \text{FUNC}(g) \wedge Dg = \text{Fv}\varphi \wedge Dg^{-1} \subseteq A(R) \wedge (\exists x \in A(R))(g \cup \{\langle x, i \rangle\} \in K_{\mathfrak{A}(R)}(\ulcorner \varphi \urcorner)).$$

We define $\mathfrak{A} \models \varphi[Y, G]$ as $\mathfrak{A} \models \varphi[G']$ where $G'^*\{0\} = Y$ and $G'^*\{i\} = G^*\{i\}$ for $i \in DG \sim \{0\}$.

THEOREM 3.11. *Let* A *and* F *be* Δ_1^{MTL} *operations and* $\mathfrak{A}(R) = [A(R), [F_R(x): x \in A(R)]]$. *Then* $\mathfrak{A}(R) \models \varphi[G]$, $\mathfrak{A}(R) \models \varphi[Y,G]$, $\mathfrak{A}(R)(X)$, *and* $\{Y: \mathfrak{A}(R)(Y) \quad \mathfrak{A}(R)(Y) \models \varphi[Y,G]\}$ *are* Δ_1^{MTL}.

PROOF:

a) $\mathfrak{A} \models \varphi[G] \leftrightarrow G$ is an assignment for φ in $\mathfrak{A}(R) \wedge (\exists x \in A(R))$ $(x \in K_{\mathfrak{A}(R)}(\varphi) \wedge G \restriction \text{Fv}\varphi = H_{\mathfrak{A}(R)}(X))$.

b) $\mathfrak{A}(R) \models \varphi[Y, G] \leftrightarrow \mathfrak{A}(R) \models \varphi[Y \times \{0\} \cup (G \sim G \restriction \{0\})]$.

c) $\mathfrak{A}(R)(X) \leftrightarrow (\exists z \in A(R))(F_R(z) = X)$.

d) $X = \{Y: \mathfrak{A}(R)(Y) \wedge \mathfrak{A}(R) \models \varphi[Y,G]\} \leftrightarrow$

$(\forall y \in X)(\mathfrak{A}(R)(y) \wedge \mathfrak{A}(R) \models \varphi[y,G]) \wedge$

$\forall y((\mathfrak{A}_R(y) \wedge \mathfrak{A}(R) \models \varphi[y,G]) \to y \in X)$.

The first part is Δ_1^{MTL} and the second is Π_1^{MTL}. But

$\forall y((\mathfrak{A}_R(y) \wedge \mathfrak{A}(R) \models \varphi[y,G] \to y \in X) \leftrightarrow$

$(\forall z \in A(R))\, \exists y((F(z) = y \wedge \mathfrak{A}_R \models \varphi[y,G]) \to y \in X)$.

Thus, the second part is also Σ_1^{MTL}, and hence Δ_1^{MTL}.

THEOREM 3.12. *If R and S are well-orderings then the following formulas and terms are* Δ_1^{MTL}: R_x, $R \cong_F S$, $R \cong S$, $R < S$, $S > R$, $R \leqslant S$, $S \geqslant R$, *"x is the last element of R"*.

PROOF:

a) $R_x = S \leftrightarrow \text{REL}(R) \wedge \text{REL}(S) \wedge S \subseteq R \wedge$
$$\forall y \in DR(y \in DS \leftrightarrow \langle y, x \rangle \in R).$$

b) $R \cong_F S \leftrightarrow \text{REL}(R) \wedge \text{REL}(S) \wedge 1\text{FUN}(F) \wedge DF = DR \wedge DF^{-1} = DS$
$$(\forall x \in DR)(\forall y \in DR)(\langle x, y \rangle \in R \leftrightarrow \langle F^{\boldsymbol{\cdot}}x, F^{\boldsymbol{\cdot}}y \rangle \in S).$$

c) $R \cong S \leftrightarrow \exists F(R \cong_F S)$
$$\leftrightarrow \forall F((1\,\text{FUN}(F) \wedge DF = DR \wedge DF^{-1} \subseteq DS \wedge$$
$$(\forall x \in DF)(\forall y \in DF)(\langle x, y \rangle \in R \leftrightarrow \langle F^{\boldsymbol{\cdot}}x, F^{\boldsymbol{\cdot}}y \rangle \in S))$$
$$\rightarrow R \cong_F S) \vee \forall F((1\text{FUN}(F) \wedge DF = DS \wedge DF^{-1} \subseteq DR \wedge$$
$$(\forall x \in DF)(\forall y \in DF)(\langle x, y \rangle \in S \leftrightarrow \langle F^{\boldsymbol{\cdot}}x, F^{\boldsymbol{\cdot}}y \rangle)) \rightarrow S \cong_F R).$$

d) $R < S \leftrightarrow S > R \leftrightarrow (\exists x \in DS)(R \cong S_x)$.

e) $R \leqslant S \leftrightarrow S \geqslant R \leftrightarrow R < S \vee R \cong S$.

f) x is the last element of $R \leftrightarrow (\forall y \in DR)(y \neq x \rightarrow \langle x, y \rangle \notin R)$. ■

THEOREM 3.13. *If* A *and* F *are* Δ_1^{MTL} *operations such that for every* $x \in A(R)$, $F_R(x)$ *is a well-ordering, then*
$$\inf[F_R(x) : x \in A(R)] \text{ is } \Delta_1^{MTL}.$$

PROOF:
$$\inf[F_R(x) : x \in A(R)] = Z \leftrightarrow (\forall x \in A(R))(Z \leqslant F_R(x)) \wedge$$
$$(\exists x \in A(R)(Z = F_R(x)).$$

THEOREM 3.14. *Let* A *and* F *be* Δ_1^{MTL} *operations such that for every*

$x \in A(R)$, $F_R(x)$ *is a well-ordering, then* $\sup[F_R(x) : x \in A(R)]$ *is* Δ_1^{MTL}.

PROOF: The rank function ρ defined by

$$\rho`x = \rho^*x \cup \bigcup\rho^*x \text{ , is } \Delta_1^{MTL}.$$

$$ds(X) = \{y: y \in X \wedge (\forall x \in X)(\rho`y \subseteq \rho`x)\}$$

$$E_R = \{\langle x,y \rangle: x,y \in A(R) \wedge F_R(x) = F_R(y)\}$$

$$B(R) = \{ds(E_R^*\{x\}): x \in A(R)\}$$

Thus, $B(R)$ is Δ_1^{MTL}.

$$T_R = \{\langle x,y \rangle: x,y \in B(R) \wedge (\exists z \in y)(\exists u \in x)(F_R(u) \leq F_R(z))\}.$$

$$G_R(y) = Z \leftrightarrow (\exists x \in y)(F_R(x) = Z).$$

T_R and G_R are Δ_1^{MTL}. Also, $[G_R(y): x \in B(R)]$ is Δ_1^{MTL}. Now,

$$S_R = \sum_{T_R} [G_R(y): y \in B(R)] = \{\langle u,v \rangle: u, v \in [G_R(y) : y \in B(R)] \wedge$$
$$[\exists y,y' \in B(R)(\exists u' \in G_R(y))(\exists v' \in G_R(y'))(u = \langle u',y \rangle \wedge$$
$$v = \langle v',y' \rangle \wedge [(y = y' \rightarrow \langle u',v' \rangle \in G_R(y)) \vee$$
$$\langle y,y' \rangle \in T_R])]\}, \text{ is } \Delta_1^{MTL}.$$

$$\sup[F_R(x): x \in A(R)] = Z \leftrightarrow ((\exists y \in DS_R)(\forall x \in A(R))((S_R)_y \geq F_R(x)) \wedge$$
$$Z = \inf[(S_R)_y: (\forall x \in A(R))(S_R)_y \geq F_R(x)])$$
$$\vee [\neg(\exists y \in DS_R)(\forall x \in A(R))((S_R)_y \geq F_R(x))$$
$$\wedge Z = S_R]. \blacksquare$$

THEOREM 3.14. *If* R *and* R' *are well-orderings, then the formulas* $L_R(X)$ *and* $TW'(R) = R'$ *are* Δ_1^{MTL}.

PROOF: Since L_R and $TW'(R)$ are defined by recursion, we have to show that the operation used to define them is Δ_1^{MTL}, and, hence, each step in the definition is Δ_1^{MTL}. In order to show that L_R is Δ_1^{MTL}, it is enough to do it for B_R and G_R.

1. $L_0(X) \leftrightarrow X = 0$.

   $TW'(0) = 0$.

2. Let $R \in V$. Then,

   $F = EN^{\bullet}R \leftrightarrow 1FUN(F) \wedge (\exists \alpha \in OR)(IN \upharpoonright \alpha \cong_F R)$

   $TW'(R) = IN \upharpoonright D(EN^{\bullet}R)$.

3. $\exists S(R \cong S+1) \leftrightarrow (\exists x \in DR)$($x$ is the last element of $R \wedge R \cong R_x + 1$).

4. $R \notin V \leftrightarrow IN \upharpoonright OR \leqslant R$.

5. Let $R \cong S+1$. Then:

   $$B_R = \bar{C}_{L_S} = \{\langle m, z\rangle : m \in \omega \wedge z \in C_{L_S}\}.$$

   $$G_R(x) = y \leftrightarrow D_{L_S}(x) = y \leftrightarrow (\exists m \in \omega)(\exists z \in C_{L_S})$$

   $$(y = \{x : L_S(x) \wedge L_S \models \varphi_m[x, H_{\mathcal{A}}(z)]\}).$$

   (Δ_1^{MTL} by 3.11).

6. Let $R \cong S+1$, $R \notin V$. Then,

   $$\begin{aligned} TW'(R) &= TW'(S) + 1 \\ &= TW'(S) \times \{0\} \cup \{\langle 0,1\rangle\}. \end{aligned}$$

7. Let $R \ncong S+1$ for all S. Then

   $$B_R = [\,B_{R_y} : y \in DR\,],$$

   $$G_R(\langle y, x\rangle) = G_{R_x}(y).$$

8. Let $R \ncong S+1$ for all S and $R \notin V$.

   a) $A(T, m, g) = \{y : L_R \models \varphi_m[y, H_{L_T}(g)]\}$

   (Δ_1^{MTL} by 3.11).

   b) $\mu(x) = T \leftrightarrow L_T(x) \wedge \forall y \in DT(\neg L_{T_y}(x))$.

c) $B(T,m,g) = \sup[\mu(y) : y \in A(T,m,g)]$.

d) $G'(T) = \sup[B(m,T,g) : m \in \omega \wedge g \in {}^{Fv(\exists v_0 \varphi_m)} B_T]$.

e) $J(T,m,g) = \inf[R(x) : x \in C]$, where

$$C = \{x : x \in DR \wedge T \leq R_x\} \text{ and}$$

$$R(x) = T' \leftrightarrow TW(R_x) = T' \wedge (\exists y \in B_R)(L_R \models \varphi_m[G_R(y), H_{L_T}(g)])$$
$$\rightarrow (\exists y \in B_{R_x})(L_R \models \varphi_m[G_{R_x}(y), H_{L_T}(g)]).$$

f) $L(T) = \sup[J(T,m,g) : m \in \omega \wedge g \in {}^{Fv(\exists v_0 \varphi_m)} B_T]$ if $T < R$;

$= T$, if $T \cong R$.

g) $M(T,0) = T$,

$M(T,m+1) = L(M(T,m))$.

h) $N(T) = \sup[M(T,m) : m \in \omega]$.

Then we have,

$$\begin{aligned} TW'(R) = R' \leftrightarrow\ & (\exists x \in DR)(G'(R_x) \cong R \wedge R_x = \inf[R_y : y \in DR \wedge \\ & G'(R_y) \cong R] \wedge R' = G'(R_x)) \vee (\neg(\exists x \in DR))(G'(R_x) \cong R) \\ & \wedge (\exists x \in DR))(N(R_x) \cong R \wedge R_x = \inf[R_y : y \in DR \wedge \\ & N(R_y) \cong R] \wedge R' = N(R_x))) \vee (\neg(\exists x \in DR)(G'(R_x) \cong R) \\ & \wedge \neg(\exists x \in DR)(N(R_x) \cong R) \wedge R' = 0). \end{aligned}$$

With this, we complete the proof of the theorem. ■

THEOREM 3.15. *$T(X)$ and $L(X)$ are Σ_1^{MTL}.*

PROOF:

(i) Define $\overline{TW}$ over all classes X, by,

$\overline{TW}(X) = y \leftrightarrow TW'(X) = y \vee$ (X is not a well-ordering $\wedge\ y = 0$).

Since, X is a well-ordering is Π_1^{MTL}, $\overline{TW}$ is Σ_1^{MTL}.

Now, $T(X) \leftrightarrow \overline{TW}(X)=X \wedge (\forall x \in DX)(\text{IN}\upharpoonright\text{OR} \leqslant X_x \to L_{X_x} \not\models \text{MT})$.

(ii) $L(X) \leftrightarrow \exists R(T(R) \wedge L_R(X))$. ∎

4. Fundamental theorems.

We define, as in [4], the class of sets hereditarily of cardinality less than κ:

$$H(\kappa) = \{x: \overline{\overline{TC(x)}} < \kappa\}.$$

The following theorem is an adaptation for MT of a theorem by Lévy. The proof is based on that appearing in [6], 7.4.

THEOREM 4.1. *Let* φ *be a* Σ_1^{MTL} *formula with at most* $v, v_0, \ldots, v_{n-1}$ *free,* κ *a cardinal,* $\kappa > \omega$. *Then*

$$MTL \vdash (\forall v_0 \ldots v_{n-1} \in H(\kappa))(\exists v\, \varphi(v, v_0, \ldots, v_{n-1}) \to$$
$$(\exists v \in H(\kappa))\, \varphi(v, v_0, \ldots, v_{n-1})).$$

(*We may replace* MTL *by* MT + WO_Q *everywhere.*)

PROOF: 1) First we shall show that it is enough to prove the theorem for Δ_0^{MTL}-formulas. Let φ be Σ_1^{MTL}. Then,

$$MTL \vdash \varphi(v, v_0, \ldots, v_{n-1}) \leftrightarrow \exists u\, \psi(v, u, v_0, \ldots, v_{n-1})$$

where ψ is Δ_0^{MTL}. Define,

$$\theta(w, v_0, \ldots, v_{n-1}) \leftrightarrow \psi(w * \{0\}, w * \{1\}, v_0, \ldots, v_{n-1}).$$

θ is also Δ_0^{MTL}.

But, then,

$$MTL \vdash \exists w\, \theta(w, v_0, \ldots, v_{n-1}) \leftrightarrow \exists v\, \varphi(v, v_0, \ldots, v_{n-1}).$$

Since $[X,Y] \in H(\kappa)$ iff $X,Y \in H(\kappa)$, we have

$\mathrm{MT}L \vdash (\exists w \in H(\kappa))\,\theta(w,v_0,\ldots,v_{n-1}) \leftrightarrow (\exists v \in H(\kappa))\,\varphi(v,v_0,\ldots,v_{n-1})$.

2) It is enough to show the theorem for κ a sucessor cardinal:

Suppose κ is a limit cardinal; then

$$H(\kappa) = \cup\{H(\lambda):\lambda<\kappa\}.$$

If $x_0,\ldots,x_{n-1} \in H(\kappa)$, there is a $\lambda^+<\kappa$ such that $x_0,\ldots,x_{n-1} \in H(\lambda^+)$.

3) Let, then, $\varphi(v,v_0,\ldots,v_{n-1})$ be $\Delta_0^{\mathrm{MT}L}$ and κ^+ a sucessor cardinal. Let $x_0,\ldots,x_{n-1} \in H(\kappa^+)$ and let X be such that $L(X)$ and $\varphi^L(X,x_0,\ldots,x_{n-1})$. Then, there is a well-ordering type R such that

$$L_R(X) \wedge L_R(x_0) \wedge \ldots \wedge L_R(x_{n-1}).$$

By reflection (10.2 of [1]), there is a w.o. type S, $S > R$, such that,

$$\varphi^L(y,y_0,\ldots,y_{n-1}) \leftrightarrow \varphi^{L_S}(y,y_0,\ldots,y_{n-1})$$

for every $y,y_0,\ldots,y_{n-1}$ with $L_S(y), L_S(y_0),\ldots,L_S(y_{n-1})$.

Since $L_S(X) \wedge L_S(x_0) \wedge \ldots \wedge L_S(x_{n-1})$ and $\varphi^L(X,x_0,\ldots,x_{n-1})$, we have, $\varphi^{L_S}(X,x_0,\ldots,x_{n-1})$, i.e. $L_S \models \varphi[X,x_0,\ldots,x_{n-1}]$.

Now, let, $a \notin \mathrm{TC}(\{x_0,\ldots,x_{n-1}\})$, $a \in V$, and $c = \mathrm{TC}(\{x_0,\ldots,x_{n-1}\}) \cup \{a\}$. Let F be the operation defined by $F(y) = y$ for $y \in \mathrm{TC}(\{x_0,\ldots,x_{n-1}\})$ and $F(a) = X$. Since $x_0,\ldots,x_{n-1} \in H(\kappa^+)$, $\overline{\overline{c}}<\kappa^+$.

Let $\mathfrak{C} = [c,[F(y): y \in c]]$. Then $\overline{\overline{\mathfrak{C}}} < \kappa^+$. Since L_S is transitive, $\mathfrak{C} \subseteq L_S$. Using 2.1, we can find a structure $\mathfrak{A}'$, $\mathfrak{A}' \prec L_S$, and $\mathfrak{C} \subseteq \mathfrak{A}'$. We have, $\mathfrak{A}'(x_0),\ldots,\mathfrak{A}'(x_{n-1})$, and $\mathfrak{A}'(X)$; but, since $\mathfrak{A}' \prec L_S$, we obtain,

$$\mathfrak{A}' \models \varphi[X,x_0,\ldots,x_{n-1}].$$

$\mathfrak{A}'$ satisfies the axiom of extensionality. By 2.2 we find a transitive structure $\mathfrak{A}$ and an isomorphism Π, with $\mathfrak{A}' \cong_\Pi \mathfrak{A}$ we have, since

$TC(\{x_0,\ldots,x_{n-1}\}) \subseteq \mathfrak{A}$, that $\Pi(x_i) = x_i$ for $i = 0,\ldots,n-1$. Also $\overline{\overline{\mathfrak{A}}} \leqslant \overline{\overline{\mathfrak{A}'}} \leqslant \kappa < \kappa^+$ and $\mathfrak{A} \models \varphi[\Pi(X), x_0,\ldots,x_{n-1}]$.

But $\mathfrak{A}$ is transitive; hence $\Pi(X) \subseteq \mathfrak{A}$, and thus $\overline{\overline{TC(\Pi(X))}} < \kappa^+$, i. e. $\Pi(X) \in H(\kappa^+)$.

Since φ is Δ_0, it is absolute for transitive models. Thus

$$\varphi^{L}(\Pi(X), x_0,\ldots,x_{n-1}).$$

Therefore,

$$((\exists y \in H(\kappa^+))\varphi(y, x_0,\ldots,x_{n-1}))^{L}.$$

This proves the theorem. ∎

Let us define $L_\kappa = \{x: L_{IN\restriction\kappa}(x)\}$ for κ an ordinal and $L = \cup\{L_\kappa: \kappa \in OR\}$.

THEOREM 4.2. *If κ is an infinite cardinal, then*

$$MT \vdash \forall x(x \in H(\kappa) \wedge L(x) \rightarrow x \in L_\kappa).$$

PROOF: Since $L(x)$ is a Σ_1^{MTL}-formula and

$$MT \vdash L(x) \leftrightarrow \exists R(T(R) \wedge L_R(x)),$$

we have by 4.1,

$$\forall x \in H(\kappa)(L(x) \rightarrow (\exists R \in H(\kappa))(T(R) \wedge L_R(x)).$$

If $R \in H(\kappa)$, then $R = IN\restriction\alpha$ for an $\alpha < \kappa$. Therefore

$L_R = L_\alpha \subseteq L_\kappa$. Hence, $x \in L_\kappa$. ∎

THEOREM 4.3. $MT \vdash V \simeq L \rightarrow V = L$.

PROOF: Let $V \simeq L$, i. e. $\forall X(L(X))$. It is clear that $L \subseteq V$. We shall prove that $V \subseteq L$. Let $x \in V$ with $L(x)$. Then $x \in H(\lambda)$ for some cardinal λ. Thus, by 4.2, $x \in L_\lambda \subseteq L$. ∎

The next theorem is proved in the same way, as [3], Lemma 2.6.

THEOREM 4.4. $MT + V \simeq L \vdash (\forall\sigma \geqslant \omega)\, \overline{\overline{L_{IN\restriction\sigma}}} = \overline{\overline{\sigma}}$.

THEOREM 4.5. MT $\vdash V \simeq L \rightarrow$ GCH.

(GCH *is the generalized continuum hypothesis*).

PROOF: Let κ be a cardinal $\geqslant \omega$. We have, $\kappa^+ \leqslant \overline{\overline{S(\kappa)}}$ (by Cantor's Theorem). Now, if $x \subseteq \kappa$, then $TC(x) \subseteq \kappa$, and hence $x \in H(\kappa^+)$. Since we have $L(x)$, $x \in L_{\kappa^+}$. Thus, $S(\kappa) \subseteq L_{\kappa^+}$. Therefore,

$$\overline{\overline{S(\kappa)}} \leqslant \overline{\overline{L_{\kappa}}} = \kappa^+. \blacksquare$$

THEOREM 4.6. *Let* $\mathfrak{A}$ *be a model of* MT *such that* $\mathfrak{A} \subseteq L$ *and suppose that* $\mathfrak{A}(\alpha)$ *for every ordinal* α. *Then* $\mathfrak{A} \simeq L$.

PROOF: We shall prove by induction over well-orderings that $\mathfrak{A}(TW'(R))$ and $L_R \subseteq \mathfrak{A}$ for every well-ordering R such that $L(R)$.

1) If $R \in V$, then $TW'(R) = IN \restriction \alpha$ for some ordinal α. Since IN is absolute we have, $\mathfrak{A}(TW'(R))$.

2) Let $R \cong S+1$, $L_S \subseteq \mathfrak{A}$, and $\mathfrak{A}(TW'(S))$. Let X, be such that $L_R(X)$. Then

$$X = \{y\colon L_S(y) \wedge L_S \models \varphi_m[y, H_{L_S}(x)]\}.$$

But since $\mathfrak{A} \models$ MT, we have, by absolutness,

$$\{y\colon L_S(y) \wedge L_S \models \varphi_n[y, H_{L_S}(x)]\}^{\mathfrak{A}} = X.$$

Thus, $\mathfrak{A}(X)$, i. e. $L_R \subseteq \mathfrak{A}$.

On the other hand, $TW'(R) = TW'(S) + 1$, if $R \notin V$. By absolutness of TW' and since $\mathfrak{A} \models$ MT, we have

$$(TW(S) + 1)^{\mathfrak{A}} = TW(R).$$

Hence $\mathfrak{A}(TW(R))$.

3) Suppose $R \not\cong S+1$ for every S and $L_S \subseteq \mathfrak{A}$ for every $S < R$. Let $L_R(X)$; then $X = G_{R_x}(y)$, for $y \in B_{R_x}$. Thus, $L_{R_x}(X)$ and hence $\mathfrak{A}(X)$.

Therefore, $L_R \subseteq \mathfrak{A}$.

If $R \notin V$, the formula that defines $TW'(R)$ is absolute. The definition of $TW'(R)$ uses the w. o. types $T < R$. Since all types $T < R$ are in $\mathfrak{A}$ by induction hypothesis and $\mathfrak{A}$ is a model of MT, we obtain $\mathfrak{A}(TW'(R))$.

Thus, $L \subseteq \mathfrak{A}$, i. e. $L = \mathfrak{A}$.

REFERENCES.

[1] R. Chuaqui, *Internal and forcing models for the impredicative theory of classes*, to appear in Dissertationes Mathematicae.

[2] R. Chuaqui, *Bernays' class theory*, this volume.

[3] F. R. Drake, Set Theory, North-Holland Pub. Co., Amsterdam, 1974.

[4] K. Gödel, *The consistency of the continuum hypothesis*, Ann. Math. Studies, Princeton Univ. Press, Princeton, 1940.

[5] A. Lévy, *A hierarchy of formulas in set theory*, Mem. Amer. Math. Soc. 57 (1965).

[6] A. Mostowski, Constructible sets with applications, North-Holland Pub. Co., Amsterdam, 1969.

Instituto de Matemática
Universidade Católica de Chile
and
Centro de Lógica, Epistemologia e
História da Ciência
Universidade Estadual de Campinas.

ADDED IN PROOF:

(1) The proof for $T \subseteq L$ which appears in Section 3 after the definition of L is incomplete. We have to show that if R is a minimal w. o. such that $\neg L(S)$ for all $S \cong R$, we have $L_R \models \mathrm{MT}$. Since for all $x \in DR$, $L(\mathrm{TW}'(R_x))$, we have by absoluteness, that $L_{\mathrm{TW}'(R_x)} \subseteq L$. Hence, $L_R \subseteq L$. From $\neg L(R)$ we get $L^L \simeq L_R$. Since $(\mathrm{MT}^L)^L$, we obtain $L_R \models \mathrm{MT}$. From this we prove that $L \simeq L_R$ and, hence $T \subseteq L$.

From Total to Partial Algebras.

by *IRENE MIKENBERG.*

ABSTRACT. In this paper we study algebraic systems that consist of a set and partial operations on this set. For this purpose, we introduce a neutral element, that does not belong to the original universe, as the value of the operations where they are not defined.

In Section 2 we give a logical system for partial algebras, very similar to the one of A. Tarski introduced in *A simplified formalization of predicate logic with identity*, Archiv fur Mathematische Logik und Grundlagen forschung, vol. 7, 1965. We also prove in this Section that this system is complete. Then, we extend two results for total algebras to partial algebras, the first one is given in Section 5, and is an extension of the characterization of the universal classes UC_Δ, that appears in A. Tarski in *Contributions to the theory of models I*, Indagationes Mathematicae, vol. 16, 1954. The second one is given in Section 6, and is a generalization of the preservation problem for Horn's sentences that appears in H. I. Keisler, *Reduced products and Horn classes*, Trans. Am. Math. Soc. 117, 1965.

Introduction.

In many parts of mathematics, algebraic systems that consist of a set and partial operations on this set are studied. These systems are not al-braic structures in the usual model-theoretic sense. There are two ways to resolve this problem, the operations may be treated as relations or they may be defined arbitrarily outside of their domain of definition. In this work we study systematically the second procedure, introducing a neutral element, that does not belong to the original universe, as the value of the operations where they are not defined.

This work is related to the so-called *Free-Logic Systems* as those of [6] and [8]. In both of these works, an individual constant may designate something that is not the value of a variable. This is possible in [8], introducing besides the internal domain $\mathcal{D}$, an external domain $\mathcal{D}'$, disjoint from $\mathcal{D}$, whose elements are also assigned to individual constants, and in [6], we have three possible truth-values, that is, true, false, and not-designated. The present work, is more similar to [8], except that the external domain $\mathcal{D}'$, consists of only one element.

We begin this work with a modified semantics of a logical system very similar to the one of A. Tarski in [10]. This system is proved to be complete by the usual method of Henkin's extensions with some modifications. Then, we extend two results for total algebras to partial algebras. The first one appears in Section 5 and is an extension of the characterization of the universal classes UC_{Δ} that appears in [11]. An announcement of this part of the work appears in [9]. The second one is given in Section 6, and is a generalization of the preservation problem for Horns's sentences that appears in [7]. Besides this, in this section, we extend also the fundamental theorem for ultraproducts to partial algebras.

1. SEMANTICS.

We shall use German letters to denote partial algebras, and capital Latin letters to denote their universes, using the notation of [5]. We shall also suppose that X represents an element that does not belong to the universe of any partial algebra.

By a *partial algebra* we shall understand a system:

$$\mathfrak{A} = \langle A, p_i, f_j, c_k \rangle_{\substack{i \in I \\ j \in J \\ k \in K}}$$

where $A \neq 0$, $X \notin A$, c_k is a distinguished element of A for each $k \in K$, $p_i \subseteq {}^n A$ for each $i \in I$, and f_j are operations of rank ρf_j for each $j \in J$ such that,

$$f_j : (A \cup \{X\})^{\rho f_j} \to (A \cup \{X\}), \quad \rho f_j > 0$$

and

$$(*) \quad f_j(\tau_{\nu_0}, \ldots, \tau_{\nu_{j-1}}) \in A \text{ implies } \tau_{\nu_i} \in A \text{ for every } i \in \{0, \ldots, j-1\}.$$

The notion of *satisfaction* is the usual one, except that the variables are restricted only to elements of A, in particular, we have,

Let $\simeq$ *denote the identity symbol, then,*

$$\mathfrak{A} \models \exists x(x \simeq \tau)\,[a_1, \ldots, a_n] \text{ iff } \tau^{\mathfrak{A}}[a_1, \ldots, a_n] \in A$$

and

$$\mathfrak{A} \models (\tau \simeq \tau')\,[a_1, \ldots, a_n] \text{ iff } \begin{cases} \tau^{\mathfrak{A}}[a_1, \ldots, a_n] \in A \textit{ and } \tau'^{\mathfrak{A}}[a_1, \ldots, a_n] \in A \\ \textit{and } \tau^{\mathfrak{A}}[a_1, \ldots, a_n] = \tau'^{\mathfrak{A}}[a_1, \ldots, a_n] \\ \textit{or} \\ \tau^{\mathfrak{A}}[a_1, \ldots, a_n] \notin A \textit{ and } \tau'^{\mathfrak{A}}[a_1, \ldots, a_n] \notin A. \end{cases}$$

We decided to choose this notion of partial algebra, because one of the fundamental concepts of our work, is the idea of preservation of identities. Here we obtain that an identity is satisfied when both sides of the identity are defined and equal, or when none of them is defined,and therefore, also equal, because there is only one element outside of the domain that is assigned to the terms that are not defined.

2. The logical system.

We define formulas and terms in the usual way, and let Greek letters to vary through them. We use $\sim$, $\to$, $\wedge$, $\forall$, $\exists$, and $\simeq$ as the "Logical Symbols". We shall also use the expression *the highest terms that occur in* ϕ ($HT(\phi)$) to mean the following:

Given a formula ϕ, let τ_ϕ be the set of all terms that occur in ϕ, then:

$$HT(\phi) = \{\tau \in \tau_\phi : \tau \textit{ is not a subterm of any } \tau' \in \tau_\phi\}.$$

We shall also need the following definitions:

i) $OC(\phi)$ = the set of all variables that occur in ϕ.

(ii) $S.I.$ = the set of all individuals symbols (i. e. variables and individual constants)

(iii) $P(\phi,\psi,\tau,\sigma)$ = the atomic formula ψ is obtained from the atomic formula ϕ substituting some occurrences of the term τ for σ.

(iv) $P'(\tau_1,\sigma_1,\tau,\sigma)$ = the term σ_1 is obtained from the term τ_1 substituting some occurrences of the term τ for σ.

We have the following logical axioms,

A1) $(\phi \to \psi) \to ((\psi \to \chi) \to (\phi \to \chi))$.

A2) $(\sim \phi \to \phi) \to \phi$.

A3) $\phi \to (\sim \phi \to \psi)$.

A4) $\forall u(\phi \to \psi) \to (\forall u\phi \to \forall u\psi)$.

A5) $\forall u\phi \to \phi$.

A6) $\phi \to \forall u\phi$ where $u \notin OC(\phi)$.

A7) $\sim \forall u \sim u \simeq \tau$ where $u \neq \tau$, $\tau \in S.I.$.

A8) $(\forall u \sim u \simeq \tau_1) \wedge (\forall u \sim u \simeq \tau_2) \to \tau_1 \simeq \tau_2$.

A9) $\tau \simeq \sigma \to (\phi \to \psi)$ where $P(\phi, \psi, \tau, \sigma)$.

A10) $\tau \simeq \sigma \to (\tau_1 \simeq \sigma_1)$ where $P'(\tau_1,\sigma_1,\tau,\sigma)$.

A11) $\exists x (x \simeq f\tau_1 \dots \tau_n) \to \exists x_1(x_1 \simeq \tau_1) \wedge \dots \wedge \exists x_n(x_n \simeq \tau_n)$

Derivation Rules:

$$\frac{\phi \qquad \phi \to \psi}{\psi} \qquad\qquad \frac{\phi}{\forall \alpha \phi}$$

This system is a modification of the one that appears in [10] in such a way that it includes terms. For example, Axiom A7 in this system is restricted only to individual symbols, instead of arbitrary terms. Axiom A8 serves to assure that all the terms that are not defined are equal to the same element, and Axiom A11 serves to assure condition (*) of the definition of partial algebra (Section 1). From this system we obtain the usual logical consequences, see [2], some of them with a few modifications. For example:

1) $\vdash \tau \simeq \tau$ *for every term* τ.

In this case, we have to prove this first for every $\tau \in S.I.$ as in [2], and then make an induction over terms using A10.

2) $\vdash \forall x \phi \wedge \exists x(x \simeq \tau) \rightarrow \phi(\frac{x}{\tau})$, *if none of the variables of* x, τ *occur bounded in* ϕ *and* $x \notin OC(\phi)$.

To prove this we use:

(*) $\vdash \tau \simeq \omega \rightarrow (\phi(\frac{\tau}{\omega}) \rightarrow \phi)$ where none of the variables of τ, ω occur bounded in ϕ.

This consequence is proved in the usual way.

Suppose that $\vdash \forall x \phi \wedge \exists x(x \simeq \tau)$ and that none of the variables of x, τ occur bounded in ϕ, then from A5 we have ϕ, and from (*) we have $\vdash x \simeq \tau \rightarrow (\sim\phi(\frac{x}{\tau}) \rightarrow \sim\phi)$; then

$\vdash \sim((\sim\phi(\frac{x}{\tau})) \rightarrow \sim\phi) \rightarrow \sim x \simeq \tau$,

$\vdash \forall x \sim(\sim\phi(\frac{x}{\tau}) \rightarrow \sim\phi) \rightarrow \forall x \sim x \simeq \tau$,

$\vdash \sim \forall x \sim x \simeq \tau \rightarrow \sim\forall x \sim(\sim\phi(\frac{x}{\tau}) \rightarrow \sim\phi)$, and, because $\exists x(x \simeq \tau)$, we have, $\vdash \exists x(\sim\phi(\frac{x}{\tau}) \rightarrow \sim\phi)$.

We need now to prove the following:

LEMMA 1. $\exists x(\phi \rightarrow \psi) \rightarrow (\phi \rightarrow \exists x \psi)$, *where* $x \notin OC(\phi)$.

We prove this lemma in two parts:

(i) $(\phi \wedge \forall x \psi) \rightarrow \forall x(\phi \wedge \psi)$, where $x \notin OC(\phi)$. Suppose $\vdash \phi \wedge \forall x \psi$, then we have $\vdash \phi$, $\vdash \forall x \psi$. But $\vdash \forall x \psi \rightarrow \psi$ by A5; then $\vdash \phi$, $\vdash \psi$. Therefore $\vdash \phi \wedge \psi$, and $\vdash \forall x(\phi \wedge \psi)$.

(ii) From (i) we have $\vdash (\phi \wedge \forall x \sim\psi) \rightarrow \forall x(\phi \wedge \sim\psi)$,

$\vdash (\phi \wedge \exists x \psi) \rightarrow \forall x \sim(\sim\phi \vee \psi)$,

$\vdash \sim(\sim\phi \vee \exists x \psi) \rightarrow \exists x(\phi \rightarrow \psi)$,

$\vdash \exists x(\phi \rightarrow \psi) \rightarrow (\phi \rightarrow \exists x \psi)$.

Now we can finish the proof of (2); we already had proved that:

$$\vdash \exists x(\sim\phi(\tfrac{x}{\tau}) \rightarrow \sim\phi).$$

By Lemma 1, we get:

$$\vdash \sim\phi\binom{x}{\tau} \to \exists x \sim\phi,$$
$$\vdash \sim \exists x \sim\phi \to \phi\binom{x}{\tau},$$
$$\vdash \phi\binom{x}{\tau}. \quad \blacksquare$$

3) *If* $\tau \neq \omega$ *and none of the variables of* ω, τ *occur in* ϕ, *then*
$$\vdash \exists\omega(\tau \simeq \omega) \wedge \forall\omega\,\phi\binom{\tau}{\omega} \to \phi.$$

To prove this again we make use of (*):

$$\vdash \tau \simeq \omega \to (\phi\binom{\tau}{\omega} \to \phi); \text{ then}$$
$$\vdash \phi\binom{\tau}{\omega} \to (\sim\phi \to \sim\tau \simeq \omega),$$
$$\vdash \forall\omega\phi\binom{\tau}{\omega} \to (\sim\phi \to \forall\omega \sim \tau \simeq \omega),$$
$$\vdash \sim\forall\omega\sim\tau \simeq \omega \to (\forall\omega\,\phi\binom{\tau}{\omega} \to \phi),$$
$$\vdash \exists\omega(\tau \simeq \omega) \wedge \forall\omega\,\phi\binom{\tau}{\omega} \to \phi. \quad \blacksquare$$

4) *Let* c *be a constant that does not occur in* ϕ *and* $\vdash \phi\binom{u}{c}$; *then* $\vdash \forall u\phi$ *in a language without* c.

To prove this property we proceed by induction over the length of the derivation in the usual way (see [2]).

3. COMPLETENESS.

Now we have the tools to prove that the theory is complete; for this we make the usual Henkin's extension, that is, we define:

$T_0 = Dr\,T$, *where* $Dr\,T$ *is the set of sentences derivable from* T;

$G_0 = \{\phi : \phi$ *is a sentence of the language of* T *of the form* $\exists x\,\phi\}$;

$C_0 =$ *the set of new constants such that* $C_0 \sim_c G_0$.

Suppose T_n, G_n, C_n defined. Then

$$T_{n+1} = Dr(T_n \cup \{\exists u\,\phi \to \phi\binom{u}{c_{\exists u\phi}} : \exists u\,\phi \in G_n\});$$

$G_{n+1} = \{\phi : \phi$ *is a sentence of the language of* T_{n+1} *of the form* $\exists u\, \phi\}$;

$C_{n+1} =$ *set of new constants such that* $G_{n+1} \sim_C C_{n+1}$,

Let be $T' = \bigcup_{n \in \omega} T_n$.

Here we can prove without problem that T' is a conservative extension of T (see [2]). And finally, applying Lindenbaum's Theorem, we extend T' to a T'' Henkin's complete theory.

We now construct the canonical structure for T''. Let τ_1, τ_2 be any terms, we define the relation $\sim$ in the following way:

$$\tau_1 \sim \tau_2 \quad \text{iff} \quad \vdash_{T''} \tau_1 \simeq \tau_2 \ .$$

Let $|\mathfrak{A}| = \{\overline{\tau} : \vdash_{T''} \exists x (x \simeq \tau)\}$ where $\overline{\tau}$ is the equivalence class of τ. Then

$$\mathfrak{A} = \langle |\mathfrak{A}|, f_i, \rho_j, d_k \rangle_{\substack{i \in I \\ j \in J \\ k \in K}} ,$$

is the partial algebra defined by:

If ρ_j is an n-ary predicate, then $\rho_j^{\mathfrak{A}} = \{(\overline{\tau}_1 \dots \overline{\tau}_n) : \vdash_{T''} \rho_j \tau_1 \dots \tau_n\}$.

If f_i is an n-ary operation symbol, then

$$f_i^{\mathfrak{A}}(\overline{\tau}_1, \dots, \overline{\tau}_n) = \begin{cases} \overline{f_i(\tau_1, \dots, \tau_n)} & \text{if } \vdash_{T''} \exists x (x \simeq f(\tau_1, \dots, \tau_n)) \\ X & \text{if } \nvdash_{T''} \exists x (x \simeq f(\tau_1, \dots, \tau_n)) \end{cases}$$

and $d_k = \overline{c}_k$.

Thus, to complete the proof, we need to show that:

$\mathfrak{A} \models \phi$ iff $\vdash_{T''} \phi$, where ϕ is a closed formula. For this, we have to make an induction over ϕ in the usual way, except that first of all we need:

$$\tau^{\mathfrak{A}} = \overline{\tau}, \quad \text{if} \quad \vdash_{T''} \exists x (x \simeq \tau),$$

for every variable free term τ. This is obvious because of Axiom A11. The rest of the induction is the usual one, except for the universal quantifier:

Let $\phi = \forall u. \psi$;

then $\mathfrak{A} \models \phi$ iff every $\tau \in |\mathfrak{A}| : \mathfrak{A} \models \psi[^{u}_{\tau}]$, but $\tau \in |\mathfrak{A}|$ iff $\tau = \bar{\tau}_1$ and $\vdash_{T''} \exists x(x \simeq \tau_1)$, and then $\tau_1^{\mathfrak{A}} = \bar{\tau}_1$. Thus,

$\mathfrak{A} \models \phi$ iff for every τ such that $\vdash_{T''} \exists x(x \simeq \tau)$, $\mathfrak{A} \models \psi[x(^{u}_{\tau}\mathfrak{A})]$,

$\mathfrak{A} \models \phi$ iff for every τ such that $\vdash_{T''} \exists x(x \simeq \tau)$, $\mathfrak{A} \models \psi(^{u}_{\tau})[x]$,

$\mathfrak{A} \models \phi$ iff for every τ such that $\vdash_{T''} \exists x(x \simeq \tau)$, $\vdash_{T''} \psi(^{u}_{\tau})$,

because of the inductive hypothesis.

We must show that $\vdash_{T''} \forall x \psi$ iff for every τ such that

$$\vdash_{T''} \exists x(x \simeq \tau), \qquad \vdash_{T''} \psi(^{u}_{\tau}).$$

From left to right, this is given by A7 and 2.

Let us see then that the other way is correct:

Suppose $\not\vdash_{T''} \forall x \psi$, so we have that $\vdash_{T''} \sim\forall x \psi$ because T'' is complete. Then $\vdash_{T''} \exists x \sim\psi$, therefore $\vdash_{T''} \exists x \sim\psi \rightarrow \sim\psi(^{u}_{c})$; for some constant c. But because of A7, $\exists u(u \simeq c)$; then $\vdash_{T''} \sim\psi(^{x}_{c})$. This means that there exists some τ (c is the one) such that $\vdash_{T''} \exists x(x \simeq \tau)$ and $\not\vdash_{T''} \psi(^{x}_{\tau})$.

4. Preliminary notions.

In this section, we modify the usual definitions for algebraic systems, in order to apply them to partial algebras. In some cases the notion for total algebras splits into several when applied to partial algebras (see [4]). We choose the concept suitable for our purposes.

1. $\mathfrak{B}$ *is a partial subalgebra of* $\mathfrak{A}$ iff $B \subseteq A$, $\mathfrak{B}$ and $\mathfrak{A}$ have the same distinguished elements and the operations and relations of $\mathfrak{B}$ are the restrictions to B of the operations and relations of $\mathfrak{A}$, that is:

Let $b_1, \ldots, b_{n_{\nu-1}} \in B$ then

$$(f_\nu | B^n)(b_0, \ldots, b_{n_{\nu-1}}) = \begin{cases} f_\nu(b_0, \ldots, b_{n_{\nu-1}}) & \text{if } f_\nu(b_0, \ldots, b_{n_{\nu-1}}) \in B \\ X & \text{if } f_\nu(b_0, \ldots, b_{n_{\nu-1}}) \notin B \end{cases}$$

2. $S(K) = \{\mathfrak{A}$: exists some $\mathfrak{B} \in K$ such that $\mathfrak{A}$ is a partial subalgebra of $\mathfrak{B}\}$.

3. $S_\omega(K) = \{\mathfrak{A} : \mathfrak{A}$ is finite and there exists some $\mathfrak{B} \in K$ such that $\mathfrak{A} \in S(\mathfrak{B})\}$.

4. $\mathfrak{A} \cong \mathfrak{B}$ iff there exists a function ϕ, one to one from $A \cup \{X\}$ onto $B \cup \{X\}$ such that:

i) $\rho^{\mathfrak{A}} a_0 \ldots a_{n-1}$ iff $\rho^{\mathfrak{B}} \phi(a_0) \ldots \phi(a_{n-1})$,

ii) $\phi(f^{\mathfrak{A}}(a\ ,\ldots, a_{n-1})) = f^{\mathfrak{B}}(\phi(a_0) \ldots \phi(a_{n-1}))$,

iii) $\phi(c_k) = c'_k$ for every distinguished element c_k, c'_k of A and B respectively.

5. $I(K) = \{\mathfrak{A} : \exists \mathfrak{B} \in K,\ \mathfrak{A} \cong \mathfrak{B}\}$.

6. Let ϕ be an universal formula, i. e., $\phi: \forall v_0 \ldots \forall v_{n-1} \psi$; ψ open and $HT(\phi) = \{\tau_1 \ldots \tau_m\}$ then we define

$$\phi^*: \forall v_0 \ldots \forall v_{n-1} \left(\bigwedge_{i=1}^{m} \exists x_i (x_i \simeq \tau_i) \rightarrow \phi \right).$$

7. $K \in \cup C'_\Delta \leftrightarrow K = Mod\ \Sigma^*$ where $\Sigma^* = \{\phi^* : \phi \in \Sigma\}$ and Σ is a class of universal sentences.

8. *Direct product of partial algebras.*

Let $\mathfrak{A}_i$, $i \in I$, be partial algebras where $\mathfrak{A}_i = \langle A_i, R_{(i)}, F_{(i)}, c_{(i)} \rangle$ with

$R_{(i)} = \langle R_{(i)k} : k \in K \rangle$ relations,

$F_{(i)} = \langle F_{(i)j} : j \in J \rangle$ partial operations,

$c_{(i)} = \langle c_{(i)m} : m \in M \rangle$ constants.

Then $\mathfrak{B} = \prod_{i \in I} \mathfrak{A}_i$ (*direct product*) iff $\mathfrak{B} = \langle \prod_{i \in I} A_i, S, G, C \rangle$ where $\mathfrak{B}$ is the partial algebra defined by:

Let $\rho_0, \ldots, \rho_{n_{\nu-1}} \in \prod_{i \in I} A_i \cup \{X\}$, then

(a) For each $j \in J$

(i) If $F_{(i)_j}(\rho_0(i),\ldots,\rho_{n_{\nu-1}}(i)) \in A_i$ for each $i \in I$, then

$G_\nu(\rho_0,\ldots,\rho_{n_{\nu-1}})(i) = F_{(i)_j}(\rho_0(i),\ldots,\rho_{n_{\nu-1}}(i))$,

(ii) If $F_{(i)_j}(\rho_0(i),\ldots,\rho_{n_{\nu-1}}(i)) \notin A_i$ for some $i \in I$, then

$G_j(\rho_0,\ldots,\rho_{n_{\nu-1}}) = X$;

(b) for each $k \in K$, $S_k = \{\langle \rho_0,\ldots,\rho_{n_{\nu-1}}\rangle : \langle \rho_0(i),\ldots, \rho_{n_{\nu-1}}(i)\rangle \in R^{\mathfrak{A}_i}_{(i)}$ for each $i \in I\}$;

(c) for each $m \in M$, $c_m = \langle c_{(i)m}\rangle_{i \in I}$.

OBSERVATION: $\prod_{i \in I} \mathfrak{A}_i$ *preserves the identities of* $\mathfrak{A}$, that is, if for every $i \in I$, $\mathfrak{A} \models \tau \simeq \tau'[t_1(i),\ldots,t_n(i)]$ then $\prod_{i \in I} \mathfrak{A}_i \models \tau \simeq \tau'[t_1,\ldots,t_n]$ where $t_1,\ldots,t_n \in \prod_{i \in I} A_i$.

9. *Reduced product of partial algebras.*

Let $\mathcal{D}$ be a proper filter over I, $\mathfrak{A}_i$ partial algebras defined as in (7) for each $i \in I$. Then; let $f, g \in \prod_{i \in I} A_i$, we define $f =_{\mathcal{D}} g$ iff $\{i \in I : f(i) = g(i)\} \in \mathcal{D}$ and $f^{\mathcal{D}} = \{g \in \prod_{i \in I} A_i : f =_{\mathcal{D}} g\}$; also $\prod_{\mathcal{D}} A_i = \{f^{\mathcal{D}} : f \in \prod_{i \in I} A_i\}$.

Now we define the *reduced product of partial algebras* as follows: $\mathfrak{B} = \prod_{\mathcal{D}} \mathfrak{A}_i$ ($\mathfrak{B}$ is the *reduced product of* $\mathfrak{A}_i$) iff $\mathfrak{B} = \langle \prod_{\mathcal{D}} A_i, S, G, C\rangle$ where $\mathfrak{B}$ is the partial algebra defined by:

Let $\rho_0^{\mathcal{D}},\ldots,\rho_{n_{\nu-1}}^{\mathcal{D}} \in \prod_{\mathcal{D}} A_i$,

(a) for each $j \in J$:

(i) If $\{i \in I : F_{(i)_j}(\rho_0(i),\ldots,\rho_{n_{\nu-1}}(i)) \in A_i\} \in \mathcal{D}$ then

$G(\rho_0^{\mathcal{D}},\ldots,\rho_{n_{\nu-1}}^{\mathcal{D}}) = \langle F_{(i)_j}(\rho_0(i),\ldots,\rho_{n_{\nu-1}}(i)) : i \in I\rangle_{\mathcal{D}}$

(ii) If $\{i \in I: F_{(i)_j}(\rho_0(i), \ldots, \rho_{n_{\nu-1}}(i)) \in A_i\} \notin \mathcal{D}$

then $G_j(\rho_0^{\mathcal{D}}, \ldots, \rho_{n_{\nu-1}}^{\mathcal{D}}) = X$;

(b) for each $k \in K$, $S_k(\rho_0^{\mathcal{D}}, \ldots, \rho_{n_{\nu-1}}^{\mathcal{D}})$ iff

$\{i \in I: R_{(i)_k}^{\mathfrak{A}_i}(\rho_0(i), \ldots, \rho_{n_{\nu-1}}(i))\} \in \mathcal{D}$;

(c) for each $m \in M$, $C_m = \langle c_{(i)_m} : i \in I \rangle_{\mathcal{D}}$.

10. We shall define now *Horn*$^{(*)}$ *formulas* in the following way,

(i) *Basic formulas* :

(a) The atomic formulas are basic formulas.

(b) $\sim P_\nu \tau_1 \ldots \tau_{n_\nu}$ are basic formulas (where P_ν is a relation symbol and $\tau_1 \ldots \tau_{n_\nu}$ are terms).

(c) $\exists x(x \simeq \tau_1) \wedge (\tau_1 \not\simeq \tau_2)$ are basic formulas (where τ_1, τ_2 are terms).

(ii) *Basic Horn*$^{(*)}$ *formulas:*

ϕ is a basic Horn$^{(*)}$ formula iff ϕ: $(\sigma_1 \wedge \ldots \wedge \sigma_n) \to \sigma_0$ where $n \geq 0$, each σ_i $(i: 1,\ldots,n)$ is an atomic formula, and σ_0 is a basic formula.

(iii) *Horn*$^{(*)}$ *formulas:*

A Horn$^{(*)}$ formula is built up from basic Horn$^{(*)}$ formulas with the connectives $\wedge$, $\exists$, and $\forall$.

5. CHARACTERIZATION OF THE UNIVERSAL CLASSES.

In this section, the notation that has not been previously defined is that of [11].

THEOREM 1. $K \in \cup C'_\Delta$ *iff the following conditions are simultaneously satisfied:*

(*i*) $I(K) = K$,

(*ii*) $S(K) = K$,

(*iii*) *For each model* $\mathfrak{A}$ *such that* $S_\omega(\mathfrak{A}|I'\ J') \subseteq S(K|I'\ J')$ *with* I', J' *finite subsets of* I, J *respectively, we have* $\mathfrak{A} \in S(K)$.

PROOF: Suppose $K \in \cup C'_\Delta$, i. e., $K = Mod\,\Sigma^*$ where Σ is a set of universal sentences.

(*i*) We have to prove that $I(K) = K$, so let $\mathfrak{A} \cong \mathfrak{B}$ and $\phi^* \in \Sigma^*$. If $\mathfrak{A} \models \phi^*$ and $\phi: \forall v_0 \ldots \forall v_{\nu-1}\psi$; then it is easy to see that, $\mathfrak{A} \models \psi[x]$ iff $\mathfrak{B} \models \psi[f \circ x]$ where $\mathfrak{A} \cong_f \mathfrak{B}$; thus $\mathfrak{A} \models \phi^*$ iff $\mathfrak{B} \models \phi^*$.

(*ii*) We have to show that $S(K) = K$.

Let $\mathfrak{A} \in S(K)$, then $\mathfrak{A}$ is a partial subalgebra of $\mathfrak{B}$ with $\mathfrak{B} \in K$, thus $\mathfrak{B} \in Mod(\Sigma^*)$ and we need to show that $\mathfrak{A} \in Mod(\Sigma^*)$. Let $\phi \in \Sigma$, ϕ : $\forall v_0 \ldots \forall v_{n-1}\ \psi$; then if $\mathfrak{B} \models \phi^*$, we have

$$\mathfrak{B} \models [\bigwedge_{i=0}^{r-1} \exists x_i(x_i \cong \tau_i) \to \psi][y]$$

for every $y \in {}^{Fv(\psi)}B$. Therefore,

$$\mathfrak{B} \models [\bigwedge_{i=0}^{r-1} \exists x_i(x_i \cong \tau_i) \to \psi][y]$$

for every $y \in {}^{Fv(\psi)}A$ because $\mathfrak{A}$ is a partial subalgebra of $\mathfrak{B}$.

Thus we need to show that:

$\mathfrak{B} \models \psi[y]$ iff $\mathfrak{A} \models \psi[y]$ for every $y \in {}^{Fv(\psi)}A$, and this can be shown by a simple induction over molecular formulas.

(*iii*) Let ϕ: $\forall v_0 \ldots \forall v_{n-1}\ \psi$, and suppose that the highest terms of ψ are assigned elements of A, then:

If $\sim(\mathfrak{A} \models \phi)$ then there exist $\tau_0, \ldots, \tau_{n-1} \in |\mathfrak{A}|$ such that

$$\sim(\mathfrak{A} \models \psi \begin{bmatrix} v_0 \ldots v_{n-1} \\ \tau_0 \ldots \tau_{n-1} \end{bmatrix}) .$$

Let $\mathfrak{A}' = \mathfrak{A}|\{\tau^{\mathfrak{A}}: \tau$ is a term that occurs in $\psi\}$; then $|\mathfrak{A}|$ is finite and $\sim(\mathfrak{A} \models \phi^*)$. Thus, if $\mathfrak{A} \notin K$, then there exists a finite $\mathfrak{A}'$ such that $\mathfrak{A}' \notin S(K)$ and $\mathfrak{A}'$ is a partial subalgebra of $\mathfrak{A}$.

To prove the theorem on the other direction we need the following:

LEMMA 2. *Let*

$$\mathfrak{A} = \langle A, \rho_i, f_j, c_k \rangle_{\substack{i \in I \\ j \in J \\ k \in K}}$$

be a finite partial algebra, let $K = \{\mathfrak{B} : (\mathfrak{A} \mid I'J') \notin IS(\mathfrak{B} \mid I'J')\}$ *where* $I' \subseteq I$, $J' \subseteq J$ *are finite sets, then* $K \in UC'_{\Delta}$.

PROOF: Let $|\mathfrak{A}| = \{a_0, \ldots, a_{n-1}\}$. Let ϕ be the set consisting of the following formulas:

i) $v_i = v_j$, $i < j < n$;

ii) $\rho_i(v_{k_0}, \ldots, v_{k_{n-1}})$, if $\sim \rho_i^{\mathfrak{A}} a_{k_0} \ldots a_{k_{n-1}}$, $i \in I'$;

iii) $\sim \rho_i(v_{k_0}, \ldots, v_{k_{n-1}})$, if $\rho_i^{\mathfrak{A}} a_{k_0}, \ldots, a_{k_{n-1}}$, $i \in I'$;

iv) $f_j(v_k, \ldots, v_{k_{n-1}}) = f_l(v_{k_0}, \ldots, v_{k_{n-1}})$, $j, l \in J'$, $j \neq l$, if $f_j^{\mathfrak{A}}(a_{k_0}, \ldots, v_{k_{n-1}}) \in A$ and $f_l^{\mathfrak{A}}(a_{k_0}, \ldots, a_{k_{n-1}}) \in A$ and $f_j^{\mathfrak{A}}(a_{k_0}, \ldots, a_{k_{n-1}}) \neq f_l^{\mathfrak{A}}(a_{k_0}, \ldots, a_{k_{n-1}})$.

v) $f_j(v_{\nu_0}, \ldots, v_{\nu_{n-1}}) \neq f_k(v_{\nu_0}, \ldots, v_{\nu_{n-1}})$, $j, k \in J'$, $j \neq k$, if $f_j^{\mathfrak{A}}(a_{\nu_0}, \ldots, a_{\nu_{n-1}}) = f_k^{\mathfrak{A}}(a_{\nu_0}, \ldots, a_{\nu_{n-1}}) \in A$.

Then Φ is a finite set of formulas, so, let $\langle \phi_0, \ldots, \phi_{r-1} \rangle$ be an enumeration of the elements of Φ, and let

$$\gamma^*: \forall v_0 \ldots \forall v_{n-1}((\bigwedge_{i=0}^{p-1} x_i(x_i \simeq \tau_i)) \rightarrow (\phi_0 \vee \ldots \vee \phi_{r-1}).$$

The rest of the proof of the Lemma and the Theorem is very similar to the theorem of characterization of universal classes for relational systems. (see [11].) ■

6. PRESERVATION OF HORN(*) FORMULAS.

LEMMA 3. *Let* $\mathfrak{B} = \prod_{D} \mathfrak{A}$ *with* D *a proper filter over* I, $\mathfrak{A}_i$ *partial al-*

gebras for each $i \in I$, *then, for every term* $\tau(x_1,\ldots,x_n)$ *and elements* $t_1^{\mathcal{D}},\ldots,t_n^{\mathcal{D}} \in B$,

$$\tau\,[t_1^{\mathcal{D}},\ldots,t_n^{\mathcal{D}}] = \begin{cases} \langle \tau^{\mathfrak{A}_i}[t_1(i),\ldots,t_n(i)] : i \in I\rangle_{\mathcal{D}}, \text{ if } \{i \in I : \\ \mathfrak{A}_i \models \exists x(x \simeq \tau)[t_1(i),\ldots,t_n(i)]\} \in \mathcal{D}; \\ \\ \textit{or} \\ \\ X, \text{ if } \{i \in I : \mathfrak{A} \models \exists x(x \simeq \tau)[t_1(i),\ldots,t_n(i)]\} \notin \mathcal{D}. \end{cases}$$

PROOF: We prove the Lemma by induction over terms.

(i) If τ is a constant or a variable, then is trivial.

(ii) If $\tau = G(\tau_1(x_1\ldots x_n),\ldots,\tau_m(x_1\ldots x_n))$ then,

$$\tau^{\mathfrak{B}}[t_1^{\mathcal{D}},\ldots,t_n^{\mathcal{D}}] = G(\tau_1^{\mathfrak{B}}[t_1^{\mathcal{D}},\ldots,t_n^{\mathcal{D}}],\ldots,\tau_m^{\mathfrak{B}}[t_1^{\mathcal{D}},\ldots,t^{\mathcal{D}}]) =$$

$$\begin{cases} F_{(i)}\,(\tau_1^{\mathfrak{A}_i}[t_1(i)\ldots t_n(i)],\ldots,\tau_m^{\mathfrak{A}_i}[t_1(i),\ldots,t_n(i)]) : i \in I\rangle_{\mathcal{D}}, \\ \text{if } \{i \in I : \mathfrak{A}_i \models \exists x(x \simeq \tau)[t_1(i),\ldots,\ _n(i)]\} \in \mathcal{D}; \\ \text{or} \\ X, \text{ if } \{i \in I : \mathfrak{A}_i \models \exists x(x \simeq \tau)[t_1(i)\ldots t_n(i)]\} \notin \mathcal{D}. \end{cases}$$

If $\{i \in I : \mathfrak{A}_i \models \exists x(x \simeq \tau)[t_1(i)\ldots t_n(i)]\} \in \mathcal{D}$ then applying the inductive hypothesis we have that

$$\tau_k^{\mathfrak{B}}[t_1^{\mathcal{D}}\ldots t_n^{\mathcal{D}}] = g_k^{\mathcal{D}} \text{ with } g_k = \langle \tau_k^{\mathfrak{A}_i}[t_1(i)\ldots t_n(i)] : i \in I\rangle ;$$

then

$$G(g_1^{\mathcal{D}},\ldots,g_m^{\mathcal{D}}) = \langle F_{(i)}\,(g_1(i),\ldots,g_m(i)) : i \in I\rangle_{\mathcal{D}},$$

and also

$$\tau^{\mathfrak{A}_i}[t_1^{\mathcal{D}}\ldots t_n^{\mathcal{D}}] = G(g_1^{\mathcal{D}},\ldots,g_m^{\mathcal{D}}) = \langle \tau^{\mathfrak{A}_i}[t_1(i)\ldots t_n(i)] : i \in I\rangle_{\mathcal{D}}.$$

And if $\{i \in I : \mathfrak{A}_i \models \exists x(x \simeq \tau)[t_1(i)\ldots t_n(i)]\} \notin \mathcal{D}$, then

$$\tau^{\mathfrak{B}}[t_1^{\mathcal{D}}\ldots t_n^{\mathcal{D}}] = X. \quad \blacksquare$$

We note that the fundamental theorem for ultraproducts is still valid for partial algebras in our sense.

THEOREM 2. *Let $\mathcal{D}$ be an ultrafilter over I, $\mathfrak{B} = \prod_{\mathcal{D}} \mathfrak{A}_i$, $\mathfrak{A}_i$ partial algebras for $i \in I$ and $t_1^{\mathcal{D}}, \dots, t_n^{\mathcal{D}}$ B. Then,*

$$\mathfrak{B} \models \phi[t_1^{\mathcal{D}}, \dots, t_n^{\mathcal{D}}] \text{ iff } \{i \in I : \mathfrak{A}_i \models \phi[t_1(i) \dots t_n(i)]\} \in \mathcal{D}.$$

PROOF: (i) If $\phi(x_1, \dots, x_n)$ is $\tau_1(x_1, \dots, x_n) \simeq \tau_2(x_1, \dots, x_n)$ then $\mathfrak{B} \models \phi[t_1^{\mathcal{D}}, \dots, t_n^{\mathcal{D}}]$ iff $\tau_1^{\mathfrak{B}}[t_1^{\mathcal{D}}, \dots, t_n^{\mathcal{D}}] = \tau_2^{\mathfrak{B}}[t_1^{\mathcal{D}}, \dots, t_n^{\mathcal{D}}]$ iff $\tau_1^{\mathfrak{B}}[t_1^{\mathcal{D}}, \dots, t_n^{\mathcal{D}}] = \tau_2^{\mathfrak{B}}[t_1^{\mathcal{D}}, \dots, t_n^{\mathcal{D}}]$ or ($\tau_1^{\mathfrak{B}}[t_1^{\mathcal{D}}, \dots, t_n^{\mathcal{D}}] \notin B$ and $\tau_2^{\mathfrak{B}}[t_1^{\mathcal{D}}, \dots, t_n^{\mathcal{D}}] \notin B$).

Suppose first that $\mathfrak{B} \models \phi[t_1^{\mathcal{D}}, \dots, t_n^{\mathcal{D}}]$, then: if $\tau_1^{\mathfrak{B}}[t_1^{\mathcal{D}}, \dots, t_n^{\mathcal{D}}] = \tau_2^{\mathfrak{B}}[t_1^{\mathcal{D}}, \dots, t_n^{\mathcal{D}}] \in B$ then

$\langle \tau_1^{\mathfrak{A}_i}[t_1(i) \dots t_n(i)] : i \in I \rangle_{\mathcal{D}} = \langle \tau_2^{\mathfrak{A}_i}[t_1(i) \dots t_n(i)] : i \in I \rangle_{\mathcal{D}}$;

because of Lemma 3, and therefore $\{i \in I : \phi[t_1(i), \dots, t_n(i)]\} \in \mathcal{D}$.

If $\{i \in I : \tau_1^{\mathfrak{A}_i}[t_1(i) \dots t_n(i)] \in A_i\} \notin \mathcal{D}$ and

$\{i \in I : \tau_2^{\mathfrak{A}_i}[t_1(i) \dots t_n(i)] \in A_i\} \notin \mathcal{D}$

then, because $\mathcal{D}$ is an ultrafilter we have that,

$\{i \in I : \tau_1^{\mathfrak{A}_i}[t_1(i) \dots t_n(i)] \notin A_i\} \in \mathcal{D}$ and

$\{i \in I : \tau_2^{\mathfrak{A}_i}[t_1(i) \dots t_n(i)] \notin A_i\} \in \mathcal{D}$; therefore,

$\{i \in I : \tau_1^{\mathfrak{A}_i}[t_1(i) \dots t_n(i)] \notin A_i \wedge \tau_2^{\mathfrak{A}_i}[t_1(i) \dots t_n(i)] \notin A_i\} \in \mathcal{D}$,

then

$\{i \in I : \tau_1^{\mathfrak{A}_i}[t_1(i) \dots t_n(i)] = \tau_2^{\mathfrak{A}_i}[t_1(i) \dots t_n(i)]\} \in \mathcal{D}$,

and therefore $\{i \in I : \mathfrak{A}_i \models \phi[t_1(i), \dots, t_n(i)]\} \in \mathcal{D}$.

Now, let us suppose that

$\{i \in I : \mathfrak{A}_i \models \phi[t_1(i), \dots, t_n(i)]\} \in \mathcal{D}$; then

$$\{i \in I : \tau_1^{\mathfrak{A}_i}[t_1(i) \dots t_n(i)] = \tau_2^{\mathfrak{A}_i}[t_1(i) \dots t_n(i)]\} \in \mathcal{D}. \quad (*)$$

If $\{i \in I : \tau_1^{\mathfrak{A}_i}[t_1(i) \dots t_n(i)] \in A_i\} \in \mathcal{D}$

then $\{i \in I: \tau_2^{\mathfrak{A}_i}[t_1(i)\dots t_n(i)] \in A_i\} \in \mathcal{D}$.

Thus

$\langle \tau_1^{\mathfrak{A}_i}[t_1(i)\dots t_n(i)] : i \in I\rangle_{\mathcal{D}} = \langle \tau_2^{\mathfrak{A}_i}[t_1(i)\dots t_n(i)] : i \in I\rangle_{\mathcal{D}}$

and therefore $\mathfrak{B} \models \phi[t_1^{\mathcal{D}},\dots,t_n^{\mathcal{D}}]$.

If $\{i \in I: \tau_1^{\mathfrak{A}_i}[t_1(i),\dots,t_n(i)] \in A_i\} \notin \mathcal{D}$ then

$\{i \in I: \tau_2^{\mathfrak{A}_i}[t_1(i),\dots,t_n(i)] \in A_i\} \notin \mathcal{D}$.

Thus, $\tau_1^{\mathfrak{B}}[t_1^{\mathcal{D}},\dots,t_n^{\mathcal{D}}] = X$, and because of (*), $\tau_2^{\mathfrak{B}}[t_1^{\mathcal{D}},\dots,t_n^{\mathcal{D}}] = X$ and, therefore, $\mathfrak{B} \models \phi[t_1^{\mathcal{D}},\dots,t_n^{\mathcal{D}}]$.

(ii) If $\phi(x_1,\dots,x_n)$ is $P_\nu \tau_1(x_1\dots x_n),\dots,\tau_m(x_1\dots x_n)$, then

$\mathfrak{B} \models \phi[t_1^{\mathcal{D}},\dots,t_n^{\mathcal{D}}]$ iff $P_\nu^{\mathfrak{B}}\ \tau_1^{\mathfrak{B}}[t_1^{\mathcal{D}},\dots,t_n^{\mathcal{D}}],\dots,\ \tau_m^{\mathfrak{B}}[t_1^{\mathcal{D}},\dots,t_n^{\mathcal{D}}]$

iff $\{i \in I: P^{\mathfrak{A}_i}_{(i)_\nu}\ \tau_1^{\mathfrak{A}_i}[t_1(i),\dots,t_n(i)],\dots,\tau_m^{\mathfrak{A}_i}[t_1(i),\dots,t_n(i)]\} \in \mathcal{D}$

iff $\{i \in I: \mathfrak{A}_i \models \phi[t_1(i),\dots,t_n(i)]\} \in \mathcal{D}$.

(iii) If $\phi(x_1,\dots,x_n)$ is $\sim\psi(x_1,\dots,x_n)$ and the theorem is true for ψ, then:

$\mathfrak{B} \models \phi[t_1^{\mathcal{D}},\dots,t_n^{\mathcal{D}}]$ iff it is not true that $\mathfrak{B} \models \psi[t_1^{\mathcal{D}},\dots,t_n^{\mathcal{D}}]$,

iff $\{i \in I: \mathfrak{A}_i \models \psi[t_1(i),\dots,t_n(i)]\} \notin \mathcal{D}$,

iff $\{i \in I: not(\mathfrak{A}_i \models \psi[t_1(i),\dots,t_n(i)]\} \in \mathcal{D}$,

since $\mathcal{D}$ is an ultrafilter,

iff $\{i \in I: \mathfrak{A}_i \models \phi[t_1(i),\dots,t_n(i)]\} \in \mathcal{D}$.

(iv) For $\phi(x_1,\dots,x_n) = \psi_1(x_1,\dots,x_n) \wedge \psi_2(x_1,\dots,x_n)$, we use the fact that $\mathcal{D}$ is a filter.

(v) If $\phi(x_1,\dots,x_n)$ is $\exists x_0 \psi(x_0,x_1,\dots,x_n)$, then: $\mathfrak{B} \models \phi[t_1^{\mathcal{D}},\dots,t_n^{\mathcal{D}}]$

iff there exists $t_0^{\mathcal{D}} \in B$ such that $\mathfrak{B} \models \psi[t_0^{\mathcal{D}},t_1^{\mathcal{D}},\dots,t_n^{\mathcal{D}}]$

iff there exists $t_0^{\mathcal{D}} \in B$ such that $\{i \in I: \mathfrak{A}_i \models \psi[t_0(i),\dots,t_n(i)]\} \in \mathcal{D}$. (*)

But $\mathfrak{A}_i \models \psi[t_0(i),\ldots,t_n(i)]$ implies that $\mathfrak{A}_i \models \phi[t_1(i),\ldots,t_n(i)]$, hence

$\{i\in I\colon \mathfrak{A}_i \models \psi[t_0(i),\ldots,t_n(i)]\} \subseteq \{i\in I\colon \mathfrak{A}_i \models \phi[t_1(i),\ldots,t_n(i)]\}$,

and because of (*) we have,

$\{i\in I\colon \mathfrak{A}_i \models \phi[t_1(i),\ldots,t_n(i)]\} \in \mathcal{D}$.

Now, if $\{i\in I\colon \mathfrak{A}_i \models \phi[t_1(i),\ldots,t_n(i)]\} \in \mathcal{D}$, then we can always choose $t_0 \in \prod_{i\in I} A_i$ such that satisfies (*), and therefore

$$\mathfrak{B} \models \phi[t_1^{\mathcal{D}},\ldots,t_n^{\mathcal{D}}]. \blacksquare$$

THEOREM 3. *Let $\phi(x_1,\ldots,x_n)$ be a Horn(*) formula, $\mathfrak{A}_i$ partial algebras for $i\in I$, $\mathcal{D}$ a proper filter over I and $t_1,\ldots,t_n \in \prod_{i\in I} A_i$.*
If $\{i\in I\colon \mathfrak{A}_i \models \phi[t_1(i),\ldots,t_n(i)]\}\in\mathcal{D}$ then $\prod_{\mathcal{D}}\mathfrak{A}_i \models \phi[t_1^{\mathcal{D}},\ldots,t_n^{\mathcal{D}}]$.

PROOF: (i) $\phi(x_1,\ldots,x_n)$ is atomic.

(a) $\phi(x_1,\ldots,x_n)\colon \tau_1(x_1,\ldots,x_n) \simeq \tau_2(x_1,\ldots,x_n)$

If $\{i\in I\colon \mathfrak{A}_i \models \phi[t_1(i),\ldots,t_n(i)]\}\in\mathcal{D}$, then

$\{i\in I\colon \tau_1^{\mathfrak{A}_i}[t_1(i),\ldots,t_n(i)] = \tau_2^{\mathfrak{A}_i}[t_1(i),\ldots,t_n(i)]\}\in\mathcal{D}$,

therefore:

Let $Y = \{i\in I\colon \tau_1^{\mathfrak{A}_i}[t_1(i),\ldots,t_n(i)] \in A_i\}$ and

$\{i\in I\colon \tau_2^{\mathfrak{A}_i}[t_1(i),\ldots,t_n(i)]\in A_i\} = Z$.

If $Y\in\mathcal{D}$ then $Z\in\mathcal{D}$ and $\tau_1^{\mathfrak{B}}[t_1^{\mathcal{D}},\ldots,t_n^{\mathcal{D}}] = \langle \tau_1^{\mathfrak{A}_i}[t_1(i),\ldots,t_n(i)]\colon i\in I\rangle_{\mathcal{D}} = \langle \tau_2^{\mathfrak{A}_i}[t_1(i),\ldots,t_n(i)]\colon i\in I\rangle_{\mathcal{D}} = \tau_2^{\mathfrak{B}}[t_1^{\mathcal{D}},\ldots,t_n^{\mathcal{D}}]$.

If $Y\notin\mathcal{D}$ then $Z\notin\mathcal{D}$ and $\tau_1^{\mathfrak{B}}[t_1^{\mathcal{D}},\ldots,t_n^{\mathcal{D}}] = X = \tau_2^{\mathfrak{B}}[t_1^{\mathcal{D}},\ldots,t_n^{\mathcal{D}}]$

and therefore $\mathfrak{B} \models \phi[t_1^{\mathcal{D}},\ldots,t_n^{\mathcal{D}}]$.

(b) If $\phi(x_1,\ldots,x_n)$ is $P_\nu\tau_1(x_1,\ldots,x_n),\ldots,\tau_m(x_1,\ldots,x_n)$

and $\{i\in I\colon \mathfrak{A}_i \models \phi[t_1(i),\ldots,t_n(i)]\}\in\mathcal{D}$, then

$$\{i \in I: P^{\mathfrak{A}_i}_{(i)_\nu}\, \tau_1^{\mathfrak{A}_i}[t_1(i),\ldots, t_n(i)],\ldots,\tau_m^{\mathfrak{A}_i}[t_1(i),\ldots, t_n(i)]\} \in \mathcal{D}$$

iff $P^{\mathfrak{B}}_\nu\, \tau_1^{\mathfrak{B}}[t_1^{\mathcal{D}},\ldots,t_n^{\mathcal{D}}] \ldots \tau_m^{\mathfrak{B}}[t_1^{\mathcal{D}},\ldots,t_n^{\mathcal{D}}]$ iff $\mathfrak{B} \models \phi[t_1^{\mathcal{D}},\ldots,t_n^{\mathcal{D}}]$.

(ii) If $\phi(x_1,\ldots,x_n)$ is $\sim P_\nu\, \tau_1(x_1,\ldots,x_n),\ldots,\tau_m(x_1,\ldots,x_n)$ and $\{i \in I: \mathfrak{A}_i \models \phi[t_1(i),\ldots,t_n(i)]\} \in \mathcal{D}$ then

$$Y = \{i \in I: \sim P^{\mathfrak{A}_i}_{(i)_\nu}\, \tau_1^{\mathfrak{A}_i}[t_1(i),\ldots,t_n(i)],\ldots,\tau_m^{\mathfrak{A}_i}[t_1(i),\ldots,t_n(i)]\} \in \mathcal{D}.$$

If $\mathfrak{B} \models P_\nu\, \tau_1[t_1^{\mathcal{D}},\ldots,t_n^{\mathcal{D}}] \ldots \tau_m[t_1^{\mathcal{D}},\ldots,t_n^{\mathcal{D}}]$ then

$$Z = \{i \in I: P^{\mathfrak{A}_i}_{(i)_\nu}\, \tau_1^{\mathfrak{A}_i}[t_1(i),\ldots,t_n(i)],\ldots,\tau_m^{\mathfrak{A}_i}[t_1(i),\ldots,t_n(i)]\} \in \mathcal{D}.$$

Then $Y \cap Z \in \mathcal{D}$, but $Y \cap Z = \emptyset$, but this can not be because $\mathcal{D}$ is a proper filter.

Therefore, not ($\mathfrak{B} \models P_\nu\, \tau_1[t_1^{\mathcal{D}},\ldots,t_n^{\mathcal{D}}] \ldots \tau_m[t_1^{\mathcal{D}},\ldots,t_n^{\mathcal{D}}]$); then $\mathfrak{B} \models \phi[t_1^{\mathcal{D}},\ldots,t_n^{\mathcal{D}}]$.

(iii) If $\phi(x_1,\ldots,x_n)$ is $\exists x(x \simeq \tau_1)(x_1,\ldots,x_n) \wedge (\tau_1(x_1,\ldots,x_n) \not\simeq \tau_2(x_1,\ldots,x_n))$ and if $\{i \in I: \mathfrak{A}_i \models \phi[t_1(i),\ldots,t_n(i)]\} \in \mathcal{D}$ then

$$\{i \in I: \mathfrak{A}_i \models \exists x(x \simeq \tau_1[t_1(i),\ldots,t_n(i)]) \wedge \tau_1[t_1(i),\ldots, t_n(i)] \not\simeq \tau_2[t_1(i),\ldots, {}_n(i)]\} \in \mathcal{D}.$$

Thus $L = \{i \in I: \mathfrak{A}_i \models \tau_1^{\mathfrak{A}_i}[t_1(i),\ldots,t_n(i)] \in A_i\} \in \mathcal{D}$

and $Y = \{i \in I: \tau_1^{\mathfrak{A}_i}[t_1(i),\ldots,t_n(i)] \neq \tau_2^{\mathfrak{A}_i}[t_1(i),\ldots,t_n(i)]\} \in \mathcal{D}$

Now, let $c = \langle \tau_1^{\mathfrak{A}_i}[t_1(i),\ldots,t_n(i)] : i \in I\rangle_{\mathcal{D}}$, then $c \in B$ because $L \in \mathcal{D}$; therefore $\mathfrak{B} \models \exists x(x \simeq \tau_1)[t_1^{\mathcal{D}},\ldots, t_n^{\mathcal{D}}]$.

Now, let $Z = \{i \in I: \tau_2^{\mathfrak{A}_i}[t_1(i),\ldots,t_n(i)] \in A_i\}$.

If $Z \notin \mathcal{D}$, then $\tau_2^{\mathfrak{B}}[t_1^{\mathcal{D}},\ldots, t_n^{\mathcal{D}}] = X$ and because $L \in \mathcal{D}$, we have that $\mathfrak{B} \models (\tau_1 \not\simeq \tau_2)[t_1^{\mathcal{D}},\ldots,t_n^{\mathcal{D}}]$; and therefore $\mathfrak{B} \models \phi[t_1^{\mathcal{D}},\ldots, t_n^{\mathcal{D}}]$.

If $Z \in \mathcal{D}$, then

$$\langle \tau_1^{\mathfrak{A}_i}[t_1(i),\ldots, t_n(i)] : i \in I\rangle_{\mathcal{D}} \neq \langle \tau_1^{\mathfrak{A}_i}[t_1(i),\ldots, t_n(i)] : i \in I\rangle_{\mathcal{D}},$$

because if they were equal, then as we had $Y \in \mathcal{D}$, then $\emptyset \in \mathcal{D}$, but this can not be, because $\mathcal{D}$ is a proper filter. Therefore

$$\mathcal{B} \models \phi[t_1^{\mathcal{D}},\ldots,t_n^{\mathcal{D}}] .$$

(iv) Let ϕ be $(o_1 \wedge \ldots \wedge o_n) \rightarrow o_0$ and $\{i \in I: \mathfrak{A}_i \models \phi[t_1(i),\ldots, t_n(i)]\} \in \mathcal{D}$, then if $\mathcal{B} \models (o_1 \wedge \ldots \wedge o_n)[t_1^{\mathcal{D}},\ldots, t_n^{\mathcal{D}}]$
we have $\{i \in I: \mathfrak{A}_i \models (o_1 \wedge \ldots \wedge o_n)[t_1(i),\ldots,t_n(i)]\} \in \mathcal{D}$.
Thus $X_j = \{i \in I: \mathfrak{A}_i \models o_j[t_1(i),\ldots, t_n(i)] \in \mathcal{D}$ for each $j = 1,\ldots, n$.
Therefore $X_1 \cap X_2 \cap \ldots \cap X_n \cap \{i \in I: \mathfrak{A}_i \models \phi[t_1(i),\ldots, t_n(i)]\} \in \mathcal{D}$.
Hence $\{i \in I: \mathfrak{A}_i \models o_0[t_1(i),\ldots,t_n(i)]\} \in \mathcal{D}$
and since o_0 is a basic formula, by the inductive hypothesis we have that

$$\mathcal{B} \models o_0[t_1^{\mathcal{D}},\ldots, t_n^{\mathcal{D}}] ;$$

therefore

$$\mathcal{B} \models \phi[t_1^{\mathcal{D}},\ldots, t_n^{\mathcal{D}}] .$$

The rest of the cases for the Horn(*) formulas, are the usual ones (see [1], Chapter 6). ■

THEOREM 4.

(i) *Every Horn(*) sentence is preserved under reduced products.*

(ii) *Every Horn(*) sentence is preserved under direct products.*

PROOF: (i) Let ϕ be a Horn(*) sentence such that $\mathfrak{A}_i \models \phi$ for every $i \in I$ then, $\mathfrak{A}_i \models \phi[t_1(i),\ldots,t_n(i)]$ for $t_1,\ldots,t_n \in \prod_{i \in I} A_i$ then,
$I = \{i \in I: \mathfrak{A}_i \models \phi[t_1(i),\ldots, t_n(i)] \in \mathcal{D}$ and therefore

$$\prod_{\mathcal{D}} \mathfrak{A}_i \models \phi[t_1^{\mathcal{D}},\ldots,t_n^{\mathcal{D}}] .$$

(ii) Is a particular case of (i) when $\mathcal{D} = \{I\}$. ■

The following theorem is necessary to prove the converse of Theorem 4; we only indicate the modifications that are needed in the case of partial algebras of the proof which appears in [1].

THEOREM 5. *Let* $\mathfrak{A}_i$, $i \in I$, *be partial algebras,* $|I| = \alpha$, $2^{\alpha} = \alpha^{+}$ *and* $|A_i| \leq \alpha^{+}$. *Let* $\mathfrak{B}$ *be a partial algebra that is finite or* α^{+}*-saturated. If each Horn(*) formula that is true in almost all* $\mathfrak{A}_i$ *is also true in* $\mathfrak{B}$, *then* $\mathfrak{B}$ *is isomorphic to the reduced product* $\prod_{D} \mathfrak{A}_i$ *modulo* D *over* I.

PROOF: Here, we have to be careful to modify the usual proof (see [1], Lemma 6.2.4) so that the isomorphism will be such that:

(i) $h: A \cup \{X\} \xrightarrow{over} B \cup \{X\}$

(ii) $h(X) = X$

(iii) for every Horn(*) sentence $\phi(x_1, \ldots, x_n)$ and $y_1, \ldots, y_n \in A$, if for almost all $i \in I$, $\mathfrak{A}_i \models \phi[y_1(i), \ldots, y_n(i)]$ then $\mathfrak{B} \models \phi[hy_1(i), \ldots, hy_n(i)]$. ■

Finally, we have the last theorem which gives the characterization.

THEOREM 6. *Suppose* $2^{\omega} = \omega^{+}$, *a formula* ϕ *is preserved under reduced product iff is equivalent to a Horn(*) formula.*

The proof is similar to [1] Theorem 6.2.5 with some obvious modifications.

It was shown in [3], also in [7], that the hypothesis $2^{\omega} = \omega^{+}$ can be eliminated from the usual theorem of characterization of the Horn's formulas, and the same method can be applied here to eliminate from Theorem 6 the condition $2^{\omega} = \omega^{+}$.

REFERENCES.

[1] C. C. Chang and H. J. Keisler, Model Theory, North-Holland Pub. Co., Amsterdam, 1972.

[2] R. Chuaqui, *Introducción a la Metamatemática y sus aplicationes*, Univ. Católica de Chile, 1976.

[3] F. Galvin, *Horn sentences*, Thesis, Univ. of Minnesota, 1965.

[4] G. Grätzer, Universal Algebras, D. Van Nostrand Company, Inc., Princeton, 1968.

[5] L. Henkin, D. Monk, and A. Tarski, Cylindric Algebras, Chapter 0, North-Holland Pub. Co., Amsterdam, 1971.

[6] J. M. Herring, *Equivalences of several notions of theory completeness in a free logic*, Reports on Math. Logic, 6 (1976), 87-92.

[7] H. J. Keisler, *Reduced products and Horn classes*, Trans. Am., Math. Soc., 117 (1965), 307-328.

[8] H. Leblanc and R. H. Thomason, *Completeness Theorem for some presupposition free logics*, Fundamenta Mathematica, LXII. 2 (1968), 125-164.

[9] I. Mikenberg, A *logical system for partial algebras*, to appear in The Journal of Symbolic Logic, 1977.

[10] A. Tarski, A *simplified formalization of predicate logic with identity*, Archiv. für Mathematische Logik und Grundlagenforschung, 7, (1965), 61-79.

[11] A. Tarski, *Contributions to the Teory of Models, I*, Indagationes Mathematicae, 16 (1954), 572-588.

Instituto de Matemática
Universidad Católica de Chile
Santiago, Chile.

ADDED IN PROOF: This article was written when the author was at the State University of Campinas, São Paulo, Brazil, with a fellowship from the Brazilian Government.

Cylindric Algebras with a Property of Rasiowa and Sikorski.

by *CHARLES C. PINTER.*

ABSTRACT. It was observed by Rasiowa and Sikorski in *A proof of the completeness theorem of Gödel*, Fund. Math., vol. 37 (1950), pp. 193-200, that in any Lindenbaum algebra of formulas of finitary first-order logic, the following equality holds:

$$(1.1) \qquad [(\exists v_\kappa)\,\phi(v_\kappa)] = \Sigma_{\lambda \neq \kappa}[\phi(v_\lambda)],$$

where $\phi(v_\lambda)$ is the result of validly replacing v_κ by v_λ in $\phi(v_\kappa)$, and $\Sigma_{\lambda \neq \kappa}$ is the least upper bound of the $[\phi(v_\lambda)]$ for $\lambda \neq \kappa$. In the language of cylindric algebras, this equality may be written:

$$(1.2) \qquad c_\kappa x = \Sigma_{\lambda \neq \kappa} s^\kappa_\lambda x,$$

where κ is understood to remain fixed while λ varies.

In this paper we investigate the class of cylindric algebras which satisfies (1.2), and its relationship with other important class of cylindric algebras, and present several conditions which are equivalent to (1.2).

1. Introduction.

It was observed by Rasiowa and Sikorski in [8] that in any Lindenbaum algebra of formulas of finitary first-order logic, the following equality holds:

$$(1.1) \qquad [(\exists v_\kappa)\,\phi(v_\kappa)] = \Sigma_{\lambda \neq \kappa}[\phi(v_\lambda)],$$

where $\phi(v_\lambda)$ is the result of validly replacing v_κ by v_λ in $\phi(v_\kappa)$, and $\Sigma_{\lambda \neq \kappa}[\phi(v_\lambda)]$ is the least upper bound of the $[\phi(v_\lambda)]$ for $\lambda \neq \kappa$. Equality (1.1) is an immediate consequence of the facts that

(i) $\vdash \phi(v_\lambda) \to (\exists v_\kappa)\phi(v_\kappa)$ *for every* $\lambda \neq \kappa$, and

(ii) *If* $\vdash \phi(v_\lambda) \to \psi$ *for every* $\lambda \neq \kappa$, *then this is true for some* v_λ *which is not free in* ψ, *hence* $\vdash (\exists v_\lambda)\,\phi(v_\lambda) \to \psi$.

In the language of cylindric algebras, (1.1) may be written in the form

$$(1.2) \qquad c_\kappa x = \Sigma_{\lambda \neq \kappa}\, s^\kappa_\lambda x,$$

where κ is understood to remain fixed while λ varies.

In many approaches to algebraic logic, the algebraic counterparts of quantifiers is formulated by means of equation (1.2); this is true, for example, in Rasiowa and Sikorski [7] and [8], Rieger [9], Leblanc [4], Preller [6], and many others. However, there are infinitary logics of a perfectly reasonable kind whose Lindenbaum algebras do not satisfy (1.2), and it is known that (1.2) does *not* hold in every cylindric algebra. In this paper we investigate the class of cylindric algebras which satisfy (1.2), and its relationship with other important classes of cylindric algebras. We present several conditions which are equivalent to (1.2), show that the class of **CA**'s satisfying (1.2) is not an equational nor even an elementary class, and prove that every CA satisfying (1.2) is representable, although there are representable **CA**'s where (1.2) fails.

Our notation and terminology is that of Henkin, Monk and Tarski [1].

2. Equivalent conditions.

We let $\mathbf{CA}_\alpha$ designate the class of cylindric algebras of degree α, and $\mathbf{RS}_\alpha$ the class of cylindric algebras of degree α which satisfy (1.2). (An algebra which satisfy (1.2) will sometimes be said to have the *RS property*.) If $\mathfrak{A}$ is an arbitrary member of $\mathbf{CA}_\alpha$, we consider the following conditions on $\mathfrak{A}$:

$$(2.1) \qquad \Sigma_{\lambda \neq \kappa}\, d_{\kappa\lambda} = 1,$$

where κ is understood to remain fixed while λ ranges over $\alpha - \{\kappa\}$.

(2.2) Let $\kappa < \alpha$ be fixed.

If $s^\kappa_\lambda x = 1$ for each $\lambda \neq \kappa$, then $x = 1$.

(2.3) Let $\kappa < \alpha$ be fixed.

If $s_\lambda^\kappa x \leqslant y$ for each $\lambda \neq \kappa$, then $x \leqslant y$.

We will prove, next, that each of these conditions is equivalent to (1.2).

(2.4) THEOREM. *If* $\mathfrak{A} \in \mathsf{CA}_\alpha$, *then* (1.2) $\Longleftrightarrow$ (2.1) $\Longleftrightarrow$ (2.2) $\Longleftrightarrow$ (2.3).

PROOF:

(1.2) $\Rightarrow$ (2.2): Suppose $s_\lambda^\kappa x = 1$ for every $\lambda \neq \kappa$. Then $s_\lambda^\kappa(-x) = 0$ for every $\lambda \neq \kappa$, so $0 = \Sigma_{\lambda \neq \kappa} s_\lambda^\kappa(-x) = c_\kappa(-x)$. It follows that $-x = 0$, so $x = 1$.

(2.2) $\Longleftrightarrow$ (2.1): (2.1) is clearly equivalent to the statement that if $x \geqslant d_{\kappa\lambda}$ for every $\lambda \neq \kappa$, then $x = 1$. The result follows immediately from the fact that $x \geqslant d_{\kappa\lambda}$ iff $s_\lambda^\kappa x = 1$.

(2.2) $\Rightarrow$ (2.3): Suppose that $s_\lambda^\kappa x \leqslant y$ for every $\lambda \neq \kappa$. Then $s_\lambda^\kappa x \leqslant s_\lambda^\kappa y$, so $s_\lambda^\kappa(x - y) = 0$ for every $\lambda \neq \kappa$. By (2.2), $x - y = 0$, that is, $x \leqslant y$.

(2.3) $\Rightarrow$ (1.2): Let $V = \{s_\lambda^\kappa x : \lambda \neq \kappa\}$, and show that $c_\kappa x = \Sigma V$. Clearly, $c_\kappa x$ is an upper bound of V. Let y be any upper bound of V; now, $y \geqslant s_\lambda^\kappa x$ $\Rightarrow c_\kappa^\delta y \geqslant s_\lambda^\kappa x$, hence $c_\kappa^\delta y$ is an upper bound of V. By (2.3), $x \leqslant c_\kappa^\delta y$, hence $c_\kappa x \leqslant c_\kappa c_\kappa^\delta y = c_\kappa^\delta y \leqslant y$. ■

In a number of arguments of algebraic logic (see, for example, the next Section), we require versions of (1.2) and (2.1) in which the sum ranges over a *cofinite subset* of $\alpha - \{\kappa\}$, rather than all of $\alpha - \{\kappa\}$. These apparently stronger versions of (1.2) and (2.1) may be stated as follows:

(1.2)* For each finite subset $F \subseteq \alpha$ such that $\kappa \in F$,

$$c_\kappa x = \Sigma_{\lambda \in \alpha - F}\, s_\lambda^\kappa x .$$

(2.1)* For each finite subset $F \subseteq \alpha$ such that $\kappa \in F$,

$$\Sigma_{\lambda \in \alpha - F}\, d_{\kappa\lambda} = 1 .$$

Although (2.1)* appears to be stronger than (2.1), it turns out, in fact, to be equivalent to it. Indeed, certainly (2.1)* $\Rightarrow$ (2.1). Conversely, if $F = \{\nu_1, \ldots, \nu_p, \kappa\}$ and $\mu \notin F$, then

$$1 = s_\mu^{\nu_1} \ldots s_\mu^{\nu_{\rho}} 1 = s_\mu^{\nu_1} \ldots s_\mu^{\nu_\rho} \Sigma_{\lambda \neq \kappa} d_{\kappa\lambda} = \Sigma_{\lambda \in \alpha - F} d_{\kappa\lambda}.$$

This proves that (2.1) $\Rightarrow$ (2.1)*. Now, (1.2)* $\Longleftrightarrow$ (2.1)* by the proof of Theorem 2.4, hence, (1.2) $\Longleftrightarrow$ (1.2)*. Thus,

(2.5) THEOREM. *In any* $\mathfrak{A} \in CA_\alpha$, (1.2) $\Longleftrightarrow$ (1.2)* *and* (2.1) $\Longleftrightarrow$ (2.1)*.

REMARK. If $\mathfrak{A}$ is a cylindric algebra and every sum

(2.5) $\quad \Sigma_{\lambda \neq \kappa} s_\lambda^\kappa x$

exists in $\mathfrak{A}$ for each fixed $\kappa < \alpha$ and every $x \in A$, one might be led to suppose that $\mathfrak{A}$ has the **RS** property. In fact, this is not so ! The sums (2.5) may exist in a cylindric algebra without being equal to $c_\kappa x$. To show this, let $\mathfrak{A}$ be a non-discrete CA and $\mathfrak{B}$ its completion in the sense of Jónsson-Tarski [2]. Since $\mathfrak{B}$ is complete, every sum (2.5) exists in $\mathfrak{B}$. If $\mathfrak{B}$ has the RS property then $\mathfrak{B}$ satisfy (2.1), so (by [2]; Theorem 2.15 and definition 1.19), there are $\mu_1, \ldots, \mu_p$ such that

$$d_{\kappa\mu_1} + \ldots + d_{\kappa\mu_p} = 1,$$

from which it follows that $\mathfrak{B}$ is discrete. This is a contradiction.

3. DIAGONAL AND DIMENSION-COMPLEMENTED CYLINDRIC ALGEBRAS.

Diagonal cylindric algebras were introduced by Donald Monk in [5]. Their importance lies in the fact that *every diagonal cylindric algebra is representable.*

(3.1) DEFINITION. (D. Monk) *An algebra* $\mathfrak{A} \in CA_\alpha$ *is called a diagonal cylindric algebra iff the following conditions holds:*

(3.2) $[x \cdot d_{\kappa\lambda} = 0$ for any two distinct $\kappa, \lambda \in \alpha - F] \Rightarrow x = 0$, where x is any element of A and F is any finite subset of α.

(3.3) THEOREM. (3.2) *is equivalent to:* $\Sigma_{\substack{\kappa\neq\lambda \\ \kappa,\lambda\in\alpha-F}} d_{\kappa\lambda} = 1$.

(Note that in this sum, both κ and λ range over $\alpha - F$.)

PROOF: Suppose (3.2) holds, and $x \geqslant d_{\kappa\lambda}$ for all $\kappa \neq \lambda$ in $\alpha - F$; that is, $d_{\kappa\lambda}\cdot(-x) = 0$ for all $\kappa,\lambda \in \alpha - F$ such that $\kappa \neq \lambda$. Then by (3.2), $-x = 0$, so $x = 1$. Thus the sum in the statement of the Theorem is equal to 1. Conversely, suppose this sum equals 1, and let $x.d_{\kappa\lambda} = 0$ for all distinct $\kappa,\lambda \in \alpha - F$. Then $-x \geqslant d_{\kappa\lambda}$ for all distinct $\kappa,\lambda \in \alpha - F$, so $-x = 1$. Thus $x = 0$, which proves (3.2). ∎

Let DI_α denote the class of all diagonal cylindric algebras of degree α. It follows immediately from Theorem 2.4 and Definition 3.1 that *every cylindric algebra with the RS property is a diagonal cylindric algebra.* Thus,

$$RS_\alpha \subseteq DI_\alpha .$$

In particular,

(3.4) *every algebra with the RS property is representable.*

Let DC_α denote the class of dimension-complemented CA's of degree α. It is known that every $\mathfrak{A} \in DC_\alpha$ has the RS property. (Indeed, it is easy to see that any $\mathfrak{A} \in DC_\alpha$ satisfies (2.1); for if (2.1) fails, there is some $x \neq 1$ such that $x \geqslant d_{\kappa\lambda}$ for every $\lambda \neq \kappa$. One easily verifies that $\Delta x = \alpha$). Thus, $DC_\alpha \subseteq RS_\alpha$. Next it is easily verified that RS_α is closed under the formation of subalgebras and arbitrary direct products. Thus, RS_α includes all subdirect products of dimension-complemented algebras. But it is known that for $\alpha \geqslant \omega$, DC_α is not closed under subdirect products, hence $RS_\alpha \neq DC_\alpha$ for $\alpha \geqslant \omega$.

A counter-example will suffices to show that $RS_\alpha \neq DI_\alpha$, for $\alpha \geqslant \omega$. Let $\mathfrak{L} = (B, \cup, \cap, \sim, 0, {}^\alpha F, C_\kappa, D_{\kappa\lambda})_{\kappa,\lambda<\alpha}$ be the full cylindric set algebra of dimension α with base a finite set F. (see [1], Definition 1.1.5). Let $G \subseteq \alpha$ be finite; for any $x \in {}^\alpha F$, it is obvious that $x \in D_{\kappa\lambda}$ for some pair of distinct $\kappa,\lambda \in \alpha - G$. Thus, the condition of Theorem 3.3 holds, so $\mathfrak{L} \in DI_\alpha$. However, if we let $P \in B$ consist of all the sequences $y \in {}^\alpha F$ such that

$$y_i \neq y_1 \quad \text{for all} \quad i > 1,$$

then $\mathcal{D}_{1\kappa} \subseteq \sim P$ for every $\kappa \neq 1$, yet $\sim P \neq {}^{\alpha}F$. Thus, (2.1) fails to hold.

In conclusion, we have the *strict* inclusions

$$DC_{\alpha} \subset RS_{\alpha} \subset DI_{\alpha},$$

for every $\alpha \geqslant \omega$.

In the case where α is finite, the situation is different. Theorem 1.3.12 in [1] asserts that any CA is discrete if $d_{\kappa\lambda} = 1$ for some $\kappa \neq \lambda$. Combining this with the fact that (2.1) $\Leftrightarrow$ (2.1)*, it is clear that for α finite, every algebra with the RS property is *discrete*. Since every $\mathfrak{A} \in DC_{\alpha}$ for finite α is discrete, it follows that $RS_{\alpha} = DC_{\alpha}$ when $\alpha < \omega$. Finally, if $\alpha < \omega$ then clearly (2.1) is equivalent to the condition in Theorem 3.3, hence $RS_{\alpha} = DI_{\alpha}$. To conclude, for any finite α,

$$DC_{\alpha} = RS_{\alpha} = DI_{\alpha}.$$

4. RS_{α} IS NOT AN ELEMENTARY CLASS.

If $\mathfrak{A}$ is a cylindric algebra, it follows from (2.2) that $\mathfrak{A} \in RS_{\alpha}$ iff $\mathfrak{A}$ omits the type

(4.1) $\{x \neq 1,\ x \geqslant d_{\kappa 1},\ x \geqslant d_{\kappa 2},\ \ldots\}$,

for each $\kappa < \alpha$.

Now, if K is any elementary class of cylindric algebras and every member of K has the RS property, then by compactness,

$$Th(K) \vdash x \neq 1 \rightarrow x \not\geqslant d_{\kappa\mu_1} \vee \ldots \vee x \not\geqslant d_{\kappa\mu_p}$$

for some $\mu_1, \ldots, \mu_p < \alpha$. This is the same as

$$Th(K) \vdash d_{\kappa\mu_1} + \ldots + d_{\kappa\mu_p} = 1,$$

from which it follows that all the algebras in K are discrete. We conclude:

(4.2) *If K is a class of cylindric algebras which are not all discrete, and $K \subseteq RS_{\alpha}$, then K is not an elementary class. In particular, K is not an equational class.*

It may be noted, finally, that RS_α is an $\forall\vee\exists$ class in the sense of [3], and (by [3], Corollary 2.2), *every generic cylindric algebra has the* RS *property*.

References.

[1] L. Henkin, D. Monk, and A. Tarski, Cylindric Algebras, North-Holland, Amsterdam, 1971.

[2] B. Jónsson, and A. Tarski, *Boolean algebras with operators, Part I*, Amer. J. Math., vol. 73 (1951), 891-939.

[3] H. J. Keisler, *Forcing and omitting types theorem*, in Studies in Model Theory, (M. Morley, Editor), Mathematical Assoc. of America, Providence, 1973.

[4] L. Leblanc, *Transformation algebras*, Canadian J. Math., vol. 13 (1961), 602-613.

[5] D. Monk, *On the transformation theory for cylindric algebras*, Pacific J. Math., vol. 11 (1961), 1447-1457.

[6] A. Preller, *Substitution algebras in their relation to cylindric algebras*, Arch. Math. Logik Grundlagenforsch., vol. 30 (1970), 91-96.

[7] H. Rasiowa, and R. Sikorski, The Mathematics of Metamathematics, Panstowe Wydanictwo Naukowe, Warszawa, 1963.

[8] H. Rasiowa, and R. Sikorski, *A proof of the completeness theorem of Gödel*, Fund. Math., vol. 37 (1950), 193-200.

[9] L. Rieger, Algebraic Methods of Mathematical Logic, Academic Press, New York, 1967.

Department of Mathematics
Bucknell University
Lewisburg, Pennsylvania 17837
U.S.A.

A Proof-Theoretic Analysis of da Costa's C_ω^*.

by *ANDRÉS R. RAGGIO.*

ABSTRACT. The original system C_ω^* was formulated by N. C. A. da Costa in *Calculs de prédicats pour les systèmes formels inconsistents*, C. R. Acad. Sc. Paris, 258 (1964), pp. 27-29; later we gave a new version of this system, NC_ω^*, using sequents (cf. A. R. Raggio, *Propositional sequence-calculi for inconsistent systems*. Notre Dame Journal of Formal Logic, IX, nº 4 (1965), pp. 91-100). As both axiomatizations are not suitable for a proof-theoretic analysis, here we present an axiomatization of NC_ω^* using Getzen's natural deduction.

Neither da Costa's original Hilbert-type axiomatization of C_ω^* (cf. [1]) nor my axiomatization (cf. [5]) using sequents are suitable for a proof-theoretic analysis.

Using Gentzen's natural deduction I give a new axiomatization of NC_ω^* which contains the following rules:

I introduction:

$\wedge$I. $$\frac{A \,,\, B}{A \wedge B}$$

$\vee$I. $$\frac{A\ (B)}{A \vee B}$$

E elimination:

$\wedge$E. $$\frac{A \wedge B}{A\ (B)}$$

$\vee$E. $$\frac{A \vee B \quad \begin{matrix}[A] \\ C\end{matrix} \quad \begin{matrix}[B] \\ C\end{matrix}}{C}$$

$A \vee B$ is the main premiss.

$$\supset\mathrm{I}.\quad \frac{\begin{array}{c}[A]\\ B\end{array}}{A \supset B} \qquad\qquad \supset\mathrm{E}.\quad \frac{A,\quad A \supset B}{B}$$

$A \supset B$ is the main premiss.

$$\neg\neg\mathrm{E}.\quad \frac{\neg\neg A}{A}$$

$$\bigwedge\mathrm{I}.\quad \frac{A(a)}{\bigwedge x\, A(x)} \qquad\qquad \bigwedge\mathrm{E}.\quad \frac{\bigwedge x\, A(x)}{A(a)}$$

$$\bigvee\mathrm{I}.\quad \frac{A(a)}{\bigvee x\, A(x)} \qquad\qquad \bigvee\mathrm{E}.\quad \frac{\bigvee x\, A(x) \qquad \begin{array}{c}[A(a)]\\ C\end{array}}{C}$$

$\bigvee x\, A(x)$ is the main premiss.

The premisses are either main or minor. The derivations are in tree form starting with assumptions that may be discharged by $\supset$I, $\vee$E and $\bigvee$E. A *path* in a derivation of a formula F is a string of formulas which : a) starts with an assumption which is not one discharged by an $\vee$E or $\bigvee$E; b) if it contains a premiss of a rule it contains its conclusion, with two exceptions, c) it stops at the minor premiss of an $\supset$I, d) if it contains the main premiss of an $\vee$E or $\bigvee$E it continues through one of the discharged assumption of this rule. A path containing the formula F is called an *endpath*. Every derivation contains at least one endpath.

A formula which is the conclusion of an I and a premiss of an E is called a *peak*.

Reductions.

$\wedge$ - reduction:

$$\frac{\dfrac{A \;,\; \overset{\downarrow}{B}}{A \wedge B}}{B} \quad\Longrightarrow\quad \overset{\downarrow}{B}$$

$\vee$ - reduction:

$$\dfrac{\dfrac{\overset{\downarrow}{A}}{A \vee B} \qquad \begin{matrix}[A]\\ C\end{matrix} \qquad \begin{matrix}[B]\\ C\end{matrix}}{C} \Longrightarrow \begin{matrix}\downarrow\\ A\\ [A]\\ C\end{matrix}$$

$\supset$ - reduction:

$$\dfrac{\overset{\downarrow}{A}\ , \qquad \dfrac{\begin{matrix}[A]\\ B\end{matrix}}{A \supset B}}{B} \Longrightarrow \begin{matrix}\downarrow\\ A\\ [A]\\ B\end{matrix}$$

$\bigwedge$ -reduction:

$$\dfrac{\dfrac{\overset{\downarrow}{A(a)}}{\bigwedge x A(x)}}{A(b)} \Longrightarrow \overset{\downarrow}{A(a)} \ ({}^{b}/_{a})$$

$\bigvee$ - reduction:

$$\dfrac{\dfrac{\overset{\downarrow}{A(b)}}{\bigvee x A(x)} \qquad \begin{matrix}[A(a)]\\ C\end{matrix}}{C} \Longrightarrow \begin{matrix}\downarrow\\ A(b)\\ [A(a)] \ ({}^{b}/_{a})\\ C\end{matrix}$$

A derivation is in normal form if it has no peaks.

THEOREM I. $\vdash_{C_\omega^*} F \iff A_1 \vee \neg A_1 \ldots A_n \vee \neg A_n \vdash_{NC_\omega^*} F$, $n \geqslant 0$.

THEOREM II. (Normalization Theorem for NC_ω^*.) *Every derivation in* NC_ω^*

can be reduced to a normal form applying a finite member of reductions. The derivation in normal form has the same endformula and no new undischarged assumptions.

We choose a peak of maximum degree which is not below any other peak of the same degree and apply the corresponding reduction. We repeat this operation until there are no peaks left. The $\neg\neg E$ present no difficulties because its premiss cannot be the conclusion of an I. For details and further information see [2], [3] and [4].

COROLLARY I. *A path in a normal derivation starts with assumptions, continues with* E*'s and ends with* I*'s. The* E*'s and* I*'s may be lacking. Besides* E*'s and* I*'s a path may contain repetitions corresponding to* $\vee E$ *and* $\vee E$.

Intuitively a derivation in normal form splits in an analysis and a synthesis. This is a further - the main - justification in calling such systems "natural deduction".

BASIC LEMMA. *A derivation in normal form cannot have a formula* P *such that* a) *two endpaths going through* P *contain above it the discharged assumptions of an* $\vee E$ *whose main premiss is a tertiun-non-datur, and* b) *one of these endpaths contains above* P *only* E*'s.*

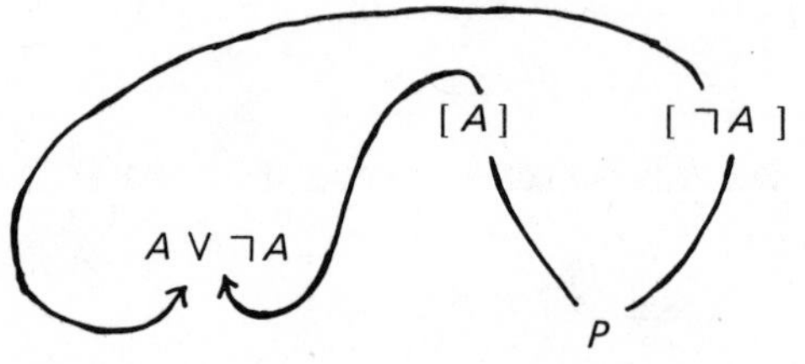

The proof is based on the fact that either A or $\neg A$ must begin with an odd number of negations.

Two path satisfying condition a) are called associated.

THEOREM III. $\nvdash_{C^*_\omega} \neg F$.

Suppose the contrary; by Theorems I and II there is normal derivation

$$A_1 \vee \neg A_1, \ldots, A_n \vee \neg A_n \vdash_{NC_\omega} \neg F ,$$

whose endpaths cannot end by an I, because no I has a negation as conclusion. Therefore all endpaths contain only E's; their assumptions must be tertium-non-datur's because otherwise they must be discharged, which is impossible applying E's. By induction we show that there must be a pair of associated endpaths. But this is impossible by the Basic Lemma.

THEOREM IV. *C_ω^* is not finitely trivializable. That is, there are no B_1, $B_2, \ldots, B_n$ such that for every D*

$$B_1, B_2, \ldots, B_n \vdash_{C_\omega^*} D.$$

Let us suppose the contrary; we may choose as D a sentencial variable p not occurring in the B_i. By Theorems I and II there is a normal derivation

$$B_1, B_2, \ldots, B_n, A_1 \vee \neg A_1, \ldots, A_n \vee \neg A_n \vdash_{NC_\omega^*} p,$$

whose endpaths cannot finish with an I. Therefore all endpaths contain only E's. They cannot start with a B_i because applying E's we get only subformulas. Therefore all endpaths start with tertiun-non-datur's. By an inductive argument we show that there must be a pair of associated endpaths. But this contradicts the Basic Lemma.

THEOREM V. $\vdash_{C_\omega^*} \bigvee x\, A(x) \implies \vdash_{C_\omega^*} A(a_1) \vee A(a_2) \vee \ldots \vee A(a_n)$.

If $\bigvee A(x)$ has no negation, then

$$\vdash_{C_\omega^*} \bigvee x\, A(x) \implies \vdash_{C_\omega^*} A(a),$$

and

$$\nvdash_{C_\omega^*} \bigvee x\, \neg A(x).$$

By Theorems I and II there is a normal derivation in NC_ω^*

$$A_1 \vee \neg A_1,\ A_2 \vee \neg A_2,\ \ldots,\ A_n \vee \neg A_n \vdash_{NC_\omega^*} \bigvee x\, A(x).$$

1) The derivation ends with an I; we are through.

2) The derivation ends with an E different from $\vee$E or $\bigvee$E. This is excluded by the Basic Lemma.

3) The derivation ends with an $\vee$E or $\bigvee$E. Let us call *endpiece* the final part of the derivation containing only $\vee$E's or $\bigvee$E's. By the Basic Lemma we see that the endpiece cannot contain $\bigvee$E's and the main premisses of the $\vee$E must be tertium-non-datur's.

The proof is trivial; see the figure below:

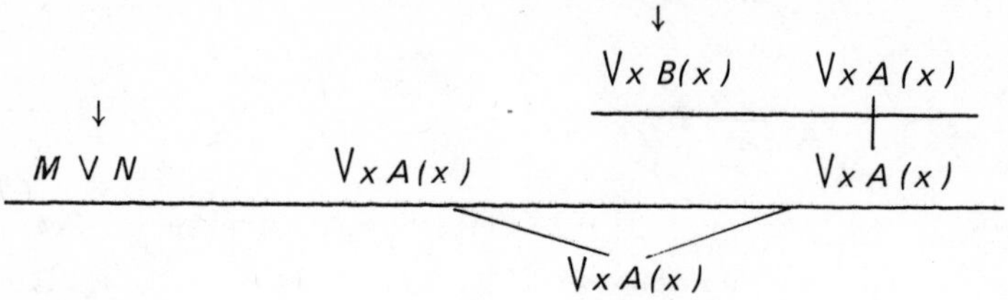

Therefore the endpiece contains only $\vee$E's whose main premisses are tertium-non-datur's. Applying again the Basic Lemma to the top formulas of the endpiece which are not main premisses, we see at once that these top formulas must be conclusions of $\bigvee$I. The derivation has the following form:

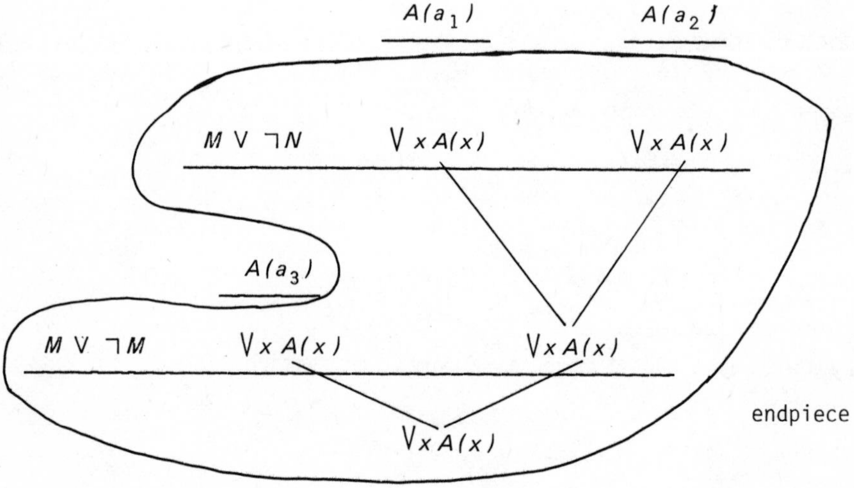

We transform it into

$$\frac{M \vee \neg M \qquad \dfrac{A(a_3)}{A(a_1) \vee A(a_2) \vee A(a_3)} \qquad \dfrac{N \vee \neg N \qquad \dfrac{A(a_1)}{A(a_1) \vee A(a_2) \vee A(a_3)} \qquad \dfrac{A(a_2)}{A(a_1) \vee A(a_2) \vee A(a_3)}}{A(a_1) \vee A(a_2) \vee A(a_3)}}{A(a_1) \vee A(a_2) \vee A(a_3)}$$

Let us assume now that $\bigvee x\, A(x)$ has no negations. Suppose N (or $\neg N$) has an odd number of negations at the beginning; the endpath going through it cannot eliminate all this negations; remenber that

$\neg\neg$E has the form $\dfrac{\neg\neg A}{A}$.

The same argument applies to M (or $\neg M$). Therefore, the derivation has no endpiece and the endformula is the conclcusion of an $\vee$I.

Using other methods N. C. A. da Costa, A. I. Arruda, A. M. Sette and A. Loparič have proved already Theorems III and IV.

References.

[1] N. C. A. da Costa, *Calculs de prédicats pour les systèmes formels inconsistants*, C. R. Acad. Sc. Paris, 258 (1964), 27-29.

[2] Dag Prawitz, Natural Deduction, Stockholm, 1965.

[3] Dag Prawitz, *Ideas and results of proof theory*, Proceedings of The Second Scandinavian Logic Symposium, edited by J. Fenstad, Amsterdam, 1971.

[4] Andrés R. Raggio, *Gentzen's Haupsatz for the systems NI and NK*,

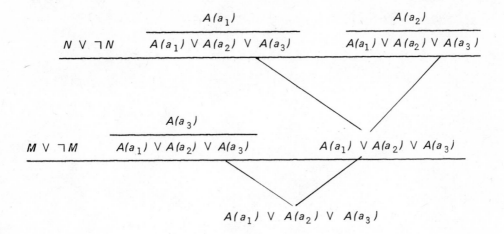

Logique et Analyse, 30 (1965), 91-100.

[5] Andrés R. Raggio, *Propositional sequence-calculi for inconsistent systems*, Notre Dame Journal of Formal Logic, vol. IX, 4 (1968) 359-366.

Departamento de Matemática
Universidade Estadual de Campinas
13.100 Campinas, S.P., Brazil

and

Philosophisches Seminar
Universitaet Erlangen
Erlangen, B.D.R.

Decision Problems: History and Methods.*

by *H. P. SANKAPPANAVAR.*

ABSTRACT. In this expository paper we try to give some glimpses into the historical origin, the development and the ramifications of the decision problems, the emphasis, however, being on the decision problems for formal systems. The paper also lists some of the decidable and the undecidable theories known since 1964, mentions a few open problems and finally includes an extensive bibliography of the work related to the decision problems for (classical) theories (including that of predicate calculus) and to the decision problems for non-classical logics, done since 1964 up to early 1977.

"All men by nature desire to know."

-Aristotle

1. Introduction.

In this expository paper we try to give some glimpses into the historical origin and the development of one of the important areas of Mathematics, namely The Decision Problems, and also we make a few remarks on some of the methods used in this area, the emphasis, however, being on the decision problems for formal systems.

Of course, there already exists in the literature an excellent survey (cf. Ershov et al.(41)), published in 1965, giving a systematic and fairly detailed account of the results and the methods in the area of the decision

---

* This is a revised version of an earlier draft which the author had used to give a lecture at the Departamento de Matemática, Universidade de Brasília, in September, 1976.

problems for classical first-order theories with an (almost) exhaustive bibliography. The present paper is, in a sense, complementary to (41); and it includes a few open problems (some being well-known) as well a list of some of the decidable and undecidable theories known since the appearance of that survey paper. Also included is an extensive bibliography of the work related to the decision problems for classical first-order theories (cf. Part Two), to solvable cases of the decision problem of the predicate calculus (cf. Part Three) and to the decision problem for non-classical logics (cf. Part Four), done since 1964 up to early 1977.

Perhaps it may not be out of place to mention here that the author's motivation to write this paper came out of his desire to give a satisfactory answer to a mathematician who once asked "What are decision problems good for ? " This paper may also be hepful to a novice to have historical perspective and to get his way around in the subject.

The author would like to thak Professors N. C. A. da Costa and A. I. Arruda for the encouragement he received during the preparation of this paper.

2. Some preliminaries.

It will be helpful if the reader has at least some rudimentary knowledge of the predicate calculus with equality, and of formal systems (say, the first chapters of Shoenfield [50]). A *first order* (or elementary) *theory* is a deductively closed set of sentences of a predicate calculus with equality (having only a finite number of non-logical symbols for our purpose). First order theories are a particular type of formal systems.

A *decision method* (or algorithm) for a formal system is a method by which, given a formula of **F**, we can decide in a finite number of steps whether or not it is a theorem of **F**. The *decision problem* for **F** is the folfowing:

Find a decision method for **F** or *prove* that no such method exists.

The decision problem for a formal system is a particular case of a more general problem which can be formulated as follows.

Suppose that A and E are two sets such that $A \subseteq E$. The *decision problem* for A in E is the following: find a decision method by which, given an element of E, we can decide in a finite number of steps whether or not it is in A or prove that no such method exists.

The decision problem for a formal system **F** is a special case in which E is the set of all formulas in **F** and A is the set of all theorems of **F**. Hilbert's tenth problem and the word problem for groups are other famous examples.

The decision problem for a formal system **F** is *unsolvable* if there does not exist a decision method for **F**; and in this case we also say that **F** is *undecidable*. An elementary theory is *essentially undecidable* if every consistent extension of it is undecidable. **F** is *decidable* (and its decision problem is *solvable*) if there is a decision method for **F**.

3. HISTORICAL ORIGIN AND DEVELOPMENT OF DECISION PROBLEMS.

The following are some brief remarks on the historical roots of growth of decision problems which, it is hoped, would bring out the fundamental importance, and the ramifications, of the area of decision problems in the continuous development of mathematics. In making these remarks I am highly influenced directly or indirectly by the works listed with asterisks in the References at the end of the paper. I gratefully acknowledge my indebtedness to the authors of these works, as well as to those mathematicians whose contributions have made the area of decision problems so important in mathematics.

The precise formulation of "decision problems" is closely related with - in fact depends on - the precise notion of "algorithm" (alias, decision procedure or effective procedure or decision method or mechanical procedure or finite combinatorial procedure, etc.). The concept of algorithm is one of the basic concepts of mathematics. By an *algorithm* is meant a finite set of instructions which can be performed machanically to give the answer, in a finite number of steps, to each of a countable number of questions of a given type. Of course, this is not a precise definition of the concept of algoritm; this only gives some intuitive feeling for it as it has evolved in a natural way over a long period of time in mathematical history. (For details on the informal description the beginner is strongly urged to read § 40 in Kleene [31] or Trakhtenbrot [55]). In fact, the precise definition evaded the mathematicians for over 2000 years, or perhaps the urgency for it did not arise for all those years ! Nevertheless, through the centuries mathematicians have indeed had a great success in discovering algorithms.

Finding algorithms to solve various problems has been an important problem in mathematics since antiquity.

The Greeks, including Euclid, sought algorithms. The recognition that we have algorithms for some mathematical questions, and that we do not know any algorithm for other questions goes far back in mathematical history. Aristotle's Topics are according to Beth [5] the source of later (unsuccessful) attempts towards the construction of an *ars inveniendi*, which was intended to provide a solution to any scientific problem; these attempts constitute, in a sense, a preamble to modern work on the decision problems.

The ancient Hindus sought algorithms. (A careful, unbiased and scholarly examination of the ancient sources, mostly in Sanskrit, of Indian mathematics seems still lacking.) Algorithmic methods to deal with algebraic problems were developed by Arabs under the influence of the Indians. The name "algorithm", by wich a mechanical procedure is commonly known, is derived from the name of the Arab mathematician and astronomer Mohammed Ibu-Musa Al Chwarizmi (or Al-Khowarizmi) wo lived during 780-850 A. D. (approximately) and wrote a treatise on algebra called Al-Jabr around 830 A. D. (The word "algebra" is derived from its name.) Among the simplest algorithms are the well-known rules, given by him, for performing the arithmetic operations on integers. Another well-known example is the Euclidean algorithm which gives a (uniform) method for solving all problems of the following type: Given two positive integers a and b, find their greatest common divisor. There is an algorithm, given by an Indian mathematician, Brahmagupta (who lived in seventh century), for deciding whether or not, for given integers a, b and c, the diophantine equation $ax + by + c = 0$ has integral solutions for x and y. This procedure consists in checking whether or not the greatest common divisor of a and b divides c. Similarly, there is an algorithm, as is well-known, for determining ehether or not an algebraic equation $a_0 + a_1 x + \ldots + a_n x^n = 0$ ($a_n \neq 0$, n a positive integer) with integral coeficients $a_0, a_1, \ldots, a_{n-1}, a_n$ has a rational root, and there is another for computing the nth prime. In fact there are plenty of examples of algorithms in practically every branch of mathematics.

The methods introduced by the Arabs inspired the Spanish Scholastic Raymond Lully (1235-1315). His Ars Magna was supposed to be a combinatorial procedure fo find out all "truth", at least so he thought. In it the idea, a splendid one, of a mechanical procedure was conceived. Lully (or Lullus) was responsible for not only the idea of mechanizing the process of logical derivation but also for constructing the first practical machine for this purpose. Not only did Lully's contemporaries not realize how revolutionary his ideas were but they even mocked at them. The Art of Lully had a great influence on the later development of mathematics. Also inspired by the Arabs was the Italian mathematician Leonardo de Pisa (about 1180-1250), more commonly known as Fibonacci. In his classic **Liber Abacaci** he describes, among other things, the algorithmic processes of arithmetic including the extraction of roots.

Later, in 16th century, the methods of solving the cubic equation given by Cardano (1501-1576) and those for the fourth degree equation by Ferrari (1522-1565) were clearly in the spirit of Lully's art. Cardano even chose the title of his work as **Artis magnae seu de regulis algebraicis liber unus**. These important contributions had a great impact on the (theoretical) research in algebra in various directions. It was natural that these methods be generalized so as to be applicable to polynomial equations of any degree and in particular to those of fifth degree. As it is well-known, the resolution of the fifth degree equation, like the three classical geometric problems of the Greeks, became one of the most famous problems of mathematics for the next centuries until the arrival of Abel and Galois. These problems were reponsible for the creation of some of the very important new ideas and methods in mathematics. Thus one can take the view that the algorithmnic considerations of Hindus and Arabs were the source for the later discovery of some important mathematics, like Galois theory. Then came the important discovery of literal algebra by François Viète (1540-1603), who initiated the systematic use of symbols (or letters) for mathematical expressions (symbols were employed in mathematics since ancient times but not systematically and in so general a way as used by him). (Since then mathematics - and other sciences - have progressed at an incredible rate.) The methods given by him in his algebra were algorithmic, for example, his method of solving equation by approximation is applicable to any polinomial equation with real coefficients and with a real root.

The predominantly algorithmic development of algebra in the 16th century gave the impression that all questions, at least in algebra, could be handled by algorithms. Furthermore, the introduction of coordinates into geometry by René Descartes (1598-1650) carried this into the realm of geometry, the only other branch of mathematics known at the time. Descartes felt that his analytic geometry made the algebraic translation of geometric problems possible and thus made them susceptible to algorithms developed in algebra. Thus, for him the decisive factor was not the mere introduction of coordinates (as is commonly believed) but the potential that lay ahead for using coordinates to deal with geometry by algorithms. Descartes thought that all algebraic problems could be solved by algorithms and thus no interesting problems were left for the creative mathematicians! Indeed he was wrong, as it was shown in this century that there are problems which cannot be solved by algorithms.

I should note here that by 1623 a friend of Kepler by the name of Wilhelm Schickard, who was a professor of Astronomy and mathematics, had built the first computing machine which demonstrated that human thought could, at least in some respects, be simulated by a machine. Of couse, this discovery shook the religious-idealistic notion of thought at that time. A little later, by about 1641, Blaise Pascal (1623-1662), a predecessor to modern axiomatic method, also built and sold some fifty calculating machines.

The German mathematician and philosopher Gottfried Wilhelm Leibniz (1646-1716), the founder of symbolic logic - under the influence of the Lully's ideas about mechanization of human thought, of Bruno's epistemology, of Pascal's ideas about axiomatic method (as well as other mathematical ideas) and of Descartes' ideas about constructing a universal method for solving scientific problems, especially mathematical problems (the so called idea of a "universal mathematics") - set forth two goals for science and philosophy: (1) the discovery of a universal characteristic (characteristica universalis) and (2) the discovery of a universal algorithm for solving any mathematical problem. In describing what he meant by a universal characteristic, Leibniz was formulating the first clear idea of a mathematical logic. Influenced by the ideas of Viète and Descartes of using symbols and variables, Leibniz proposed in (1) to extend the formal method of mathematics to the field of logic. Thus the universal characteristic, taken as a logical plan, is a precise symbolic language within which all statements of science could be expressed - that would be the source of a true logical algebra applicable to various types of thought and that would make it possible to obtain all logical consequences that necessarily follow from the initially assumed statements. In (2) he dreamed to create "a general method in which all truths of the reason be reduced to a kind of calculation". Such an "arte combinatoria" would make it possible to replace arguments by formal computations. As a consequence, philosophers would not need to involve themselves in unproductive disputes; instead they would check their conclusions by calculations. (For further details about Leibniz' methodological views, see Styazhkin [52] and Bell [04].) It should also be mentioned that Leibniz built a calculating machine far superior to Pascal's. (While Pascal's machine handled only addition and subtraction, Leibniz' could also do multiplication, division and extraction of roots.) Let us also remember that Calculus (invented by Leibniz and Newton) is full of algorithmic methods.

Unfortunatelly, the dreams of Leibniz, great as they were, fell by the wayside for a long time after Leibniz was gone. His first dream got refined into what is now called Mathematical Logic, which began to flower only in the 19th century. It sprang out of the works of De Morgan, Boole, Jevons, Schröder, Poretskiy and Peirce on the one hand, and of Frege and Peano on the other. To this study Cantor added his "paradise" (viz. set Theory). The culmination point was the appearance of the Principia Mathematica [60] (in 1910-1913) in which a large part of mathematics was deduced by the help of a logical calculus, in other words, it was presented as a formal (deductive) system.

To "save" classical mathematics from the paradoxes discovered at the end of the last century and at the beginning of this century, Hilbert in 1904 proposed the so-called Hilbert's Program (which was undertaken seriously by Hilbert's School only after 1920) to formalize mathematics and then to prove its consistency by "finitistic" methods (for details see Kleene [30] or [31] or Wang [59]). Out of this and other foundational studies emerged the precise notion of "formal system" (also called logistic system) among whose important ingredients are the notions of "formula", "proof", and "provability" and the requirements that the notions of "formula" and "proof" be "efective", i.e. that there shall be effective procedures to check whether or not a given finite string of symbols is a formula and a given finite sequence of formulas is a proof. Thus it was only natural to ask whether the notion of provability was also effective, i. e. whether there was a decision method such that, given any sentence of a formal system **F**, the method would enable us to decide (in a finite number of steps) whether or not it is a theorem of **F**. This problem came to be known as the "decision problem for the formal system **F** ".

The first recognition of the decision problems for formal systems is in 1895 due to Schröder [49] who stated the decision problem in a way adequate for rigourous treatment. Löwenheim, in his famous article of 1915 (cf. [35]) had already established perhaps the first really significant results in this area. They were in the current terminology: 1) The reduction of the decision problem for first-order logic with predicates of rank greater than 1 to the case in which only binary predicates are present, and 2) the solvability of the decision problem for first-order logic (with equality) with only unary predicates. This work opened up a whole new area of investigation. Vaught writes in [58], regarding the influence of Löwenheim's paper on model theory as follows: "...The Löwenheim's paper is truly a critical point in mathematical history, signalling a whole development to come". His observation applies equally well to the area of decision problems.

After these pioneering results of Löwenheim there followed the work [51] of Skolem. In 1919 Skolem gave an important decision method for the monadic first-order logic with equality and also a decision method for monadic second order logic by means of the so-called method of quantifier-elimination. Although Löwenheim , in 1915, had used it very sketchily, Skolem recognized quantifier-elimination as a method for decidability proofs, developed it in its full form and applied it to obtain the above mentioned decidability results. Langford [33] and [34] applied this method to obtain the decision procedure and completeness for the theory of dense orders and also for the theory of discrete orders with a first but no last element.

Ever since the appearance of Löwenheim's and Skolem's results, there has been a series of significant results published both on the reduction problem and on the problem of discovering new solvable cases of the decision problem (of the predicate calculus). As it is impossible to give the full account here I will only refer the reader to the following works where more information (or at least references) can be found: Church [09], pages 293-294, Suranyi [65], Ackermann [02], and also Part 3 of the references at the end of this paper which covers (perhaps not completely) the more recent works done during 1965-1976.

The simplest (and the earliest) of the formal systems is the propositional calculus (for its history see again Church [09]). The decision problem for the propositional calculus is, as is well-known, solvable; and the truth-table method is an algorithm for it. Although the truth-table method was first applied informally to special cases by Frege in his Begriffsschrift, the first recognition of it as a decision method is by Peirce, six years later. Its present form is due essentially to Post [41] and Łukasiewicz [36]. For the work on the desicion problems for several partial systems of the propositional calculus see Church [09], Chapter II.

The explicit formulation of the predicate calculus as a separate formal system first appeared in 1928 due to Hilbert and Ackermann (cf. [20]) and received a crowning conclusion with Gödel's completeness theorem of 1930. In the meantime, Leibniz' second dream received some refinements to be one of the most famous problems of mathematical logic, namely, the decision problem for the predicate calculus. In fact, to dramatize the importance of this problem, Hilbert and Ackermann refer to it as "the main problem" of mathematical logic.

WHY A PRECISE DEFINITION ?

If the decision problem of predicate calculus were solvable then it would mean that we would have a decision procedure for solving problems in all (finitely axiomatized) theories. Such an algorithm would in fact be a single effective method for solving almost all of the mathematical problems which have been formulated (in the predicate calculus) and have remained unsolvable. This is why construction of such an "omnipotent algorithm" is so appealing. In 1920's some progress was made, as noted above, in this direction by discovering several solvable cases of the decision problem. However, in spite of this progress and of persistent and dedicated efforts of many gifted mathematicians, the difficulties of finding such an algorithm remained unresolved and unsurmountable untill 1936.

Furthermore, similar obstacles were also felt in solving decision problems which arose in the conventional branches of mathematics at the down of this century. Among these were the famous Hilbert's tenth problem of finding *an algorithm to decide of a polynomial equation, in several variables, with integer coefficients, whether or not the equation has solutions in integers*, and the word problem which was first considered by M. Dehn and A. Thue in 1912-14. For more details on these two problems see

Davis (26), Jones (81), and Boone et al (13).

Mathematicians' attempts to construct such algorithms were also without success. All these fruitless attempts made it clear that the difficulties involved were basic and inherent, and people soon began *conjecturing*, for the first time in more than two thousand years, that *it is not possible* to construct an algorithm for *every* class of questions, thus forecasting of a new and important stage in the development of mathematics.

The claim that a certain class of problems cannot be solved algorithmically does not simply mean that no algorithm has yet been discovered; it means that such an algorithm in fact can *never* be discovered; in other words, that no such algorithm can exist. It is a statement about *all* imaginable algorithms in general. If one has to have faith in such an assertion, it must be based on some *mathematical* proof. However, to prove such a statement we must have more than just the natural feeling for the concept of an algorithm possessed by every mathematician, we must have a precise definition of "algorithm", since otherwise it is unclear what one is trying to prove impossible. Thus the formulation of such a precise definition of an algorithm turned out to be one of the major problems of this century (more specifically of the 1930's).

Similar situation existed already in the history of mathematics. For example, an exact definition of continuous curves was called for before one could answer (or even could ask) the question whether there are any space-filling continuous curves. Another perhaps more dramatic instance occurred when mathematicians were faced with the problem of finding precise definitions for "a construction using compass and straight edge" and "solving equations by radicals" before they could address thenselves to solve the three famous Greek problems, namely, squaring a circle, duplication of a cube and trisection of an angle by using compass and straight edge alone, as well as to solve an equation of degree $\geqslant 5$ by radicals (it has been noted above as to how the last problem arose). It was only in the beginning of the nineteenth century that Abel and Galois furnished the impossibility proofs for all the above problems; they did so only after being able to formulate the needed precise definitions. For an elegant treatment of these impossibility proofs, see Jacy Monteiro [62].

THE INCREDIBLE 1930'S AND THE BIRTH OF THE THEORY OF ALGORITHMS.

Thus the necessity of the precise notion of an algorithm was felt so great that mathematicians, especially mathematical logicians (perhaps with the determination to settle the decision problem of predicate calculus), in 1930's really searched for a definition. Gödel's 1931 epoch making paper on "Incompleteness Theorems" gave an impetus to this search. It should be pointed out here, as Trakhtenbrot so aptly does, that a definition of "algorithm" is not just an agreement that may be arrived at by mathematicians as to the meaning of the word "algorithm"; but more importantly, the definition must *reflect accurately* what constitutes the intuitive notion of an algorithm. In other words, it must be a precise and complete description of the feelings (however vague) of mathematicians, acquired through the specific algorithms discovered through the ages (as noted earlier in this paper). For a nice discussion of these "feelings" I urge the newcomer to read § 1.1 in Rogers [47], and Chapter I in Hermes [19]. Thus the logicians were confronted with the problem of formulating a mathematically exact definition of the notion of *algorithm* which would be complete both in form and, more importantly, in substance.

It was soon realized that from the point of view of solving decision problems it would be sufficient to consider the problem of formulating the definition of the concept of *numerical algorithm* (i. e. algorithms whose inputs are n-tuples of non-negative integers, and outputs are non-negative integers, written in some concrete notation), which was in turn equivalent to the problem of characterizing the class of effectively calculable functions with arguments as n-tuples of natural numbers and values as natural numbers (a function f is effectively calculable if there is a (numerical) algorithm which, given an argument a as input, would yield $f(a)$ as the output). For a further discussion of this point see Kleene [31], especially pages 227-230, Shoenfield [50], pages 106-108, or Rogers [47], page 1.

Gödel [15] had already defined and made extensive use of a class of number-theoretic functions, which are now called, following Kleene, primitive recursive functions. It was apparent that these functions were intuitively calculable functions, an observation which led immediately to the question: How inclusive is the class of primitive recursive functions? It was soon observed that there were calculable functions which were not primitive recursive. Already in 1928 Ackermann had made an important contribution to the clarification of the notion of calculable function by constructing a function $f(x,y,z)$, now called Ackemann's generalized exponential, which was accepted to be a calculable function, but was shown to be not primitive recursive. For details, see Ackermann [01], and Pēter [63], pages 1-69. In fact the method of diagonalization could be applied to the class of primitive recursive functions (or to any formally characterizable class of algorithmic functions for that matter) to yield an algorithmic (calculable) function ouside that class, thus showing some of the difficulties associated with this problem (see Rogers [47], § 1.4). Thus, stimulated by the discoveries of Gödel and Ackermann, the logicians of the 1930's set out to resolve these (and other) difficulties involved in finding a satisfactory characterization of algorithm and of calculable function. Soon such a definition was in the making.

In the fall of 1931 Church was giving a logic course in Princeton which Kleene and Rosser were (as students) taking notes. In this course Church introduced his system and a definition of natural numbers in it. Then arose a natural question of developing the theory of natural numbers in Church's system, and Kleene began working on it. The efforts of Kleene led Kleene and Church to a class of calculable functions called "λ-definable functions", which were intensively studied by them during 1931-1933. For an account as to how this class of functions actually came into being, read Kleene's own reminiscences in Crossley [11], p. 4.

In the spring of 1934 Gödel gave a series of lectures in Princeton (at which also Kleene and Rosser were note-takers). In this series Gödel brought out another exactly defined class of calculable functions called general recursive functions, for which he is said to have been motivated by Herbrand (see Gödel [16]). Then the question came up: Does the class of general recursive functions embrace all effectively effectively functions, and does it coincide with the class of λ-definable functions which were already developed by Kleene and Church during 1931-1933?

In the late spring of 1934 - that is when Church was talking with Gödel about general recursive functions - Church came out explicitly with the thesis (Church's Thesis) that all functions which are intuitively regarded as effectively calculable are λ-definable. As to the circumstances in which Church proposed his thesis, Kleene reminisces, "He (Church) spent some month sweating over it and saying: 'Don't you think it is so?' and I was a sceptic, and when he came out asserted the thesis I said to myself: 'He can't be right'. So I went home and I thought I would diagonalize out of the class of the λ-definable functions and get another effectively calculable function that was not λ-definable. Just in one night I realized you could not do that, and from that point on I was a convert. But until Church really came out and said so, I guess I had not really believed they would be all of them." Soon thereafter it was proved by Kleene and Church that the class of λ-definable functions was the same as that of general recursive functions (cf. Kleene [29] and Church [07]. Thus convinced, Church published his thesis in 1936 giving some very strong evidence in favor of it.

Also in 1936-37 another important paper (Turing [56]) was published in which was proposed, much before the arrival of moderm computers, a precise (and very natural) definition of "machine", and using these machines - now called Turing machines - Turing defined a class of calculable functions, namely the computable (by Turing machines) functions. In the same paper he also proposed his thesis to identify the informal notion of "effectively calculable" with (his) formal notion of "computable". Shortly after this, in 1937 he proved that the class of computable functions is the same as that of λ-definable functions (and hence coincides with that of recursive functions). Thus Church's thesis was shown to be equivalent to Turing's thesis. It is a tribute to Turing's incredibly clear presentation of his ideas that practically every logician came to believed in the truth of Church-Turing's thesis. As for Gödel's feelings about Church's thesis Kleene remarks (Crossley [11], pp. 10-11): "I think he was sceptical, and it may well be that Turing's presentation also brought Gödel around". (Read also "Postscriptum" at the end of Gödel's paper published in Davis (25).) Independently of Turing, Post in 1936 published (albeit very briefly) an essentially same formulation as Turing's. Later in 1943 Post published a fourth formulation which also turned out to be equivalent to the previous definitions. In 1951 Markov, with his theory of algorithms, provided still another equivalent formulation.

Thus several definitions, apparently different and with different logical starting points, were all proved to be equivalent - a fact which is indeed remarkable. It shows that all these definitions define the same concept and that this concept is natural, fundamental, and useful. This is the precise mathematical definition of the concept if "algorithm". This definition, it is generally agreed, came into being in 1935. As a consequence, the decision problems received the status of precise mathematical problems,with the Church-Turing's thesis helping one to feel that the precise definition "agrees" with one's intuitive definition of decision probelms.

As soon as Church formulated his thesis, it was observed that there must be functions which are not computable (or recursive) because the set of different possible algorithms is countably infinite (since algorithm, roughly speaking, is describable in a finite number of words, while the set of all number theoretic functions is uncountable (recall the proof, first given by Cantor, that the set of reals is uncountable). However, it was in 1936 that the first explicit example of undecidability results were reported by Church. He first obtained an undecidability result in connection with λ-definability and then applied it to decision problems for formal systems; more specifically, he showed that the formalized first-order number theory **N** is undecidable (**N** consisits of the (logical) axioms of first-order calculus and a few non-logical axioms expressed in predicate calculus chosen in such a way that one can, in a formal fashion develop in **N** the usual (intuitive) elementary number theory which is treated in introductory texts of number theory. For more details see Kleene ([31], § 38). Still in 1936, Church proved that the pure predicate calculus is undecidable, thus giving another example of an unsolvable (decision) problem. Thus showing for the first time in the history of mathematics that there are problems for which it is impossible to find algorithmic solutions - one of the remarkable achievements of this century.

Gödel's first incompleteness theorem states that in the formalized elementary arithmetic **N** there are statements (formulas) T true in the standard model (viz.,natural numbers) but formally undecidable (i. e. neither T not the negation of T is provable in the system **N**), and hence no adequate formalization can exist for the (informal) arithmetic (much less for all of mathematics, thus throwing cold water into Leibniz' hope of creating a Universal Charateristic). Gödel's second incompleteness theorem says (or at least implies) that we cannot prove the consistency of arithmetic by the methods of arithmetic alone! This gave a blow to the Hilbert s Program which called for proving the consistency of mathematics by "finitary" methods (for more on the implications of these celebrated theorems see for instance Kleene [30] and [31]). In view of the surprising and important implications of Gödel's theorems (Gödel proved these theorems for a modified version of the system given in Principia Mathematica [60]), logicians immediately started wondering "how general this thing is?" Some, including Church, even felt that it might perhaps be possible to escape from Gödel's theorems if one uses some other particular system having a format totally different from the one used by Gödel (see Crossley [11], p. 3, about Church's initial reaction to Gödel's theorems). Soon after the appearance of the precise definition of algorithms and the Church's thesis, it was realized that there is indeed no escape from Gödel's theorems and that the idea of computability is connected with the incompleteness phenomenon. Indeed, Kleene in 1936 showed, as an application of Church's thesis, that Gödel's theorems remain true for any formal system in which one can effectively recognize the axioms (and the proofs); thus there is no escape from them. A consequence of all this work of Gödel, Church, Kleene, etc. for model theory, as Vaught ([58], p. 68) observes, was that model theory received more emphasis for its growth as it was realized that the method of elimination of quantifiers used with such a great success by Presburger, Tarski and others (see next paragraph) would not work for many theories.

Important achievements were made in another direction. The method of quantifier-elimination was applied in proving the decidability of some important theories. The intensive study on the application of the quantifier-elimination method in the decidability proofs, begun in Tarski's seminar at Warsaw University during 1926-1928, led to the famous work of Presburger, in 1930, on the theory (in the sense of the last section) of $\langle \omega, + \rangle$ and of $\langle Z, +, < \rangle$ (ω and Z being the set of non-negative integers and the set of integers respectively) and to Tarski's classic work announced by him in 1931 and published in full in Tarski [54], giving a decision method for elementary algebra and geometry by sharpening Sturm's method for deciding the number of real roots of a given algebraic equation in one variable. For the far-reaching consequences of the quantifier-elemination method, through the just mentioned (and other) work of Tarski, consult Vaught [58] (especially p. 60) and Robinson [46].

The thirties also witnesses the important logical discoveries in a still another direction, of Herbrand [17] and Gentzen [64] on the predicate calculus. Herbrand gave a decision procedure to reduce a proof of a formula to a proof which does not use modus ponens ("Unweg") and began to use his theorem to give a uniform treatment of all known cases of the decision problem for the predicate calculus. Gentzen showed that the predicate logic can be formalized in systems called Gentzen-type systems (and in natural deduction systems), and in his "Haupsatz" or "normal form theorem" he showed that the uses of the *cut rule* can be eliminated from any given proof. As an application of his theorem, Gentzen in 1934-35 gave the first proof that the intuitionistic propositional calculus is decidable. This celebrated result was also obtained independently almost at the same time by Jaśkowski [23]. (It should be noted here that natural deduction systems were also considered by Jaśkowski at the suggestion of Łukasiewicz in 1926.) For a treatment of theorems of Herbrand and Gentzen, see § 54 and 55 in Kleene [31]. A good deal of work has been done using ideas of Herbrand and Gentzen not only in metamathematical investigations but also in the area of mechanical mathematics (Cf. Wang (59), Chap. IX, Wang (136), and Danton and Dreben (04)).

THE DECISION PROBLEMS FOR THEORIES.

Soon after the appearance of the undecidability results of Church and Rosser, namely that the Peano's arithmetic is undecidable and essentially undecidable, Tarski took over, so to speak, and developed in 1938-1939 a general method, extending the method used by them for proving the undecidability of first-order (or elementary) theories and published it in an abstract form in 1949, and in complete detail in the monograph Tarski [53]. Since 1949 a great deal of work was done in this fertile area, and Ershov (41) deals with this development in Chapter 3. Also following the other work of Tarski on the decidability via quantifier-elimination (refered to earlier) some decidability results were obtained, an account of which is also given in Ershov (41). (For the more recent work since 1965 see Part 2 of the references of this paper.) Here we would like to note that in that survey paper the important (unfortunately less known) work of Jaśkowski was left out, except for one paper. Some results of Jaśkowski (published without proof) were later verified by other people. Many other results of Jaśkowski (cf. the review of Kotas [68]), have not been fully worked out yet. In view of this we shall mention here only some of his results * :

* Most of the information in the next paragraph and elsewhere on Jaśkowski was kindly provided to the author by Professor N. C. A. da Costa.

In [24] Jaśkowski presents a general "mise au point" of decision problems. He has also shown that the decision problem for predicate calculus is reducible to the decision problem of several theories: for example: 1) to the theory of groups which has in addition to individual variables, a set variable (cf. [25]); 2) a topological theory (cf. [25]); 3) the theory of abelian groups which has at most 3 set variables (in addition to individual variables, cf. [25]); 4) a class of theorems of free groupoids, with one unary predicate variable E, of the form $\forall E \exists x_1 \exists x_2 \ldots \exists x_n \alpha$, where α is a formula of the first-order (cf. [28]); and 5) a class of ordinary differencial equations of a certain particular form having a solution (cf. [26]). He also proved in 1947 the undecidability of the theory of closure algebras. In addition to these and other reduction problems, Jaśkowski showed the decidability of the elementary theory of boolean rings and of the theory of totally additive boolean algebras (cf. [27]).

NON-CLASSICAL LOGICS.

Ever since Łukasiewicz first discovered in 1920 a system of 3-valued logic, logicians have studied new kinds of formal systems, different from the classical propositional and predicate calculi, the so-called non-classical logics, for example intuitionistic logic, modal logic, many-valued (even infinitely many-valued) logics, etc. Following Post's and Łukasiewicz' discoveries of decision procedures for classical propositional calculus, a good deal of work has been done on the decision problems of these systems. For instance, it has already been remarked that Gentzen and Jaśkowski have shown that the intuitionistic propositional calculus is decidable. Jaśkowski has also shown the decidability of several non-classical systems of propositional calculi such as Lewis **S5**. McKinsey [66] has solved the decision problem for Lewis's **S2** and **S4**. McKinsey and Tarski [67] proved Gödel's conjecture concerning the connections between the decision problems of Lewis **S4** system and the intuitionistic propositional calculus.

The work on the decision problem of non-classical logics certainly merits a systematic survey of results and methods used, perhaps in the style of Ershov (41). Such a work, to the author's knowledge, seems to be still lacking. In view of this we have included a (modest) bibliography of the papers published on this subject during 1965-1975 (see Part Four of the References at the end of this paper).

Church on May 19, 1936 (soon after proposing his thesis and obtaining his first undecidability results) wrote to Kleene of his desire that these results be used to prove the unsolvability of some mathematical problems not apparently related to logic. This desire was indeed fulfilled in 1947 when Post and Markov, independent of each other, showed that the "word problem for semigroups" is unsolvable. Then followed Turing's paper in 1950 showing the unsolvability of the "word problem for semigroups with cancellation". These works led to the 143-pages article by Novikov [40] in 1955 showing the unsolvability of the "word problem for groups". In 1958 Markov demonstrared the unsolvability of the "homeomorphism problem for four-dimensional manifolds in topology". All this work underwent, in recent years, considerable simplification and improvement. The reader could easily get an idea as to how intense the research has been in this new area by consulting the recent work (cf. Boone et al (13)). Thus the work of logicians - which of course grew out of the efforts to make logic mathematical and out of Hilbert's Formalism (proposed as a way out of the foundational crisis of mathematics) - became an indispensable part of mathematics. In this respect it is perhaps worthwhile to quote from the Editors' Introduction in Word Problems (cf. Boone et al. (13)).

" To us it seems unlikely that the argument of Novikov and Adjan for the Burnside Problem, published in 1968, and whose central ideas are explained in the present volume, or the argument of Britton for the same result in the present volume, could have been discovered by a 'pure' group theorist!" Later on they remark that "... the logic and algebra are so merged in theorems and proofs that it is impossible to separate one from the other."

4. Some Remarks on the Methods.

In the attempts at obttaining the unsolvability of the decision problems of particular theories two types of methods, namely the direct and the indirect method, have been used.

The direct method was used by Church to show the unsolvability of Peano's arithmetic and by Rosser to prove the essential undecidability of this theory. This method is quite involved and is applicable only to those theories in which the recursive functions are "definable". For the details we refer the reader to the Chapter II of the well-known monograph of Tarski [53], and to the more recent survey of Ershov et all (41), where the Tarski's treatment has been shortened and revised.

The indirect method calls for reducing the decision problem for a theory T_1 to the decision problem of another theory T_2 which is known to be undecidable.

The original form of the indirect methods was very limited in applications. It could be applied only to those theories which were at least as strong as Peano's arithmetic, while it could not be applied to theories which were "weaker" than Peano's arithmetic. For such weak theories neither the direct nor the (original) indirect method could be applied.

However, Tarski developed a general method - the method of interpretation - by extending and modifying the original indirect method. Tarski and his co-workers (and others) applied this method to prove the undecidability of a wide variety of theories. This method was described in the monograph [53] (see also (41), Chapter III, § 4) whose influence has been considerable in later studies of the decision problem for elementary theories.

In the early 60's two new and rather more efficient reduction methods for proving undecidability were published. One of these is due to Rabin (104) and the other due to Ershov (35). These methods have been applied to give (considerably) simpler proofs for the earlier undecidability results. In order to prove the undecidability of a theory T_2, via any one of these methods, one uses an undecidable theory T_1 and a certain effective transformation of T_1 into the sentences of the language of T_2. Whereas for Tarski's method one needs to pick a finitely axiomatizable and essentially undecidable thoery for T_1, any undecidable theorie will do for T_2 in Rabin's method. It is also known (Ershov (40)) that if the undecidability of a theory is proved by Tarski's interpretation method, then it is also provable by Erschov's (weak) elementary definability method. Rabin's method has been modified (cf. (16)) to prove (the more definitive information of) recursive inseparability of theories. The details of these methods will not be given here; instead the reader is asked to refer to the papers cited above.

The methods of proving the decidability results are more involved (naturally so). These methods are satisfactorily treated in (41). I only wish to make a couple of observations regarding the most recent work in this area. Rabin's method has also been used, in conjunction with Rabin's another remarkable paper (106), as a method of proving decidability of theories by Hauschild, Herre and Rautenberg (cf. (62), (65), (73) etc.). Burris and Werner (cf. (17)) have shown, using sheaf-theoretic methods and the decidability of the theory of countable boolean algebras with quantification over ideals (studied in Rabin (106)), that the first-order theory, with quantification over congruences, of the countable algebras in a residually finite discriminator variety is decidable, a result which includes the several known examples of varieties with decidable theories such as relatively complemented distributive lattice, n-valued Post algebras (both first proved by Erschov) and residually finite varieties of monadic algebras (proved by Comer) and so on.

5. CONCLUDING REMARKS.

There is a great deal more than what (if at all !) is expounded here. A few more general observations follow.

The theory of algorithms since 1936 has grown rapidly (refer Rogers [47] for instance) with its own challenging problems and exciting results. It has been also useful in clarifying some of the (vague) intuitionistic concepts. Although it is presently of limited "practical" value, some of the motivation in the development of the theory of computer programming has come from its concepts. The theory of algorithms, in the words of Rogers, can be viewed as a limiting, asymptotic version of the practical and down-to-earth computaional methods. The undecidability results have some important implications; for example, they demonstrate (conclusively !) that mathematics *is not just* constructing algorithms and that it is *impossible* for the machines, however powerful (even in future), to *completely replace* (quite contrary to the common belief) the creative faculty of the human brain ! They also made a mathematician searching for some algorithm aware that he must also consider the possibility that the sought-for-algorithm may not exist. An undecidability result, however, is not a cause for despair; it only shows that the class of questions considered is too wide to posses an algorithm and thus leads one to consider (interesting) subclasses.

At the same time it is important and useful to know how much of the thought process is mechanical (i. e. susceptible to machines); for the work in this direction see, for example, Wang [59] and (137), Davidov et al (03) and van Westrhenen (35). See also Kokorin and Kozlov [84]. The whole trend of formalization and the recent work connected with it made the possibility of using machines as an aid to mathematical research a reality.

People are also interested in computational complexity and in "measuring the difficulty" of theories (05). There are interesting connections of automata theory with decidability (Buchi (20), Rabin (105) , (106) and Thatcher and Wright (131). Work is also in progress to find more and more "efficient" and practically feasible and economical algorithms for decidable theories. For the work on abstract models of computers see (113) and (114).

The decision problems for the various restricted theories (equational theories, universal theories, Horn theories,etc.) have also been studied. (See, for instance, Ershov (38), Kshisamiev (83), Kokorin and Kozlov (84), Simmons (119), and Raimanov (123). For the work on "derived" theories (e. g. theories of congruence lattices, subalgebra lattices, lattices of varieties of algebras etc.) see Kozlov (85), Taitslin (125), (126), (127) and Burris and Sankappanavar (16), and Sankappanavar (112).

Undecidability results have also appeared in topology (Boone, Kaen and Poēnaru (12)), analysis, game theory - practically every branch of mathematics (see Jones (81), and Kleene [31], p. 265, for more references).

Burris and Werner (17), Grzegorczyk (53), and Schmerl (112) seem to open up the work towards the unification of the known results on decidability (probably by using model-theoretic, universal algebraic and/or category theoretical conditions).

Let me conclude by invoking a conviction of Hilbert and Ackermann:

"The task of extending the class of formulas for which the decision problem is solved remains rewarding and significant."

APPENDIX 1.

DECIDABLE THEORIES.

THEORY OF	WHO PROVED AND WHEN
I. SEMIGROUPS.	
1.1. Finitely presented (defined) commutative semigroups with cancellation.	Taĭclin, 1966 (123)
1.2. A semigroup prevariety (quasi-variety, variety) has a decidable theory iff ir is one of the following: 1) G x S, where G is a prevariety (quasivariety, variety) of periodic groups with a decidable theory, and S is one of the classes: a) left zero semigroups, b) right zero semigroups, or c) rectangular semigroups; 2) $xy = zt$; 3) $xy = x^2$; 4) $xy = y^2$.	Zamjatin, 1973 (140)
II. ABELIAN GROUPS.	
2.1. Torsion-free abelian groups with a predicate that distinguishes a subgroup.	Kozlov, Kokorin, 1969 (86)
2.2. Extended universal theory of lattice-ordered abelian groups.	Kozlov, Kokorin, 1968 (84)
III. GRAPHS.	
3.1. i) finite trees. ii) finitely generated trees. iii) finite symmetric graphs of width $\leqslant n$ (given n). iv) finitely generated, n-separated graphs.	Hauschild, Herre, Rautenberg, 1972 (66)
3.2. Monadic second-order theory of n-separated graphs.	Hauschild, Herre, Rautenberg, 1972 (67)

3.3. Ordered trees with minimum property and having k-length structure ($k < \omega$).	Hauschild, Herre, 1971 (73)
3.4. n-separated symmetric graphs of valency at most k.	Hauschild, Herre, 1972 (74)
3.5. Universal theory of graphs.	Hauschilf, Herre, 1971 (62)
3.6. Graphs with orbits having length at most $n+2$.	Rautenberg, 1972 (65)
IV. LINEAR ORDER (new proof).	Läuchli, Leonard, 1966 (88)
V. INTEGERS.	
5.1. i) Systems of integers with natural numbers as distinguished elements, addition, subtraction, less than and congruence modulo m (as non-logical notions).	Folk, Sestopal, 1968 (43)
ii) Natural numbers with the same signature as above.	
5.2. System of integers with addition, order, and multiplication by a single number.	Penzin, 1973 (98)
VI. WEAK SECOND-ORDER THEORY of $\langle\alpha,<\rangle$ for any ordinal α.	Buchi, 1964 (20)
VII COUNTABLE BOOLEAN ALGEBRAS quantification over ideals.	Rabin, 1969 (106)
VIII. ($x^n = x$)-RINGS.	Comer, 1974 (141)
IX. RESIDUALLY FINITE VARIETIES of monadic algebras.	Comer, 1975 (142)
X. COUNTABLE ALGEBRAS in a redually finite discrimiantor variety, with quantification over congruences.	Burris, Werner, 1976 (18)
XI. SUBGROUP LATTICES of	Taĭclin, 1970 (127)
i) finite abelian groups with r	

generators.

ii) Finite abelian p-groups with r generators.

iii) Direct sums of r isomorphic finite cyclic groups.

XII. LATTICE OF VARIETIES of monounary algebras.	Burris, Sankappanavar, 1975 (16)
XIII. IDEAL LATTICES of a Dedekind domain.	Taĭclin, 1969 (125)
XIV. EVERY $\aleph_0$-CATEGORICAL TREE.	Schmerl, 1977 (112)

Appendix 2.

UNDECIDABLE THEORIES.

THEORY OF	WHO PROVED AND WHEN
I. FIELDS.	
1.1. The field of formal power series over a field whose theory is undecidable.	Ax, 1965 (04)
1.2. Pythagorian fields	Hauschild, 1974/75 (64)
1.3. Field of rational functions of one variable over an arbitrary finite field of characteristic $\neq 2$.	Ershov, 1965 (35)
1.4. The same for characteristic $= 2$.	Penzin, 1973 (97)
II. GROUPS, SEMIGROUPS, etc.	
2.1. Metabelian p-groups satisfying the identity $x^{p^2}=1$ $(p \neq 2)$ or $x^8=1$ $(p=2)$.	Gurevič, 1966 (55)
2.2. The general linear groups, sym-	Ershov, 1966 (36)

	plectic groups, the unimodular groups, their projective groups.	
2.3.	The free product of two free groups with an amalgamated subgroups.	Rabin, 1964 (104)
2.4.	i) Abelian groups with four predicates for subgroups. ii) Torsion-free abelian groups with five predicates for subgroups.	Bauer, 1975 (08)
2.5.	Grupoids with a unary predicate selecting a substructure which is an infinite cancellative groupoid.	Garfunkel, Schmerl, 1974 (48)
2.6.	The subsemigroups of natural numbers under addition.	McKenzie, 1971 (94)
2.7.	Every variety of groups containing a non-abelian finite group.	Ershov, 1974 (40)
2.8.	Involutorially generated groups.	Hauschild, Rautenberg, 1977 (68)
2.9.	Semifree groups.	
2.10.	Abelian groups with a unary predicate specifying a subgroup.	Slobodskor, Fridman, 1975 (118) Bauer, 1975 (08) Baudish, 1974 (06)
2.11.	Abelian groups with a unary predicate specifying a finite subgroup.	Baudish, 1974 (06)
III.	RINGS.	
3.1.	Finite commutaitve rings.	Rabin, 1974 (104)
3.2.	A ring of formal power series with more than one variable over an arbitrary non-commutative ring with unity.	Ershov, 1966 (36) (rec. insep.)
3.3.	Commutative semisimple Banach algebras over C (in the ring-signature), in particular, Banach algebras.	Macintyre, 1971 (91)
3.4.	Universal theories of i) rings of characteristic zero; ii) torsion free rings of characteristic zero.	Simmons, 1970 (118)
IV.	LATTICES, etc.	
4.1.	Finitely generated free distibutive lattices.	Ershov, 1966 (36) (rec. insep.)

4.2. Atomistic distributive lattices.	Rabin, 1964 (104)
4.3. Heyting lattices.	Sankappanavar, 1973 (111) (rec. insep.)
4.4. Atomless and coatomless distibutive lattices.	Hauschild, Rautenberg, 1970 (68)
4.5. Monadic algebras.	Rabin, 1975 (109)
4.6. i) Partition lattices ii) Congruence lattices of semigroups (excluding group varieties) and of unary algebras. iii) Subalgebra lattices of rings, algebras over $\mathbb{Z}_p$ and boolean algebras.	Burris, Sankappanavar, 1975 (16) (rec. insep.)
4.7. Congruence lattices of congruence-distributive pseudocomplemented semillatices.	Sankappanavar, 1973 (111)
V. PLANAR GRAPHS.	Herre, 1973 (75)
VI. THE LATTICE OF IDEALS in any polynomial ring of more than one variable over an arbitrary field.	Taǐclin, 1968 (126)
VII. PASCH-FREE GEOMETRY.	Szczerba, 1971 (122)
VIII. METRIC SPACE of signature $\langle 0, \rho, +, \leqslant \rangle$ (rec. insep.)	Bondi, 1973 (02)
IX. SYSTEM OF INTEGERS with addition, order, and multiplication by power of a single number.	Penzin, 1973 (98)

APPENDIX 3.

SOME CONJECTURES AND OPEN PROBLEMS.

1) Every non-abelian variety of groups has an undecidable theory (Ershov 1974).

2) Is there a finitely axiomatized decidable theory (i.e. a first-order theory) such that the theory of its finite models is undecidable? (Dyson 1964.) More specifically:
2.1) Ask the above question for the theory of abelian groups.

3) Decidability of the theory of free algebras in a (popular) variety of algebras; for instance, modular lattices, Stone algebras, Heyting algebras etc. (The problem for groups was posed by Tarski. It has been solved for semigroups, lattices, Lie rings, soluble rings, etc.)

4) Is the theory of real closed ordered fields with an additional unary predicate $C(x)$ characterizing the field of constructible numbers decidable ? (Tarski)(Cf. Robibson (108).)

5) Is the theory of real closed ordered fields with the additional function 2^x decidable ? (Tarski. Cf. Robinson (108).)

6) Describe the decidable $\aleph_0$-categorical extensions of the theory of a partial ordering (theories of η, $1+\eta$, $1+\eta+1$, $\eta+1$, atomless Boolean algebras, and any $\aleph_0$-categorical tree are some such extensions known).

7) Is the theory of distributive lattices for which the set of meet-irreducible elements forms a tree decidable? (The decidability of the theory of the finite models of this theory is due to Gurevich.)

8) Is the theory of congruence lattices of Boolean algebras decidable? (If we restrict to countable Boolean algebras the decidability is known.)

9) Is the theory of orthomodular lattices recursively inseparable?

REFERENCES: PART ONE.

[01] W. Ackermann, *Zum Hilberstschen der reelem*, Zahler, Mathematische Annalen, 99 (1928), 118-133.

[02] W. Ackermann, Solvable cases of the decision problem, North-Holland, 1954.

[03]* E. T. Bell, Mathematics: Queen and Servants of Sciences, McGraw-Hill, New York, 1951.

[04]* E. T. Bell, Men of Mathematics, Simmon and Schuster, 1965.

[05]* E. W. Beth, The Foundations of Mathematics, Harper & Row, New York, 1966.

[06]* C. B. Boyer, History of Mathematics, John Willey & Sons, 1968.

[07] A. Church, *An unsolvable problem of elementary number theory*, Amer. J. Math., 58 (1936), 345-363. Reprinted in The Undecidable (Davis Ed.) 1965, 88-107.

[08] A. Church, *A note on the Entscheidungsproblem*, J. Symbolic Logic, 1 (1936), 40-41. Reprinted in The Undecidable (Davis, Ed.), 1965, 108-115.

[09] A. Church, Introduction to Mathematical Logic, Vol.I, Princeton, R. J., 1956.

[10] N. C. A. da Costa, Introdução aos Funadamentos da Matemãtica, Editora Hucitec, São Paulo, 1977.

[11]* J. N. Crossley, *Reminiscence of a logician*, in Algebra and Logic, Lecture Notes in Mathematics, vol. 450, Springer-Verlag, 1975, 1-16.

[12] M. Dehn, *Über unendliche diskontinuierliche Gruppen*, Math. Ann., 71 (1912), 144-166.

[13] H. Delong, A Profile of Mathematical Logic, Addison-Wesley, 1970.

[14] K. Gödel, *Die Vollstandigkeit der Axiome des logischen Funktionenkalkulus*, Monatsh., Math. Phys., 37 (1930), 349-360. English translation in van Heijenoort (1967).

[15] K. Gödel, *Über formal unentscheidbare Satze der Principia Mathematica und verwander System I*, Ibid. (1931), 173-198. English transla-

tion in Davis (1965), 4-38, and in van Heijenoort (1967).

[16] K. Gödel, *On undecidable propositions of formal mathematical systems*, Notes by Kleene and Rosser (1934). Reprinted (with a postscript) in Davis (1965), 39-74.

[17] J. Herbrand, *Recherches sur la théorie de la démonstration*, Travaux de la Société des Sciences et des lettres de Varsovie, Classe III, Sciences mathématiques et physiques, nº 33, 128 pp.

[18] J. van Heijenoort, From Frege to Gödel: A source book in Mathematical Logic, Harvar Univ. Press, Cambridge, 1967, 1874-1931. MR. 35 # 15.

[19]*H. Hermes, Enumerability, Decidability, Computability, Springer-Verlag, 1965.

[20]*D. Hilbert and W. Ackermann, Principles of Mathematical Logic, Chelsea, New York, 1950.

[21] D. Hilbert, *Mathematical problems* (Lecture delivered before the International Congress of Mathematicians at Paris in 1900). English translation: Bull. Math. Soc., 8 (1901-2), 437-479.

[22] D. Hilbert, *Axiomatic Denken*, Math. Ann., 78 (1918), 405-415.

[23] S. Jaśkowski, *Recherches sur la logique intuitioniste*, Actes du Congress Intern. de Philosophie Scientifique 6, Hermann, 1936, 58-61.

[24] S. Jaśkowski, *Quelques problèmes actuels concernant les fondements des mathématiques*, Casopis pro pestovani matematiky, 75 (1949), 74-78.

[25] S. Jaśkowski, *Sur le problème de la décision de la topologie et de la théorie des groupes*, Colloquium Mathematicum, 1 (1948), 176-178.

[26] S. Jaśkowski, *Examples d'une classe de systèmes d'equations differentielles ordinaires n'ayasat pour d'algorithme de solubilité des problèmes d'existence* (in Russian), Bull. de l'Académie Polonaise des Sciences, 2 (1954), 153-155.

[27] S. Jaśkowski, *Investigations into the dicidability of extended boolean algebras*, Casopis pro pestovani matematiky a physiki, 75 (1949), 136-137.

[28] S. Jaśkowski, *Undecidability of first-order sentences in the theory of free groupoids*, Fund. Math., 43 (1956), 36-45.

[29] S. C. Kleene, *λ-definability and recursiveness*, Duke Math. J., 2 (1936), 340-353.

[30]* S. C. Kleene, Introduction to Metamathematics, North-Holland, Amsterdam, 1952.

[31]* S. C. Kleene, Mathematical Logic, John Willey & Sons, New York, 1967.

[32] S. C. Kleene, *General recursive functions of natural numbers*, Math. Ann., 112 (1936), 727-742. Reprinted in Davis (1956), 236-253.

[33] C. H. Langford, *Some theorems on deducibility*, Ann. of Math., 28 (1926) 16-40.

[34] C. H. Langford, *Theorems on deducibility*, Ann. of Math., 29 (1927), 459-471.

[35] L. Löwenheim, *Über Möglichkeiten im Relativkalkül*, Math. Ann., 76 (1915), 447-479.

[36] J. Łukasiewicz, *O logice trojwartos ciowej* (*On three-valued logic*), Ruch Pilozoficzny (Lwow), 5 (1920), 169-171.

[37] A. A. Markov, *On the impossibility of certain algorithm in the theory of associative systems*, Comptes Rendus (Doklady) de l'Acadēmie des Sciences de l'URSS, 55 (1947), 583-586.

[38] A. A. Markov, *The theory of algorithms*, American Mathematical Society Translations, Ser. 2 , 15 (1960), 1-14. MR 22 # 5572.

[39] A. A. Markov, *Unsolvability of the problem of homeomorphy*, Proc. Internat. Congress Math., Edinburg, 1958.

[40] P. S. Novikov, *On the algorithmic unsolvability of the word problem in group theory*, Am. Math. Soc. Translations, Ser. 2, 9 (1958), 1-122. Russian original 1955.

[41] E. Post, *Introduction to a general theory of elementary propositions*, Am. J. Math., 43 (1921), 163-185. Reprinted in van Heijenoort 1967.

[42] E. Post, *Finite combinatory processes - formulation 1*, J. Symbolic Logic, 1 (1936), 103-105. Reprinted in Davis 1965, 288-291.

[43] E. Post, *Recursively enumarable sets of positive integers and their decision problems*, Bull. Am. Math. Soc. 50 (1944), 284-316. Reprinted

in Davis 1965, 404-337.

[44] M. Presburger, *Über die Vollstandigkeit eines gewissen systems der Arithmetik ganzer Zahlen*, in Welchen die Addition als einzige Operation hervoraitt (Comptes Rendus du I Congrès des Mathématician des Pays Slaves), Warsaw 1930, 92-101, 395.

[45] H. Putnam, *Recursive functions and hierarchy*, Am. Math. Monthly, 80 (1973), nº 6, part II, 68-86. MR 50 # 71.

[46] A. Robinson, *A decision method for elementary algebra and geometry-revisited*, Proceedings of Tarski Symposium, AMS Publication XXV, Providence, 1974, 139-152.

[47]* Jr. H. Rogers, Theory of Recursive Functions and Effective Computability, McGraw-Hill, 1967.

[48] J. B. Rosser, *Extensions of some theorems of Gödel and Church*, J. Symbolic Logic, 1 (1936), 87-91.

[49] E. Schröder, Vorlesungen über die Algebra der Logik, Vol.1, Leipzeg, 1890; Vol. II, Leipzig, 1891; Vol. III, Algebra und Logik der Relative, part 1, Leipzig 1895.

[50]* J. R. Shoenfield, Mathematical Logic, Addison-Wesley, 1967.

[51] Th. Skolem, *Untersuchungen über die Axiome des Klassenkalkuls und über Productions und Summations probleme welche gewisse Klassen von Aussagen betreffen*, Skr. Vid. Krist. Math-Naturaid. Kl. nº 3 (1919) 37 pp.

[52]* N. I. Styazhkin, History of Mathematical Logic from Leibniz to Peano (translated from Russian), the M.I.T. Press, 1969.

[53]* A. Tarski, A. Mostowski, and R. M. Robinson, Undecidable Theories, North-Holland, Amsterdam, 1953.

[54] A. Tarski, A decision method for elementary algebra and geometry, The Rand Corporation Santa Monica L948. MR 10 # 400.

[55]* B. A. Trakhtenbrot, Algorithms and automatic computing machines, D. C. Heath and Company, 1963. (Translated from Russian edition 1960)

[56]* A. Turing, *On computable numbers, with an application to the Entscheidungsproblem*, Proc. London Math. Soc. Ser. 2, 42 (1936-1937),

230-265. A correction, ibid. 43 (1937), 544-546. Reprinted in Davis (1965). 115-154.

[57] A. Turing, *Computability and λ-definablity*, J. Symbolic Logic 2 (1937), 153-163.

[58]* R. L. Vaught, *Model theory before 1945*, in Proceedings of the Tarski Symposium, AMS Publication XXV, Providence, 1974, 153-172.

[59]* H. Wang, A Survey of Mathematical Logic, Science Press, Peking; North-Holland Publ. Co, Amsterdam, 1963. MR 27 # 2394.

[60] A. N. Whitehead and B. Russell, Principia Mathematica, Vols. 1, 2 and 3, Cambridge University Press, 1910. 1912, 1913.

[61] C. C. Chang and H. J. Keisler, Model Theory, North-Holland Publ. Co., Amsterdam, 1973.

[62] L. H. Jacy Monteiro, Teoria de Galois, 7º Coloquio Brasileiro de Matemática, 1969.

[63] R. Pēter, Recursive Functions, Academic Press, 1967.

[64] G. Gentzen, *Untersuchungen über das logische Schliessen*, Math. Zeitshr. 39 (1934-35), 176-210, 405-431.

[65] J. Suranyi, Reductionstheorie des Entscheidungsproblems, Budapest, 1959.

[66] J. C. C. Mckinsey, *A solution of the decisiom problem for the Lewis systems S2 and S4, with application to topology*, J. Symbolic Logic 6 (1941).

[67] J. C. C. McKinsey and A. Tarski, *Some theorems about the sentencial calculi of Lewis and Heyting*, J. Symbolic Logic 13 (1948).

[68] J. Kotas (Editor), Stanislaw Jaškowski's Achievements in Mathematical Logic (Proc. of the 20º Conference on the History of Logic, Cracow, 1974), Studia Logica 34 (1975). MR 51 # 7800.

REFERENCES: PART TWO.

(01) V. N. Agafonov, *Complexity of computation of pseudorandom sequences*, Algebra i Logika 7 , nº 2 (1968), 4-19. MR 41 # 66.

(02) Z. A. Almagambetov, *Solvability of the elementary theory of certain classes of free nilpotent algebras*, Algebra i Logika Sem. 4, nº 6 (1965), 5-14. MR 34 # 59.

(03) C. Ash, *Undecidable $\aleph_0$-categorical theories*, Notices Am. Math. Soc. 18 (1971), 243. Ab 71T-E10.

(04) J. Ax, *On the undecidability of power series fields*, Proc. Am. Math. Soc. 16 (1965), 846. MR 31 # 2148.

(05) G. Ausiello, *Difficult logical theories*, Société Mathématique de France, Astérisque, 38-39 (1976), 3-21.

(06) A. Baudisch, *Theorien abelscher Gruppen mit einen einstellingen Predikat*, Fundamenta Mathematica 83 (1974), 121-127.

(07) A. Baudisch, *Elementare Theorien von Halbgruppen mit Kurzungsregeln mit einem einstelligen Predikat*, Bull. Acad. Polonaise des Sciences Math. Astr. et Phys. 23 (1975), 107-109. MR 51 # 10067.

(08) W. Bauer, *Decidability and undecidability of theories of abelian groups with predicates for subgroups*, Compositio Mathematica 31 (1975), 23-30.

(09) I. L. Bondi, *The decision problem for metric spaces*, Izv. Vyss. Ucebn. Zaved. Matemática 1973, nº 1 (128), 24-27. MR 47 # 6464.

(10) I. L. Bondi, *The decision problem for normed linear spaces and for Hilbert spaces* (Russian), Izv. Vyss. Ucebn. Zaved. Matemática 1973, nº 5 (132), 3-10. MR 48 # 8218.

(11) G. Boolos and R. Jaffrey, Computability and Logic, Cambridge University Press, London, 1974.

(12) W. W. Boone, W. Kaen, and V. Poénaru, *On recursively insolvable problems in topology and their classification*, in Contributions to Mathematical Logic (Proceedings of the Logic Colloquim, Hanover, 1966), Ed. by H. A. Schimidt, K. Schütte and H. J. Thiele, North-Holland, Amsterdam, 1968.

(13)* W. W. Boone, F. B. Cannonito, and R. C. Lyndon, Word Problems, North-Holland, Amsterdam, 1973.

(14) W. W. Boone, Between Logic and Group Theory, Springer-Verlag, Lecture Notes 372, 1974, 90-102.

(15) R. E. Bradford, *Cardinal addition and the axiom of choice*, Ann. Math. Logic 3, nº 2 (1971), 111-196. MR 46 # 38.

(16) S. Burris and H. P. Sankappanavar, *Lattice-theoretic decision problems in universal algebra*, Algebra Universalis 5 (1975), 163-177. MR 51 # 13359.

(17) S. Burris and H. Werner, *Decidable theories*, Notices Am. Math. Soc. 22 (1975), A-475 (Ab 75T-E42).

(18) S. Burris and H. Werner, *Sheaf constructions and their elementary properties*, to appear in Trans. Am. Math. Soc.

(19) J. R. Büchi, *Decision methods in the theory of ordinals*, Bull. Am. Math. Soc. 71 (1965), 767-770. MR 32 # 7413.

(20) J. R. Buchi, *Transfinite automata recursion and weak second-order theory of ordinals*, in Logic, Methodology and Philosophy of Science, North-Holland , L965, 3-23. MR 35 # 1480.

(21) J. R. Buchi and L. H. Landweber, *Definability in the monadic second order theory of sucessor*, J. Symbolic Logic 34 (1969), 166-170. MR 42 # 4387.

(22) L. L. Cinman, *Certain algorithm in a formal arithmetic system*, Doklady Akad. Nauk. SSRR 189 (1969), 489-490. MR 41 # 64.

(23) P. J. Cohen, *Decision procedures for real and p-adic fields*, Communications on Pure and Applied Mathematics 22 (1969), 131-151. MR 39 # 5342.

(24) S. D. Comer, *Finite inseparability of some theories of cylindrification algebras*, J. Symbolic Logic 34 (1969), 17-176. MR 42 # 7508.

(25) M. Davis, The Undecidable, Raven Press, New York, 1965.

(26) M. Davis, *Hilbert's thenth problem is unsolvable*, Am. Math. Monthly 80 (1973), 233-269. MR 47 # 6465.

(27) M. Davis, Computability, Notes by Barry Jacobs from lectures given

during 1973-74; Courant Institute of Mathenatical Sciences, New York University, 1974. MR 50 # 77.

(28) M. Dalla Chiara Scabia, Modeli Sintattici e Semantici delle Teorie Elementari, Feltrinelli Editore, Milano, 1968. MR 39 # 5308.

(29) M. Dulac, *Décidabilité et opérations entre théories*, C. R. Acad. Sc. Paris, ser. A-B, 273 (1971), 1113-1114. MR 45 # 3204.

(30) V. H. Dyson, *On the decision problem for theories of finite models*, Israel J. of Math. 2 (1964), 55-70. MR 31 # 2149.

(31) V. H. Dyson, *On the decision problem for extensions of a decidable theory*, Fundamenta Mathematica 64 (1969), 7-40. MR 40 # 5431.

(32) A. Ehrenfeucht, A *finitely axiomatizable complete theory with atomless* $F_1(T)$, Arch. Math. Logik Grundlagenforsch. 14 (1971), 162-166.

(33) S. Eilenberg and C. C. Elgot, Recursiveness, Academic Press, New York, 1970. MR 42 # 2939.

(34) C. C. Elgot and M. O. Rabin, *Decidability and undecidability of extensions of second (first) order theory of (generalized) successor*, J. Symbolic Logic 31 (1966), 169-181.

(35) Y. L. Ershov, *The undecidability of certain fields*, Doklady Akad. Nauk. SSRR 161, nº 1 (1965), 27-29.

(36) Y. L. Ershov, *New examples of undecidable theories*, Algebra i Logika Sem. 5, nº 5 (1966), 37-47. MR 34 # 7375.

(37) Y. L. Ershov, *Fields with a solvable theory*, Soviet Math. Dokl. 8 (1967), 575-576.

(38) Y. L. Ershov, *Restricted theories of totally ordered sets*, Algebra i Logika 7, nº 3 (1968), 38-47. MR 41 # 44.

(39) Y. L. Ershov, *Elementary theories of groups*, Doklady Akad. Nauk. SSRR 203 (1972), 1240-1243.

(40) Y. L. Ershov, *Theories of non-abelian varieties of groups*, Proceedings of the Tarski Symposium, AMS Publication, XXV, 255-264. MR 51 # 5280.

(41)* Y. L. Ershov, I. A. Lavrov, A. D. Taimanov, and M. A. Taitstin, *Elementary theories*, Uspechi Mat. Nauk 20, nº 4 (124), (1965), 37-108. MR 32 # 4012.

(42) J. Ferrante and C. Rackoff, *A decision procedure for the first-order theory of real addition with order*, SIAM J. Compt. 4 (1975), 66-76. MR 52 # 10403.

(43) N. F. Folk and G. A. Šestopal, *The solvability of elementary theories of integral and natural numbers with addition*, Rjazansk. Gos. Ped. Inst. Učen. Zap. 67 (1968), 141-155. MR 41 # 1522.

(44) H. Friedman, *Algorithmic procedures, generalized Turing algorithms and elementary recursion theory*, Logic Colloquium' 69, North-Holland, 1971, 361-389. MR 46 # 3275.

(45) H. Friedman, *One hundred and two problems in mathematical Logic*, J. Symbolic Logic 40 (1975), 113-129.

(46) Utz. Friedrich, *Entscheidbarkeit der monadischen Nachfolgerarithmetic mit endlichen Automaton ohne Analysetheorem*, Wiss Z. Humboldt Univ. Berlin Math. Natur. Reihe 21 (1972), 503-504. MR 48 # 3724.

(47) F. Galvin, *Reduced products, Horn sentences, and decision problems*, Bull. Am. Math. Soc. 73 (1967), 59-64. MR 35 # 39.

(48) S. Garfunkel and J. H. Schmerl, *The undecidability of theories of groupoids with an extra predicate*, Proc. Am. Math. Soc. 42 (1974), 286-289. MR 48 # 3725.

(49) S. Garfunkel and H. Shank, *On the undecidability of finite planar cubic graphs*, J. Symbolic Logic 37 (1972), 595-597. MR 47 # 4781.

(50) K. Gödel, *Zum Entscheidungsproblem des logischen Funktionenkalküls*, Monatschefte fur Math. und Physik 40 (1973), 433-443.

(51) I. J. Grilliot, *Inductive definitions and computability*, Trans. Am. Math. Soc. 158 (1971), 309-317.

(52) A. Grzegorczyk, *Logical uniformity by decomposition and categoricity in* $\aleph_0$, Bull. Acad. Polonaise des Sciences Math. Astr. et Phys. XVI (1968), 687-692.

(53) A. Grzegorczyk, Decision Procedures for Theories categorical in $\aleph_0$, Lecture Notes 125, Springer-Verlag 1970. MR 42 # 7496.

(54) Ju. Š. Gurevich, *Existential interpretation*, Algebra i Logika 4, nº 4 (1965), 71-85. MR 34 # 46.

(55) Ju. S. Gurevich, *Certain algorithmic questions for the theory of classes of algebraic systems*, Interuniv. Sc. Sympos. General Algebra, Tartu. Gos. Univ. Tartu , 1966, 38-41. MR 34 # 29.

(56) Ju. S. Gurevich, *On the elementary theory of lattice ordered abelian groups and k-limeals*, Dokl. Akad. Nauk SSSR 175 (1967), 1213-1215. MR 36 # 1373.

(57) Ju. S. Gurevich, *Hereditary undecidability of one class of lattice ordered abelian groups*, Algebra i Logika 6, nº 1 (1967), 45-62.

(58) W. Hanf, *The boolean algebra of lògic*, Bull. Am. Math. Soc. 81 (1975), 587-589. MR 52 # 10404.

(59) W. Hanf, *Model-theoretic methods in the study of elementary logic*, in Theory of Models, North-Holland, 1965, 132-145. MR 35 # 1457.

(60) K. Hauschild, *Über die unentscheidbarkeit nichtassoziativer Schiefringe*, Wiss. Z. Humboldt-Univ. Berlin Math. Natur.- Reihe 15 (1966), 681-683. MR 36 # 2501.

(61) K. Hauschild, *Equivalence of models in respect to special classes of sentences*, Bull. Acad. Polon. Sci. Sēr. Sci. Math. Astronom. Phys. 17 (1969), 609-610. MR 42 # 4383.

(62) K. Hauschild, *Entscheidbarkeit in der Theorie tewerteter Graphen*, Elektron, Informationveratbeit. Kybernetik 7 (1971), 485-492.

(63) K. Hauschild, *Universalität in der Ringtheorie*, Wiss. Z. Humboldt-Univ. Berlin Math. Natur. Reihe 21 (1972), 505-506. MR 48 # 8221.

(64) K. Hauschild, *Rekursive Unentscheidbarkeit der Theorie der pythagoräischen Körper*, Fundamenta Mathematica 82, nº 3 (1974/75), 191-197.

(65) K. Hauschild, H. Herre and W. Rautenberg, *Interpretierbarkeit und Entscheidbarkeit in der Graphen Theorie II*, Zeitschr. fur Math. Logic und Grundlagen 18 (1972), 457-480. MR 48 # 3737.

(66) K. Hauschild, H. Herre and W. Rautenberg, *Entscheidbarkeit der monadischen Theorie 2. Stufe der n-separierten Graphen*, Wiss. Z.

Humboldt Univ. Berlin Math. Natur. Reihe, 21 (1972), 507-511. MR 50 # 79.

(67) K. Hauschild, H. Herre and W. Rautenberg, *Entscheidbarkeit der elementaren Theorie der endlich Baüme und verwandter Klassen endlicher Strukturen;* Wiss. Z. Humboldt Univ. Math. Natr. Reihe 21 (1972), 497-502. MR 50 # 6818.

(68) K. Hauschild and W. Rautenberg, *Universelle Interpretierbarkeit in Verbänden*, Wiss. Z. Humboldt Univ. Berlin Math. Natur. Reihe 19 (1970), 575-577. MR 48 # 8220.

(69) K. Hauschild and W. Rautenberg, *Interpretierbarkeit in der Gruppen-theorie*, Algebra Universalis 1 (1971), 136-151. MR 46 # 39.

(70) K. Hauschild and W. Rautenberg, *Entscheidungsprobleme der Theorie zweir Aquivalenzrelationen mit beschänkter Zahl von Elementen in der Klassen*, Fundamenta Mathematica 81, nº 1 (1973), 35-41. MR 49 # 2338.

(71) I. J. Heath, *Decidable classes of number-theoretic sentences*, Z. Math. Logik Grundlagen Math. 15 (1969), 411-420. MR 42 # 72.

(72) H. Hecker, *Recursive Untrennbarkeit bei elementen Theorien*, Z. Math. Logic Grundlagen Math. 17 (1971), 443-464. MR 48 # 3726.

(73) H. Herre, *Entscheidungsproblem in der elementaren Theorie einer zweistelligen Relation*, Zeitschr. f. Math. Logic und Grundlagen der Math. 17 (1971), 301-313. MR 45 # 1749.

(74) H. Herre, *Die Entscheidbarkeit der elementaren theorie der n-separierten symmetrischen Graphen endlicher valez*, Zeitschr. f. Math. Logik und Grundlagen der Math. 18 (1972), 249-254. MR 46 # 3289.

(75) H. Herre, *Unentscheidbarkeit in der Graphentheorie*, Bull. Acad. Polon. Sci. Ser Sci. Math. Astr. Phys. 21 (1973), 201-208. MR 50 # 1865.

(76) H. Herre and W. Rautenberg, *Das Basistheorem und einige Anwendungen in der Modeltheorie*, Wiss. Z. der Humboldt Univ. Berlin Math. Nat. Reihe 19 (1970), 579-583.

(77) H. Herre and H. Wolter, *Entscheidbarkeit von Theorien in Logiken mit verallgeimeinerten Quantoren*, Zeitschr. f. Math. Logic und

Grundlagen der Math. 21 (1975), 229-246. MR 53 # 2623.

(78) K. Hirose and I. Shigeaki, *A proof of negative answer to Hilbert's 10 th problem*, Proc. Japan Acad. 49 (1973), 10-12. MR 49 # 2325.

(79) N. G. Hisamiev, *Unsolvability of the elementary theory of a free lattice*, Algebra i Logika 6, nº 5 (1967), 45-48. MR 37 # 3921.

(80) N. G. Hisameiv, *Strongly constructive models of a decidable theory*, Izv. Akad. Nauk. Kazah. SSRR Ser. Fiz-Mat. nº 1 (1974), 83-84. MR 50 # 6824.

(81) J. P. Jones, *Recursive undecidability - an exposition*, Am. Math. Monthly 81 (1974), 724-738. MR 50 # 9568.

(82) S. Kenžebaev, *Undecidability of the elementary theory of a certain ring*, Kazah. Gos. Ped. Inst. Učen. Zap. 23 (1966), 24-25. MR 37 # 2604.

(83) N. G. Kshisamiev, *Universal theory of lattice-ordered abelian groups*, Algebra i Logika 5 (1966), 71-76.

(84) A. I. Kokorin and G. T. Kozlov, *Extended elementary and universal theories of lattice-ordered abelian groups with a finite number of threads*, Algebra i Logika 7, nº 1 (1968), 91-103. MR 38 # 4298.

(85) G. T. Kozlov, *Unsolvability of the elementary theory of lattices of subgroups of finite abelian P-groups*, Algebra i Logika 9 (1970), 167-171. MR 43 # 7329.

(86) G. T. Kozlov and A. I. Kokorin, *An elementary theory of torsion-free abelian groups, with a predicate that distinguishes a sub-group*, Algebra i Logika 8 (1969), 320-334. MR 41 # 3263.

(87) A. H. Lachlan, *The elementary theory of recursively enumerable sets*, Duke Math. J. 35 (1968), 123-146. MR 37 # 2593.

(88) H. Läuchli and J. Leonard, *On the elementary theory of linear order*, Fundamenta Mathematica 59 (1966), 109-116.

(89) V. A. Livishits, *Some classes of reduction and undecidable theories*, Notes of the Science Seminar of Mathematics Department of the Sketlov Mathematical Institute, Leningrad, 4 (1967).

(90) M. H. Loeb, *Decidability of the monadic predicate calculus with unary function symbols*, J. Symbolic Logic 32 (1967), 563 (abstract).

(91) A. Macintyre, *On the elementary theory of Banach algebras*, Ann. Math. Logic 3, nº 3 (1971), 239-269. MR 48 # 87.

(92) V. I. Mart'yanov, *The theory of abelian groups with predicates specifying a subgroup, and with endomorphism operations*, Algebra and Logic 14 (1971), 330-334.

(93) R. Mckenzie, *Negative solution of the decision problem for sentences true in every subalgebra of* ⟨N,+⟩, J. Symbolic Logic 36 (1971), 607-609. MR 45 # 6625.

(94) R. Mckenzie, *On spectra, and the negative solution of the decision problem for identities having a finite nontrivial model*, J. Symbolic Logic 40 (1975), 186-196. MR 51 # 12499.

(95) B. Meltzer, *The use of symbolic logic in proving mathematical theorems by means of a digital computer*, Foundations of Mathematics (Symp. Commemorating Kurt Gödel, Columbus, Ohio, 1966), Springer-Verlag, 1969, 39-55. MR 41 # 3279.

(96) A. R. Meyer, *Weak monadic second order theory of sucessor in not elementary-recursive*, Logic Colloquium (Boston, Mass. 1972-73) Lecture Notes in Math. 453, Springer-Verlag, 1975, 132-154. MR 52 # 13358.

(97) Yu. G. Penzin, *The undecidability of fields of rational functions over fields of characteristic 2*, Algebra and Logic 12 (1973), 116-119.

(98) Ju. G. Penzin, *Decidability of the theory of integers with addition order and multiplication by an arbitrary number*, Mat. Zametki 13 (1973), 667-675 (Math. Notes 13 (1973), 401-405). MR 48 # 1907.

(99) Ju. G. Penzin, *The decidability of certain theories of integers*, (Russian) Sibirsk. Mat. Z. 14 (1973), 1139-1143, 1160. MR 48 # 8222.

(100) M. G. Peretjat'kin, *Complete theories with a finite number of Countable models*, Algebra i Logika 12 (1974), 550-576, 618. MR 50 # 6827.

(101) P. Perkins, *Unsolvable problems for equational theories*, Notre Dame J. Formal Logic 8 (1967), 175-185. MR 38 # 4310.

(102) A. Ju. Pljuskevicene, *Extension of the inverse method to axiomatic theories with equality*, Investigations in constructive Mathematics and Mathematical Logic, V. Zap. Naucn. Sem. Leningrad, Otdel. Mat. Inst. Steklov. (LOMI) 32 (1972), 108-115. MR 52 # 5376.

(103) H. Putnam, *Recursive functions and hierarchies*, Am. Math. Monthly 80, nº 6 (1973), part II, 68-86. MR 50 # 71.

(104)* M. O. Rabin, *A simple method for undecidability proofs and some applications*, in Logic, Methodology and Philosophy of Science (Proceedings of the 1964 International Congress) Ed. Bar Hillel, North-Holland, 1965, 58-68. MR 36 # 4976.

(105) M. O. Rabin, *Decidability of second-order theories and automata on infinite trees*, Bull. Am. Math. Soc. 74 (1968), 1025-1029. MR 38 # 44.

(106)* M. O. Rabin, *Decidability of second-order theories and automata on infinite trees*, Trans. Am. Math. Soc. 141 (1969), 1-35. MR 40 # 30.

(107) W. Rautenberg and K. Hauschild, *Interpretirbarkeit und Entscheidbarkeit in der Graphentheorie I*, Zeits. f. math. Logic und Grundlagen d. Math. (1971), 47-55.

(108) A. Robinson, *A decision method for elementary algebra and geometry revisited*, Proceedings of the Tarski Symposium, AMS Publication XXV (1974), 139-152.

(109) M. Rubin, *Boolean algebras: undecidability with distinguished subalgebras and interpretation in automorphism groups*, Notices AMS, 75T-E63.

(110) T. K. Sajahmetov, *Undecidability of certain theories with a supplemental predicate*, Vestnik. Akad. Nauk. Kazah SSR 24 (1968), nº 3, (275), 48-50. MR 37 # 1245.

(111) H. P. Sankappanavar, *On the decision problem of the congruence lattices of pseudocomplemented semilattices*, Non-Classical Logics, Model Theory and Computability, Eds. A. I. Arruda, N. C. A. da Costa, and R. Chuaqui, North-Holland, 1977, 255-266.

(112) J. H. Schmerl, *On $\aleph_0$-categoricity and the theory of trees*, Fundamenta Mathematica XCIV (1977), 122-128.

(113) A. Schurmann, *Functions computable by a computer*, Studia Logica 27 (1971), 59-72. MR 47 # 6458.

(114) J. C. Schepherdson and H. E. Sturgis, *Computability of recursive functions*, J. Assoc. Compt. Mach. 10 (1973), 217-255. MR 27#1359.

(115) J. R. Shoenfield, *A theorem on quantifier elimination*, Symposia Mathematica 5 (1971), 173-176.

(116) J. R. Shoenfield, *Quantifier elimination in fields*, Non-Classical Logics, Model Theory and Computability, Eds. A. I. Arruda, N. C. A. da Costa and R. Chuaqui, North-Holland, 1977, 243-252.

(117) D. Siefkes, *Undecidable extensions of monadic second-order successor arithmetic*, Z. math. Logik und Grundlagen der Math. 17 (1971), 385-394. MR 45 # 1763.

(118) H. Simmons, *The solution of a decision problem for several classes of rings*, Pacific J. Math. 34 (1970), 547-557. MR 42 # 1658.

(119) A. M. Slobodskoi and E. I. Fridman, *Theories of abelian groups with predicates specifying a subgroup*, Algebra and Logic 14 (1975), 253-255.

(120) M. E. Szabo (Ed.), The Collected Papers of Gerhard Gentzen, North-Holland, 1969. MR 41 # 6660.

(121) L. W. Szczerba, *Undecidability of elementary Pasch-free geometry*, Bull. Acad. Polonaise Sc. Ser. Sci. Math. Astr. et Phys. 19 (1971), 469-474. MR 46 # 1580.

(122) A. D. Taĭmanov, *Systems with a solvable universal theory*, Algebra i Logika 6, nº 5 (1967), 33-43. MR 37 # 6173.

(123) M. A. Taĭclin, *On elementary theories of commutative semigroups with cancellation*, Algebra i Logika 5 (1966), 51-69. MR 33#5486.

(124) M. A. Taĭclin, *Some further examples of undecidable theories*, Algebra i Logika 6 (1967), 105-111. MR 37 # 69.

(125) M. A. Taĭclin, *The elementary theories of lattices of ideals in polinomial rings*, Algebra i Logika 7, nº 2 (1968), 94-97. MR 38 # 4312.

(126) M. A. Taĭclin, *Prime ideals in polynomial rings*, Algebra i Logika 7, nº 6 (1968), 64-66. MR 40 # 5607.

(127) M. A. Taĭclin, *Elementary theories of lattices of subgroups*, Algebra i Logika 9 (1970), 473-483. MR 44 # 1739.

(128) R. L. Tenney, *Second-order Ehrenfeucht games and the decidability of the second-order theory of an equivalence relation*, J. Austral. Math. Soc. Ser. A 20, nº 3 (1975), 323-331. MR 52 # 2861.

(129) J. W. Thatcher, *Decision problems for multiple sucessor arithmetics*, J. Symbolic Logic 31 (1966), 182-190. MR 34 # 4139.

(130) J. W. Thatcher and J. B. Wright, *Generalized finite automata theory with an application to a decision problem of second-order logic*, Math. Systems Theory 2 (1968), 57-81. MR 37 # 75.

(131) E. S. Vasil'ēv, *The elementary theories of complete trosion-free abelian groups with p-adic topology*, Mat. Zametki 14 (1973), 201-208. (Math. Notes 14 (1973), 673-677 (1974)). MR 49 # 2329.

(132) Ju. M. Vazenin, *The elementary theory of symmetric groups and semigroups*, Izv. Vyss. Ucebn. Zaved. Matematika 140, nº 1 (1974), 15-20. MR 50 # 4271.

(133) S. Vinner, *A generalization of Ehrenfeucht's game and some applications*, Israel J. Math. 12 (1972), 279-298. MR 47 # 3142.

(134) E. G. Wagner, *Uniformly refelexive structures: on the nature of gödelizations and relative computability*, Trans. Am. Math. Soc. 144 (1969). 1-41. MR 40 # 2543.

(135) E. G. Wagner, *Bounded action machines: toward an abstract theory of computer structure*, J. Comput. System Sci. 2 (1968), 13-75. MR 43 # 5764.

(136) H. Wang, *Remarks on machines, sets, and the decision problem*, in Formal Systems and Recursive Functions (Proc. Eight Colloq. Oxford, 1963), North-Holland, 1965, 304-320. MR 39 # 6729.

(137) V. Weispfenning, On the theory of Hensel fields, Ph. D. thesis, Heilderberg, 1971.

(138) H. Wolter, *Entscheidbarkeit der Arithmetik mit Addition und Ordnung in Logiken mit verallgemeiten Quantoren*, Z. Math. Logik Grundlagen der Math. 21 (1975), 321-330. MR 51 # 7816.

(139) A. P. Zamjatin, *Prevarieties of semigroups whose elementary theory*

is decidable, Algebra and Logic, 12 (1973) 233-241 (1975). MR 52 # 10401.

(140) A. P. Zamjatin, *Varieties of associative rings whose elementary theory is decidable*, DoKl. Akad. Nauk. SSRR 229 (1976), 276-279.

(141) S. Comer, *Elementary properties of structures of sections*, Bol. Soc. Mat. Mexicana 19 (1974), 78-85.

(142) S. Comer, *Monadic algebras with finite degree*, Algebra Universalis 5 (1975), 315-329.

REFERENCES: PART THREE.

(01) S. O. Aanderaa, *On the decision problem for formulas in which all disjunctions are binary*, Proceedings of the Second Scandinavian Logic Symposium (Oslo 1970), North-Holland, 1971, 1-18. MR 50 # 1864.

(02) E. Börger, *Eine entscheidbare Klasse von Kromformeln*, Z. Math. Logik Grundlagen Math. 19 (1973), 117-120. MR 49 # 2327.

(03) G. V. Davydov, S. Ju. Maslov, G. E. Minc, V. P. Orekov, and A. O. Slisenko, *A machine algotithm for establishing deducibility on the basis of the inverse method*, Zap. Naucn. Sem. Leningrad Otdal. Mat. Inst. Steklov (LOMI) 16 (1969). 8-19. MR 41 # 3278.

(04) B. Dreben and J. Denton, *A supplement to Herbrand*, J. Symbolic Logic 31 (1966), 393-398. MR 34 # 4107.

(05) B. S. Drebben, *Solvable Sranyi subclasses: An introduction to Herbrand theory*, Proc. Harvard Sympos. Digital Computers Appl. (Cambridge, 1961), Harvard Univ. Press, 1962, 32-47. MR 36 # 1318.

(06) Ju. Š. Gurevič, *The problem of reduction for the logic of predicates and operations*, Algebra i Logika 8 (1969), 284-308. MR 41 # 8205.

(07) Ju. Š. Gurevič, *On the decision problem for pure narrow calculus of predicates*, Dokl. Akad. Nauk. SSSR 166 (1966), 1032-1034. MR 34 # 47.

(08) Ju. Š. Gurevič, *Solvability problem for a restricted predicate*

calculus, Dokl. Akad. Nauk. SSSR 168 (1966), 510-511. MR 34 # 60.

(09) Ju. Š. Gurevič, *Effective recognition of realizability of formulae of the restricted predicate calculus*, Algebra i logika 5, nº 2 (1966), 25-55.

(10) R. Goldberg, *On the solvability of a subclass of the Suranyi reduction class*, J. Symbolic Logic 28 (1963), 237-244. MR 35 # 23.

(11) H. Hermes, *A simplified proof for the unsolvability of the decision problem in the case* $\wedge \vee \wedge$, in Logic Colloquium 69, Edited by R. O. Gandy and C. E. M. Yates, North-Holland, 1971.

(12) R. B. Jensen, *Ein newer Beweis für die Entscheidbarkeit des einstelligen Predicatenkalküls mit Identitat*, Arch. Math. Logik Grundlagenforsch. 7 (1965), 128-138. MR 34 # 1193.

(13) V. F. Kostyrko, *The reduction class* $\forall \exists^n \forall$, Algebra i Logika Sem. 3 (1964), 5-6, 45-55. MR 31 # 27.

(14) V. F. Kostyrko, *On the* $\forall \exists \forall$ *reduction class*, Kibernetika)Kiev) nº 1 (1966), 17-22. MR 34 # 4136.

(15) V. F. Kostyrko, *The reduction class* $\forall x \forall y \exists z\ F(x, y, z) \wedge \forall^m A(F)$, Kibernetika (Kiev), nº 5 (1971), 1-3. MR 46 # 1550.

(16) M. R. Krom, *The decision problem for segregated formulas in first-order logic*, Math. Scand. 21 (1967), 233-240. MR 39 # 1286.

(17) M. R. Krom, *A decision procedure for a class of formulas of first-order predicate calculus*, Pacific J. Math. 14 (1964), 1305-1319. MR 31 # 3316.

(18) V. F. Kostyrko, *On the decidability problem for Ackermann's case*, Sibirsk. Math. Z. 6 (1965), 342-363. MR 31 # 2150.

(19) M. R. Krom, *The decision probelm for a class of first-order formulas in which all disjunctions are binary*, Z. Math. Logik Grundlagen Math. 13 (1967), 15-20. MR 35-24.

(20) H. Lewis and W. D. Golfarb, *The decision problem for formulas with a small number of atomic subformulas*, J. Symbolic Logic 38 (1973), 471-480. MR 49 # 2328.

(21) V. A. Lifsic, *Certain reduction classes and undecidable theories*, Zap. Naucn. Sem. Leningrad. Otdel. Inst. Steklov. (LOMI) 4 (1967), 65-68. MR 38 # 5603.

(22) V. A. Lifsic, *Deductive general validity and reduction classes*, Zap. Naučn. Sem. Leningrad. Otdel. Mat. Inst. Steklov (LOMI) 4 (1967), 69-77. MR 38 # 5604.

(23) S. Ju. Maslov, *An inverse method of establishing deducibility in the classical predicate calculus*, Dokl. Akad. Nauk. SSSR 159 (1964) 17-20. MR 30 # 3005.

(24) S. Ju. Maslov, *Application of the inverse method for establishing deducibility to theory of decidable fragments in the classical predicate calculus*, Soviet. Math. Dokl. 7 (1966), 1653-1657. MR 35 # 25.

(25) S. Ju. Maslov, *Inverse method of establishing deducibility for non-prenex formulas of predicate calculus*, Dokl. Akad. Nauk. SSSR 172 (1967), 22-25. MR 35 # 19.

(26) S. Ju. Maslov, *An inverse method of establishing deducibility for logical calculi*, Trudy. Mat. Inst. Steklov 98 , 26-87. MR 40# 5416.

(27) S. Ju. Maslov, *The connection between tactics of the inverse method and the resolution method*, Zap. Naučn. Sem. Leningrad Otdel. Mat. Inst. Steklov (LOMI) 16 (1969), 137-146. MR 42 # 1645.

(28) S. Ju. Maslov and V. P. Orevkov, *Decidable classes that reduce to a single quantifier class*, Trudy Mat. Inst. Steklov 121 (1972), 59-66, 165. MR 50 # 12678.

(29) Z. Mijajlovich, *On decidability of one class of Boolean formulas*, Mat. Vesnik 11 (26), (1974), 48-54. MR 50 # 80.

(30) V. P. Orevkov, *Two unsolvable classes of formulas of the classical predicate calculus*, Zap. Naucn. Sem. Leningrad Otdel. Mat. Inst. Steklov (LOMI) 8 (1968), 202-210. MR 42 # 5776.

(31) A. Pieczkowski, *Undecidability of the homogeneous formulas of degree 3 of the predicate calculus*, Studia Logica 22 (1968), 7-16. MR 38 # 4311.

(32) J. Reichbach, *On characterizations and undecidablity of the first-order function calculus*, Yokohama Math. J. 13 (1965), 11-30. MR 34 # 4108.

(33) A. B. Slomson, *An undecidable two sorted predicate calculus*, J.

Symbolic Logic 34 (1969), 21-23. MR 39 # 2636.

(34) T. V. Turasvili, *A reduction of the decidability problem of first-order predicate logic to a class with asymmetric and irreflexive two-places predicates,* Sakharth. SSR Mecn. Akad. Moambe 75 (1974), 297-300. MR 50 # 6821.

(35) S. C. van Westrhenen, *A computer programme for the first-order predicate calculus without identity,* Automation in Language Translation and Theorem Proving, Commision of the European Communities, Brussels, 1968, 69-83. MR 39 # 5310. (also MR 38 # 5606, and MR 38 # 5607.)

REFERENCES: PART FOUR.

(01) J. G. Anderson, *Superconstructive propositional calculi with extra axiom schemes containing one variable,* Z. math. Logik Grundlagen Math. 18 (1972), 113-130. MR 46 # 8815.

(02) D. I. Bakerelov, *Extensional logics* (Russian), C. R. Acad. Bulgare Sci. 25 (1972), 1609-1612. MR 48 # 55.

(03) A. Bayart, *On truth tables for M, B, S4 and S5*, Logique et Analyse (N. S.) 13 (1970), 335-375. MR 45 # 1734.

(04) N. D. Belnap and J. R. Wallace, *A decision procedure for the system EL of entailment with negation,* Z. Math. Logic Grundlagen Math. 77 (1965), 277-289. MR 31 # 4714.

(05) D. Bollman and M. Tapia, *On the recursive unsolvability of the provability of the deduction theorem in partial propositional calculi,* Notre Dame J. Formal Logic 13 (1972), 124-128. MR 46 # 3288.

(06) R. A. Bull, *A modal extension of intuitionistic logic,* Notre Dame J. Formal Logic 6 (1965), 142-146. MR 36 # 2475.

(07) J. Czermak, *Ein Vollstandigkeitsbeweis für die aussagenlogischen Modalitätensysteme M, S4, Br. und S5,* Arch. Math. Logik Grundlagenforsch. 17 (1975), 45-49. MR 52 # 13331.

(08) M. Fidel, *The decidability of the calculi* C_n , to appear.

(09) K. Fine, *The logics containing S4.3*, Z. Math. Logik Grundlagen Math. 17 (1971), 371-376. MR 45 # 1735.

(10) K. Fine, *Logics containing S4 without the finite model property*, Conference in Mathematical Logic - London'70, Lecture Notes 255, Springer-Verlag, 1972. MR 48 # 10764.

(11) D. M. Gabbay, *Decidability results in non-classical logics III, Systems with stability operators*, Israel J. Math. 10 (1971), 135-146. MR 46 # 1553.

(12) D. M. Gabbay, *On decidable, finitely axiomatizable, modal and tense logics without the finite model property, I, II*, Israel J. Math. 10 (1971), 478-495; ibid. 10 (1971), 496-503. MR 45 # 4961.

(13) D. M. Gabbay, *Sufficient conditions for the undecidability of intuitionistic theories with application*, J. Symbolic Logic 37 (1972) 275-284. MR 47 # 8278.

(14) D. M. Gabbay, *Decidability of some intuitionistic predicate theories*, J. Symbolic Logic 37 (1972), 579-587. MR 48 # 86.

(15) D. M. Gabbay, *A general filtration method for modal logics*, J. Philos. Logic 1, nº 1 (1972), 29-34. MR 50 # 12657.

(16) D. M. Gabbay, *The undecidability of intuitionistic theories of algebraically closed fields and real closed fields*, J. Symbolic Logic 38 (1973), 86-92. MR 47 # 6466.

(17) D. M. Gabbay, *On 2nd. order intuitionistic propositional calculus with full comprehension*, Arch. Math. Logik Grundlagenforsch. 16 (1974), 177-186. MR 50 # 12672.

(18) D. M. Gabbay and D. H. J. de Jongh, *A sequence of decidable finitely axiomatizable intermediate logics with the disjunction property*, J. Symbolic Logic 39 (1974), 67-78. MR 51 # 10038.

(19) D. M. Gabbay, *The decision problem for some finite extensions of the intuitionistic theory of abelian groups*, Studia Logica 34, nº1, (1975), 59-67. MR#12500.

(20) D. M. Gabbay, *Decidability results in non-classical logics, I*, Ann. Math. Logic 8 (1975), 237-295. MR 52 # 68.

(21) R. I. Goldblatt, *Decidability of some extensions of J*, Z. Math. Logik Grundlagen Math. 20 (1974), 203-205.

(22) R. Harrop, *Some structure results for propositional calculi*, J. Symbolic Logic 30 (1965), 271-292. MR 33 # 2523.

(23) S. A. Kripke, *The undecidability of monadic modal quantification theory*, Z. Math. Logik Grundlagen Math. 8 (1962), 113-116. MR 28 # 2975.

(24) E. J. Lemmon, *Algebraic semantics for modal logics I and II*, J. Symbolic Logic 31 (1966), 46-65; and 191-218. MR 34 # 5660 and MR 34 # 5661.

(25) V. A. Lifsic, *The decision problem for certain constructive theories of equality*, Zap. Naucn. Sem. Leningrad Otdel. Mat. Inst. Steklov (LOMI) 4 (1967), 78-85. MR 39 # 65.

(26) A. Loparič, *Une ētude semantique de quelques calculs propositionnels*, C. R. Acad. Sc. Paris 284 A (1977), 835-838.

(27) E. G. K. Lopez-Escobar, *A decision method for the intuitionistic theory of sucessor*, Nederl. Akad. Wetensch. Proc. Ser. A 71, Indag. Math. 30 (1968), 466-467. MR 40 # 4081.

(28) S. Ju. Maslov, G. E. Minc and V. P. Orevkov, *Unsolvability of the constructive calculus of predicates of certain classes of formulae containing only one-place predicate variables*, Soviet Math. Dokl. 6 (1966), 918-920. MR 33 # 51 and # 52.

(29) G. C. Mckay, *The decidability of certain intermediate propositional logics*, J. Symbolic Logic 33 (1968), 258-264. MR 39 # 1300.

(30) R. K. Meyer, *An undecidability result in the theory of relevant implication*, Z. Math. Logik Grundlagen Math. 14 (1968), 255-262. MR 38 # 4287.

(31) G. E. Minc, *Some calculi of modal logic*, Trudy Mat. Inst. Steklov 98 (1968), 88-111. MR 39 # 40.

(32) G. E. Minc, *Solvability of the problem of deducibility in LJ for a class of formulas which do not contain negative occurrences of quantors*, (Russian) Trudy Mat. Inst. Steklov 98 (1968), 121-130. MR 41 # 6660.

(33) A. Nakamura, *On a propositional calculus whose decision problem is recursively unsolvable*, Nagoya Math. Journal 38 (1970), 145-152. MR 41 # 3277.

(34) J. Okee, *A semantical proof of the undecidability of the monadic intuitionistic predicate calculus of the first-order*, Notre Dame J. Formal Logic 16 (1975), 552-554. MR 52 # 5370.

(35) V. P. Orevkov, *Unsolvability in constructive predicate calculus of a class of formulas of type* $\neg\neg\forall\exists$, Dokl. Akad. Nauk. SSRR 163 (1965), 581-583.

(36) V. P. Orevkov, *Certain reduction classes and solvable classes of sequents for the constructive calculus of predicates*, Dokl. Akad. Nauk. SSRR 163 (1950), 30-32. MR 33 # 51.

(37) V. P. Orevkov, *Unsolvability in the modal predicate calculus of a class of formulae which contain only one-place predicate variable*, Zap. Naucn. Sem. Leningrad Otdel. Inst. Steklov (LOMI) 4 (1967), 168-173. MR 41 # 1514.

(38) V. P. Orevkov, *Undecidable classes of formulas for the constructive predicate calculus I*, (Russian) Trudy Mat. Inst. Steklov 121 (1972), 100-108), 165. MR 49 # 7132.

(39) R. Routley, *The decidability and semantical incompleteness of Lemmon's system S0.5*, Logique et Analyse (N.S.) 11 (1968), 413-421. MR 38 # 4290.

(40) R. Routley, *Decision procedures and semantics for C1, El and* $S0.5^0$, Logique et Analyse (N.S.) 11 (1968), 468-471. MR 40 # 1267.

(41) K. Segerberg, *Decidability of four modal calculi*, Theoria 34 (1968), 21-25. MR 39 # 1309.

(42) K. Segerberg, *Decidability of S4.1*, Theoria 34 (1968), 7-20. MR 39 # 1308.

(43) C. Smorynski, *Elementary intuitionistic theories*, J. Symbolic Logic 38 (1973), 102-134. MR 48 # 5842.

(44) R. H. Thomason, *A decision procedure for Fitch's propositional calculus*, Notre Dame J. Formal Logic 8 (1967), 101-117.
MR 38 # 3139.

(45) D. Ulrich, *Some results concerning finite models for sentential calculi*, Notre Dame J. Formal Logic 13 (1972), 363-368.
MR 46 # 8808.

Instituto de Matemática
Universidade Federal da Bahía
40.000 Salvador, BA., Brazil.

ADDED IN PROOF: The author has learned that the argument of Britton refered to on page 258 has been found to be incorrect; however the original proof is, of course, valid.

Functorialization of First-Order Language with Finitely Many Predicates.

by A. M. SETTE and J. S. SETTE.

ABSTRACT. In this paper we are concerned with the notion of language and its relation to category theory. Our motivations has been Farïssé's treatment of formulas as "operators". We show that the formulas of first-order language with finitely many predicates can be (naturally) characterized as a class of functors from one category to another.

Introduction.

In this paper we are concerned with the notion of language and its relation to category theory. Our motivation has been Fraïssé's treatment of formulas as "operators" which as we shall show, can be very simply described in the context of categories, formulas being treated as functors. In fact, this is implicit in [1].

This approach is quite different from the one of Lawvere and the people currently dealing with topoi, although we believe that it may possible to link the two points of view.

In a crude outline, the fundamental philosophy is the following: languages arise in connection with one's capacity to group "objects" in the "world", comparing them, creating categories and finally relating these categories, bringing out some invariants of the "objects" by means of functors.

We shall show that first-order languages with finitely many predicates can be analysed within this approach. We also claim that this will be the correct approach to study the analogies between homology theory and logic.

1. Basic concepts.

We will consider first-order languages $\mathcal{L}$ with a finite number of predicate symbols $\rho_1^{m_1}, \ldots, \rho_k^{m_k}$, an equality symbol $=$, denumerably many variables $x_1, x_2, \ldots, x_n, \ldots$, and the logical symbols $\wedge, \vee, \neg, \exists, \forall$. An *structure* (or *multirelation*) *adequate to* $\mathcal{L}$ is a $(k+1)$-tuple $\langle |M|, R^{n_1}, \ldots, R_k^{n_k} \rangle$ where $|M|$ is a non-void set and $R_i^{n_i}$, $i = 1, \ldots, k$, are n_i-ary relations defined on $|M|$, i.e., are functions from $|M|^{n_i}$ to $\{0,1\}$. The n-tuple $\sigma = \langle n_1, \ldots, n_k \rangle$ will be called the *type* of M (or $\mathcal{L}$).

Given a multirelation M and a formula $\phi(x_{i_1}, \ldots, x_{i_r})$ (with the free variables $x_{i_1}, \ldots, x_{i_r}$) we can define, in a natural way, an r-ary relation $\phi(\boldsymbol{M})$ on $|\boldsymbol{M}|$ as follows: *for every* $(a_1, \ldots, a_r) \in |\boldsymbol{M}|^r$, $\phi(\boldsymbol{M})(a_1, \ldots, a_r) = 1$ *iff* $M \models \phi(a_1, \ldots, a_r)$.

With any formula ϕ of $\mathcal{L}$ we now associate a pair of integers $\langle k, p \rangle$, $k \leq p$, called the *characteristic* of ϕ (in symbols $char(\phi)$), as follows:

(i) if ϕ is quantifier-free, we put $char(\phi) = \langle 0, 0 \rangle$.

(ii) if ϕ is one of the formulas $\phi_1 \wedge \phi_2$, $\phi_1 \vee \phi_2$ and $char(\phi_1) = \langle k_1, p_1 \rangle$, $char(\phi_2) = \langle k_2, p_2 \rangle$, then $char(\phi) = \langle \max\{k_1, k_2\}, \max\{p_1, p_2\} \rangle$.

(iii) $char(\neg\phi) = char(\phi)$.

(iv) Let ϕ be the formula $\forall x_1, \ldots, x_n \phi_1$ and $char(\phi_1) = \langle k, p \rangle$. Then we have two cases:

a) ϕ_1 is of the form $\forall y_1, \ldots, y_n \phi_2$; in this case, $char(\phi) = \langle k, p+n \rangle$;

b) ϕ_1 is not of the form $\forall y_1, \ldots, y_n \phi_2$, then $char(\phi) = \langle k+1, p+n \rangle$.

(v) For the formulas $\exists x_1, \ldots, x_n \phi_1$ the definition of $char(\phi)$ is analogous to that in (iv).

With this definition it is easy to verify that if ϕ is a formula in prenex normal form and $char(\phi) = \langle k, p \rangle$ then p is the number of quantifiers in ϕ and k the numbers of blocks of these quantifiers.

A *category* C consists of

(i) A class (of *objects*) denoted by $Ob(C)$;

(ii) For each pair A, $B \in Ob(C)$, a set $Mor_e(A,B)$. An element of $Mor_e(A,B)$ will be called an *arrow* and denoted by $f: A \longrightarrow B$ or $A \xrightarrow{f} B$.

(iii) For any triple A, B, C in $Ob(C)$, a function
$o: Mor_e(B,C) \times Mor_e(A,B) \longrightarrow Mor_e(A,C)$
$(g,f) \longmapsto g \circ f$
satisfying the followina conditions:

(a) given $A \xrightarrow{f} B \xrightarrow{g} C \xrightarrow{h} D$ then $h \circ (g \circ f) = (h \circ g) \circ f$,

(b) for every $A \in Ob(C)$ there is $I_A \in Mor_e(A,A)$ such that: if $A \xrightarrow{f} B$ and $A \xrightarrow{g} C$ are arrows then $I_A \circ f = f$ and $g \circ I_A = g$.

A *functor* $F: C \longrightarrow D$ from the category C to the category $\mathcal{D}$ is a function such that:

(i) to each object A of C os associated an object $F(A)$ of D,

(ii) to each arrow $A \xrightarrow{f} B$ of is associated an arrow $F(A) \xrightarrow{F(t)} (B)$ of D, satisfying the folowing conditions:

(a) $F(g \circ f) = F(g) \circ F(t)$,

(b) $F(I_A) = I_{F(A)}$.

2. Some important categories.

Let $M_0 = \langle |M_0|, R^n, \ldots, R_k^{n_k} \rangle$, $M_1 = \langle |M_1|, S_1^{n_1}, \ldots, S_k^{n_k} \rangle$ be multirelations. A *local-isomorphism* from M to M is a bijection f defined on a subset F of $|M_0|$ onto a subset E of $|M_1|$ such that:

$$R_i^{n_i}(a_1,\ldots,a_{n_i}) = S_i^{n_i}(f(a_1),\ldots, f(a_{n_i}))$$

for $i = 1,\ldots,k$ and all $a_1,\ldots,a_{n_i} \in F$.

We will associate with each local-isomorphism f a pair of integers $\langle k,p \rangle$, recursively, as follows:

Every local-isomorphism is a $\langle 0,p\rangle$-isomorphism. for any $p \geq 0$. For $k \geq 1$ we say that $f\colon F \subset |M_0| \longrightarrow E \subset |M_1|$ is a $\langle k,p\rangle$-isomorphism if for every $a_1,\dots,a_q \in |M_0|$, $(a_1,\dots,b_q \in |M_1|)$, $q \leq p$, there is an extension $\bar{f}\colon F \cup \{a_1,\dots,a_q\} \longrightarrow |M_1|$ ($\overline{f^{-1}}\colon E \cup \{b_1,\dots,b_q\} \longrightarrow |M_0|$ which is a $\langle k-1,p-1\rangle$-isomorphism. (This notion is connected with the notion of game; see [2]).

Now, let M_0, M_1 and M_2 be multirelations of the same type and $M_0 \xrightarrow{f} M_1 \xrightarrow{g} M_2$, $\langle k,p\rangle$-isomorphisms, i. e., $f\colon F \subset |M_0| \longrightarrow E \subset |M_1|$, $g\colon G \subset |M_1| \longrightarrow H \subset |M_2|$. We define the composition of f and g by:

$$g\circ f = \begin{cases} \emptyset \text{ (the empty function) if } E \cap G = \emptyset \\ g\circ f\colon f^{-1}(E \cap G) \subset |M_0| \to g(E \cap G) \subset |M_1| \text{ if } E \cap G \neq 0. \end{cases}$$

The following properties are easily verified:

P1. Every $\langle k,p\rangle$-isomorphism is $\langle k',p'\rangle$-isomorphism for $k' \leq k$ and $p' \leq p$.

P2. Every local-isomorphism is a $\langle k,0\rangle$-isomorphism for $k \geq 0$.

P3. Every $\langle p,p\rangle$-isomorphism is a $\langle k,p\rangle$-isomorphism for all k.

P4. If $f\colon M_0 \longrightarrow M_1$ is a $\langle k,p\rangle$-isomorphism, then every restriction of f is also a $\langle k,p\rangle$-isomorphism.

P5. Every (global) isomorphism is a $\langle k,p\rangle$-isomorphism for all k and p.

P6. If $M_0 \xrightarrow{f} M_1 \xrightarrow{g} M_2$ are $\langle k,p\rangle$-isomorphisms, then $g\circ f$ is a $\langle k,p\rangle$-isomorphism.

If we take the class of all multirelations for $Ob(M^{\tau}_{\langle k,p\rangle})$, the set of all $\langle k,p\rangle$-isomorphisms from M_0 to M_1 for $Mor_e(M_0,M_1)$, and consider the composition $g\circ f$ of $\langle k,p\rangle$-isomorphisms as defined above, the properties P1 - P6 imply in particular that we have a category. This category will be denoted by $M^{\tau}_{\langle k,p\rangle}$. In particular, $M^{\langle n\rangle}_{\langle 0,0\rangle}$ is the category of n-ary relations with the local-isomorphisms as arrows.

3. FORMULAS AS FUNCTORS.

In this section we will show that each formula ϕ of $\mathcal{L}$ with n free variables defines a functor from the category $M^{\tau}_{\langle k,p\rangle}$ in the category $M^{\langle n\rangle}_{\langle 0,0\rangle}$. More precisely:

PROPOSITION 1. *If ϕ is a formula of $\mathcal{L}$ with* n *free variables and Char $(\phi) \leq \langle k,p\rangle$ then ϕ defines a functor F_{ϕ}: $M^{\tau}_{\langle k,p\rangle} \longrightarrow M^{\langle n\rangle}_{\langle 0,0\rangle}$ making the diagram commutative:*

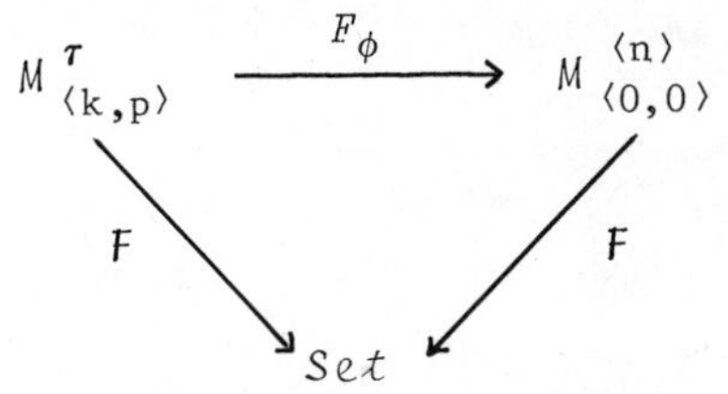

where τ is the type of the multirelations adequate to $\mathcal{L}$, Set is the category of sets with the local-isomorphisms and F *is the forgetful functor.*

PROOF: Define F_{ϕ} as follows: $F_{\phi}(M) = \langle |M|, \phi(M)\rangle$ and $F_{\phi}(f) = f$. We must show that F_{ϕ} is a functor, i. e., if $M_0 \xrightarrow{f} M_1$ is a $\langle k,p\rangle$-isomorphism then $F_{\phi}(M_0) \xrightarrow{f} F_{\phi}(M_1)$ is a local-isomorphism. ($F_{\phi}(I_M) = I_{F_{\phi}(M)}$ and $F_{\phi}(g o f) = F_{\phi}(g) o F_{\phi}(f)$ obviously.) If ϕ is atomic or a negation of a formula, the proof is straightforward.

Suppose the Proposition is valid for formulas ϕ_1 and ϕ_2 such that $Char(\phi_1)$ and $Char(\phi_2) \leq \langle k, p\rangle$. For the inductive hypothesis, if f: $F \subset |M_0| \longrightarrow |M_1|$ is a $\langle k, p\rangle$-isomorphism then f is a local- isomorphism from $F_{\phi_i}(M_0)$ to $F_{\phi_i}(M_1)$, i = 1, 2. Thus $(\phi_1 \wedge \phi_2)(M_0)(a_1, \ldots, a_n) = 1$ iff $\phi_1(M_0)(a_1, \ldots, a_n) = 1$ and $\phi_2(M_0)(a_1, \ldots, a_n) = 1$, and this holds iff $(\phi_1 \wedge \phi_2)(M_1)(f(a_1), \ldots, f(a_n)) = 1$.

Now, take a formula ϕ of the form $\forall x_{n+1}, \ldots, x_{n+r}\, \phi_1(x_1, \ldots, x_n, x_{n+1}, \ldots, x_{n+r})$ and suppose that $F \subset |M_0| \xrightarrow{f} |M_1|$ is a $\langle k, p\rangle$-isomorphism, $a_1, \ldots, a_n \in F$, $\phi(M_0)(a_1, \ldots, a_n) = 1$, and that ϕ_1 defines the functor F_{ϕ_1} : $M^{\tau}_{\langle k-1, p-q\rangle} \longrightarrow M^{\langle n+r\rangle}_{\langle 0,0\rangle}$. $\phi(M_0)(a_1, \ldots, a_n) = 1$ iff for all $b_1, \ldots, b_r \in |M_0|$, $\phi_1(M_0)(a_1, \ldots, a_n, b_1, \ldots, b_r) = 1$ (by the definition of $\phi_1(M_0)$). But then $\phi(M_1)(f(a_1),\ldots,f(a_n)) = 1$. If not, there are $b'_1,\ldots,b'_r \in |M_1|$ such that $\phi_1(M_1)(f(a_1),\ldots,f(a_r), b'_1,\ldots,b'_r) = 0$. On the other hand, $\phi_1(M_1)(f(a_1),\ldots,f(a_n), b'_1,\ldots, b'_r)$ must be equal to $\phi_1(M_0)(a_1,\ldots,a_n,\overline{f^{-1}}(b'_1),\ldots,\overline{f^{-1}}(b'_n))$ for some $\langle k-1,p-r\rangle$-isomorphism $\overline{f^{-1}}$, extension of f^{-1}. This is a contradiction. It is easy to show that if $\phi(M_0)(a_1,\ldots,a_n) = 1$ then $\phi(M_1)(f(a_1),\ldots,f(a_n)) = 1$.

The case where ϕ is a formula of the form $\exists x_{n+1},\ldots,x_{n+r}\ \phi_1(x_1,\ldots,x_n)$ can be treated analogously.

DEFINITION. *We note by* S^{τ}_n *the class of all pairs* $\langle M,a\rangle$ *where* M *is a multirelation of type* τ *and* a *belongs to the set* $|M|^n$. *If* $\langle M,a\rangle$ *and* $\langle M',a'\rangle \in S^{\tau}_n$, *we say that they are* $\langle k,p\rangle$*-equivalent, in symbols,* $\langle M,a\rangle \equiv_{\langle k,p\rangle} \langle M',a'\rangle$, *iff the function* $a_i \longmapsto a'_i$ $i = 1,\ldots,n$, is a $\langle k,p\rangle$-isomorphism ($a = (a_1,\ldots,a_n)$; $a' = (a'_1,\ldots,a'_n)$).

LEMMA 1. $\langle k,p\rangle$*-equivalence is an equivalence relation.*

The proof is straightforward.

LEMMA 2. *The number of* $\langle k,p\rangle$*-equivalence classes is finite (i. e.,* $S^{\tau}_n/\equiv_{\langle k,p\rangle}$ *is finite).*

PROOF: $k = 0$. In that case the number of $\langle k,p\rangle$-equivalence classes will be the same as the number of all multirelations of type τ defined on a subset of a set with n elements.

Suppose that the Lemma holds for $k-1$, $q \leqslant p$, and all n, and show that it still holds for k,p and all n.

Let σ_q be a function $\sigma_q\colon S^{\tau}_n \longrightarrow \mathcal{P}(S^{\tau}_{n+q}/\equiv_{\langle k-1,p-q\rangle})$

$$\langle M,a\rangle \longmapsto \{\overline{\langle M,ab\rangle} : b \in |M|^q\}$$

Since $\langle M,a\rangle \equiv_{\langle k,p\rangle} \langle M',a'\rangle$ implies $\sigma_q(\langle M,a\rangle) = \sigma_q(\langle M',a'\rangle)$, σ_q may be factored as follows:

$$\begin{array}{ccc} S^{\tau}_{n} & \xrightarrow{\sigma_q} & \mathcal{P}(S^{\tau}_{n+q}/\equiv_{\langle k-1,p-q\rangle}) \\ \Pi \downarrow & \nearrow \bar{\sigma}_q & \\ S^{\tau}_{n}/\equiv_{\langle k,p\rangle} & & \end{array}$$

where Π is the canonical projection.

Let $\bar{\sigma} : S^{\tau}_{n}/\equiv_{\langle k,p\rangle} \longrightarrow \prod_{q=0}^{p} \mathcal{P}(S^{\tau}_{n+q}/\equiv_{\langle k-1,p-q\rangle})$

be the product of the family $\{\bar{\sigma}_q\}_q$.

By the definition of $\equiv_{\langle k,p\rangle}$, $\sigma_q(\langle M,a\rangle) = \sigma_q(\langle M',a'\rangle)$, $q = 0$ $1,\ldots,p$, implies $\langle M,a\rangle \equiv_{\langle k,p\rangle} \langle M',a'\rangle$. Thus $\bar{\sigma}$ is an injection. By the inductive hypothesis, $\mathcal{P}(S^{\tau}_{n+q}/\equiv_{\langle k-1,p-q\rangle})$ is finite; therefore $S^{\tau}_{n}/\equiv_{\langle k,p\rangle}$ is finite. ∎

LEMMA 3. *Given* $\langle M,a\rangle \in S^{\tau}_{n}$ *and the integers* k, p, *there is a formula* ϕ *such that:*

(i) $Char(\phi) \leqslant \langle k,p\rangle$;

(ii) $\langle M,a\rangle \equiv_{\langle k,p\rangle} \langle M',a'\rangle$ *iff* $\phi(M')(a') = 1$.

PROOF: For $k = 0$ consider the variables $x_1,\ldots,x_n$ of $\mathcal{L}$ and let

$$\rho^{n_i}_{i}(x_{j_1},\ldots,x_{j_{n_i}}) = \begin{cases} \rho^{n_i}_{i}(x_{j_1},\ldots,x_{j_{n_i}}) & \text{if } R^{n_i}_{i}(a_{j_1},\ldots,a_{j_{n_i}}) = 1 \\ & \text{and} \\ \neg\rho^{n_i}_{i}(x_{j_1},\ldots,x_{j_{n_i}}) & \text{if not.} \end{cases}$$

Take for ϕ the conjunction of all $\rho^{n_i}_{i}(x_{j_1},\ldots,x_{j_{n_i}})$ $i = 1,\ldots,k$, $x_{j_1} \in \{x_1,\ldots,x_n\}$ with the formula ϕ' asserting that all $x_i = x_j$ or $x_i \neq x_j$ according as $a_i = a_j$ or $a_i \neq a_j$; $i,j = 1,\ldots,n$.

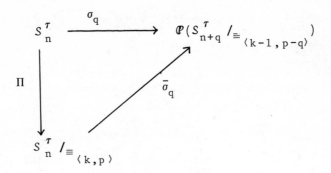

Suppose $k \geq 1$ and observe that $\bar{\sigma}(\overline{\langle M,a\rangle}) = \bar{\sigma}(\overline{\langle M', a'\rangle})$ iff $\bar{\sigma}_q(\overline{\langle M,a\rangle}) = \bar{\sigma}_q(\overline{\langle M', a'\rangle})$, $q = 0,\dots,p$.

Let $\{\overline{\langle M, c_s\rangle} : s = 0,\dots, r_q\}$ be an enumeration of

$$S^{\tau}_{n+q} /\equiv_{\langle k-1,p-q\rangle}$$

such that $\bar{\sigma}_q(\overline{\langle M,a\rangle}) = \{\overline{\langle M, ab_s\rangle} : s = 0,\dots,\ell_q\}$, i. e., the elements of $\sigma_q(\overline{\langle M,a\rangle})$ occur in the first ℓ_q places. By the induction hypothesis, for every $s = 0,\dots, r_q$ there is a formula ϕ^s_q of characteristic $\leqslant \langle k-1, p-q\rangle$ such that

$$\langle M', c'\rangle \equiv_{\langle k-1,p-q\rangle} \langle M, ab_s\rangle$$

iff

$$\phi^s_q(M')(c') = 1.$$

If we denote the formula $\exists x_{n+1},\dots, x_{n+q}\ \phi^s_q$ by ψ^s_q and the formula $\psi^0_q \wedge \dots \wedge \psi^{\ell_q}_q \wedge \neg\psi^{\ell_q+1}_q \wedge \dots \wedge \neg\psi^{r_q}_q$ by ϕ_q we can see that $\phi_q(M')(a')=1$ iff $\sigma_q(\overline{\langle M,a\rangle}) = \sigma_q(\overline{\langle M',a'\rangle})$, for if $\phi_q(M')(a') = 1$, we have $\psi^s_q(M')(a')=1$ for $s = 0,\dots,\ell_q$, i. e., there is $b' \in |M|^q$ such that $\phi^s_q(M')(a'b') = 1$. But this means that $\overline{\langle M', a'b'\rangle} = \overline{\langle M, ab_s\rangle}$, thus $\sigma_q(\overline{\langle M',a'\rangle}) \supset \sigma_q(\overline{\langle M,a\rangle})$. Now, if $\langle N, c\rangle \in \bar{\sigma}_q(\overline{\langle M', a'\rangle}) - \sigma_q(\overline{\langle M,a\rangle})$ then $\overline{\langle N,c\rangle} = \overline{\langle M',a'b'\rangle}$ for some b'. Since $\overline{\langle M', a'b'\rangle} \notin \bar{\sigma}_q(\overline{\langle M,a\rangle})$, there is $\psi^{\ell_q+i}_q(M')(a') = 1$. This is a contradiction. Thus $\bar{\sigma}_q(\overline{\langle M',a'\rangle}) = \bar{\sigma}_q(\overline{\langle M,a\rangle})$. (The other side of of the implication is easy to prove). The conjunction $\phi_0 \wedge \dots \wedge \phi_p$ is the required formula. ■

PROPOSITION 2. *A functor* $F\colon M^{\tau}_{\langle k,p\rangle} \longrightarrow M^{\langle n\rangle}_{\langle 0,0\rangle}$ *is defined by a formula of* $\mathcal{L}$ *iff the following diagram commutes:*

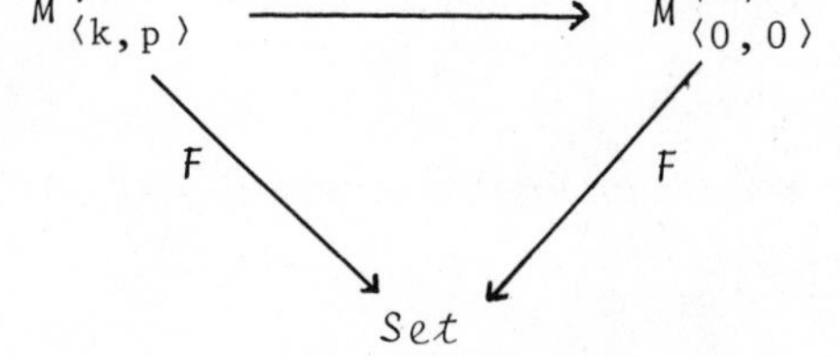

where Set and F are as Proposition 1.

PROOF: One half of the Proposition is exactly Proposition 1. For the other half, let $F\colon M^{\tau}_{\langle k,p\rangle} \longrightarrow M^{\langle n\rangle}_{\langle 0,0\rangle}$ be a functor making the above diagram commute. Then, we know that:

(i) $F(M) = \langle |M|, F(M)\rangle$,

(ii) $F(f) = f$.

Thus, the function

$$\dot{F}\colon S^{\tau}_{n} \longrightarrow \{0,1\}$$

$$\langle M, a\rangle \longmapsto F(M)(a),$$

induced by the functor F, can be factored as follows:

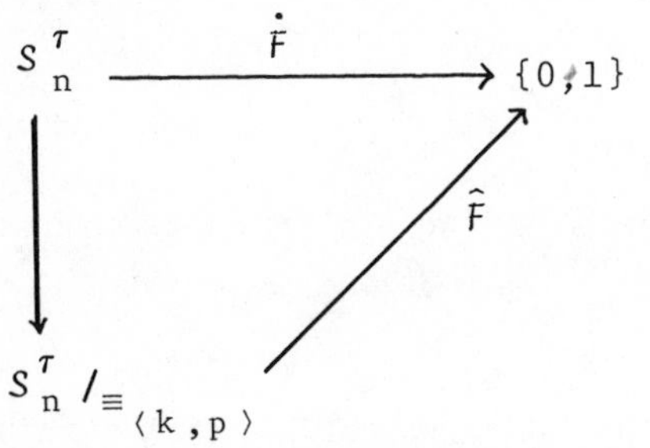

Take $\{C_1,\ldots,C_r\} = (\hat{F})^{-1}(1) \subset S^{\tau}_{n}/_{\equiv_{\langle k,p\rangle}}$, by Lemma 3 there are formulas $\phi_1,\ldots,\phi_r$ such that $\phi_i(M)(a) = 1$ iff $\langle M,a\rangle \in C$, $i = 1, \ldots, r$. If we let ϕ the disjunction $\phi_1 \vee \ldots \vee \phi_r$, ϕ defines F. For $\phi(M)(a) = 1$ iff there is a $i \in \{1,\ldots,r\}$ such that $\phi_i(M)(a) = 1$, i.e., iff $\overline{\langle M,a\rangle} \in (\hat{F})^{-1}(1)$ and this holds iff $\hat{F}(\overline{\langle M,a\rangle}) = 1$ or, iff $F(M)(a) = 1$. ∎

Problems.

1- For what (usual) logical languages can we obtain results analogous to Proposition 2?

2- The same question can be put for the homological functors. More explicitly: can we introduce isomorphisms in the category of topological spaces in such a way as to characterize the functors H_n in Proposition 2 ?

3- What is the link between logic viewed from this point of view and the topoi theory ?

OBSERVATION.

We have now an obviously general definition of language. Namely: a language $\mathcal{L}$ on a category C and taking values in a category D is a triple

$$\langle \{C_i\}_{i\in I}, \{F_{ij}\}_{i\in I}^{j\in J}, D \rangle$$

such that:

(i) $Ob(C_i) = Ob(C)$

(ii) F_{ij} are functors from C_i to D.

In the case treated in this paper, we can take for D a category a little larger than $\bigsqcup_n M_{\langle 0,0\rangle}^{\langle n\rangle}$, introducing morphisms between relations of different aryties. This can be done in such a way that if ϕ and ψ are formulas of $\mathcal{L}$, $\phi \vdash \psi$ iff the function that associates the arrow

$$F_\phi(M) \xrightarrow{i_{|M|}} F_\psi(M)$$

to each multirelation M is a natural transformation from F_ϕ to F_ψ.

This will be better explained and analysed in a forthcoming work.

REFERENCES.

[2] A. Ehrenfeucht, An application of games to the completeness problem for formalized theories, Fundamenta Mathematica, XLIV (1961).

[1] R. Fraïssé, Course of Mathematical Logic, vol. II, D. Reidel Publishing Company, Dordrecht - Boston.

Instituto de Matemática
Universidade Federal de Pernambuco
50.000 Recife, PE., Brazil.